The Boundary Element Method for Plate Analysis

The Boundary Element Method for Plate Analysis

by

John T Katsikadelis

National Technical University of Athens, Athens, Greece

ELSEVIER

AMSTERDAM • BOSTON • HEIDELBERG • LONDON
NEW YORK • OXFORD • PARIS • SAN DIEGO
SAN FRANCISCO • SINGAPORE • SYDNEY • TOKYO

Academic Press is an imprint of Elsevier

AP

Academic Press is an imprint of Elsevier
The Boulevard, Langford Lane, Kidlington, Oxford OX5 1GB, UK
225 Wyman Street, Waltham, MA 02451, USA
Radarweg 29, PO Box 211, 1000 AE Amsterdam, The Netherlands
525 B Street, Suite 1800, San Diego, CA 92101-4495, USA

British Library Cataloguing in Publication Data
A catalogue record for this book is available from the British Library

Library of Congress Catalog Number
A catalog record for this book is available from the Library of Congress

ISBN–13: 978-0-12-416739-1

For information on all Academic Press publications
visit our website at http://store.elsevier.com/

14 15 16 17 18 10 9 8 7 6 5 4 3 2 1

To my wife Efi for her loving patience and support

Contents

Companion Website for this book: http://booksite.elsevier.com/9780124167391

Foreword

At first glance this book may appear to describe yet another highly specialised method applied to the solution of plate problems, namely the one the author calls Analog Equation Method (AEM). Nothing could be further from the truth. Professor Katsikadelis has instead presented for the first time a generalised and consistent BEM for all types of plate analysis. This has been possible only because of his brilliant interpretation of the principle of virtual work.

The first two introductory chapters set the basis for the subsequent treatment.

After having set up the basic principles of boundary elements (BEM) in an elegant and consistent manner in the first chapter, the reader acquires the necessary knowledge to understand how these principles can be employed in subsequent chapters to solve many different problems.

This basic theory is then used to formulate the direct BEM for the analysis of thin plates. The benefits of having previously described the fundamentals of the method in a clear manner then become evident. Once the basic integral equations are derived, the author then demonstrates how they can be applied to write a computer programme, which results are validated through a series of comparisons.

The beauty of the approach followed by the author is that it describes how the mathematical process gives rise to equations which can be reduced to computational form for solving realistic engineering problems.

The above introductory chapters, important as they are, pale into insignificance in comparison with the rest of the book, where a series of most novel concepts are described. The author starts by describing the analysis of plates under membrane and bending forces, which leads to the equations governing buckling, large deflections and post buckling of plates. Important as these cases are, the most significant aspect is that they are solved using an original methodology based on the author's Analog Equation Method, which leads to the full analysis of a wide range of plate problems [1,2].

Few developments in Boundary Elements have been as significant as this idea of Professor Katsikadelis' and, as with all truly original ideas, it is striking in its simplicity and elegance.

To understand the AEM properly we have to refer to the basic idea behind the principle of virtual work as defined by Aristotle who stated that the behaviour of physical systems could be expressed in terms of "potentiality" and "actuality". In other words, Aristotle set up the principle of virtual "potentialities" or what we now call the principle of virtual work. While an "actual" field function is to satisfy the equations giving the problem, a "virtual" function can be more general. Usually we assume that the virtual function satisfies the same equation as those governing the actual field, or sometimes a reduced version of those equations as in the case of the Dual Reciprocity Method [3]. Professor Katsidadelis instead gave a much wider interpretation to the virtual functions – one that would have pleased Aristotle – by stating that they do not necessarily need to satisfy the same type of governing equations of the actual problem, provided that they have the necessary degree of continuity (in the case of plate bending fourth order for instance).

The resulting Method (AEM) when combined with the use of the localised particular solutions proposed by the Dual Reciprocity Method, opens up a huge range of possibilities to Boundary Elements, some of which are presented in this book.

The part dealing with the time and non-linear analysis of plates for instance leads to a series of original formulations based on the AEM. The possibility of solving problems with membrane as well as bending forces can now be fully exploited for cases like dynamic buckling, including flutter instability and a series of applications of fundamental importance in aerospace engineering for instance. Extensions to the case of membranes, non linear materials and large deformations follow effortlessly.

Throughout the book the reader will find a clarity of exposition and consistency which allows the progression from simple to more complex problems in a stepwise fashion. This results in obtaining a full comprehension of the basic principles and how they are applied to obtain practical solutions in a way that is frequently missing in the current engineering sciences literature.

The fact that this book centres on the concept of the AEM developed by the author does not imply any restrictions as the AEM can be interpreted to be the most general version of the principle of virtual work developments ever presented in science and engineering.

Professor Katsikadelis' Method effortlessly transforms a series of complex problems into alternative problems which can be solved in BEM form using simple fundamental solutions.

An added advantage of the AEM is that it allows for the solution of a given set of problems, in the case of plates for instance, using the same type of computer programme. This generality will allow boundary elements to become more widely used for plates and shells, types of problems for which the method still lags behind the less accurate but more versatile finite element method.

Those interested in knowing more about the many contributions of Professor Katsikadelis to the solution of a wide variety of engineering problems and the development of many different ideas, ought to refer to my own appraisal of his work in reference [4]. It was while compiling that paper that I came to fully appreciate his work, including his many contributions to the analysis of plates.

The contents of the present book represent without doubt, a major development in engineering sciences.

Carlos A. Brebbia

REFERENCES

[1] J.T. Katsikadelis, The analog equation method – a powerful BEM-based solution technique for solving linear and non-linear engineering problems, in: C.A. Brebbia (Ed.), Boundary Element Methods XVI, Computational Mechanics Publications, Southampton and Boston, 1994, pp. 167–182.

[2] Katsikadelis, J.T., Nerantzaki, M.S., The boundary element method for nonlinear problems. Eng. Anal. Bound. Elem. 23 (5), 365–373.

[3] D. Nardini, C.A. Brebbia, New approach to free vibration analysis using boundary elements, in: C.A. Brebbia (Ed.), Boundary Element Methods in Engineering, Springer Verlag, Berlin and Computational Mechnics Publications, Southampton and Boston, 1982, pp. 312–326.

[4] C.A. Brebbia, In praise of John Katsikadelis, in: E.J. Sapountzakis (Ed.), Recent Developments in Boundary Element Methods, WIT Press, Southampton and Boston, 2010, pp. 1–16.

Preface

This book presents the Boundary Element Method, BEM, for the static and dynamic analysis of plates and membranes. It is actually a continuation of the book *Boundary Elements: Theory and Applications* by the same author and published by Elsevier in 2002. The latter was well received as a textbook by the relevant international scientific community, which is ascertained by the fact that it was translated into three languages, Japanese by the late Prof. Masa Tanaka of the Shinshu University, Nagano (Asakura, Tokyo 2004), in Russian by the late Prof. Sergey Aleynikov of the Voronezh State Architecture and Civil Engineering University (Publishing House of Russian Civil Engineering Universities, Moscow 2007), and in Serbian by Prof. Dragan Spasic of the University of Novi Sad (Gradjevinska Knjiga, Belgrade 2011).

The success of the first book on the BEM encouraged me to prepare this second book on the BEM for plate analysis. Though there is extensive literature on BEM for plates published in journals, there hasn't been any book published on this subject to date, either as a monograph or as a textbook. To my knowledge, there are only two edited books with contributions of various authors on different plate problems. These books are addressed to researchers and are not suitable for introducing students or even scientists to the subject. Some books on BEM contain the application of the method to plates as a concise chapter aiming, rather, on the completeness of their book, than the presentation of material necessary to understand the subject.

The main reasons for not writing a book on plates at an earlier time include the following:

1. The basic plate problem, i.e., the problem for thin Kirchhoff plates, is described by the biharmonic differential operator whose treatment with the BEM requires special care, both in deriving the boundary integral equations and in obtaining their numerical solution. Thus, a comprehensive presentation of the material to the student is a tedious task and demands a great effort from the author.
2. Different plate problems (e.g., plates on elastic foundation, plates under simultaneous membrane loads, anisotropic plates, etc.) are described by different fourth-order partial differential equations (PDEs) that require the establishment of the fundamental solution, in general not possible, and, thus, different formulations for the derivation of the boundary integral equations and special numerical treatment is needed to obtain results.
3. The difficulties in applying the conventional BEM become insurmountable when plates with variable thickness and dynamic or nonlinear plate problems must be treated.

The above reasons have discouraged potential authors from writing a book on plates. Many have envisioned it as a digest of BEM formulations for plate problems rather than as an efficient computational method for practical plate analysis and design.

During the last 20 years, intensive research has been carried out in an effort to overcome the above shortcomings, especially to alleviate the BEM from establishing a fundamental solution for each plate problem. Several techniques have been developed to cope with the problem. The DRM (Dual Reciprocity Method) has enabled the BEM to efficiently solve static and dynamic engineering problems. Although this method is quite general, it produces boundary-only solutions for those cases where a linear operator with a well-known fundamental solution could be extracted from the full governing equation. However, this is not always possible. The AEM introduced in 1994 overcomes all restrictions of the DRM and enables the BEM to efficiently solve any problem. It is based on the concept (principle) of the analog equation according to which a problem governed by a linear or nonlinear differential equation of any type (elliptic, parabolic, or hyperbolic) can be converted into a substitute problem described by an equivalent linear equation of the same order as the original equation having a simple known fundamental solution and subjected to a fictitious source, unknown in the first instance. The value of this source can be established using the BEM. By applying this idea, coupled linear or nonlinear equations can be converted into uncoupled linear ones. This method is employed to solve all plate problems discussed in the present book.

As any plate problem is described by a single fourth-order PDE or coupled with two second-order PDEs in the presence of membrane forces, the classical plate equation and two Poisson's equations serve as substitute equations. Both types of equations have simple known fundamental solutions and can be readily solved by the conventional direct BEM. A major advantage of the AEM is that the computer program for the classical plate problem can be used to solve any particular plate problem. The research of the author has highly contributed to this end. Most of the material presented in this book can be found in the journal articles written by the author and his colleagues. The AEM renders the BEM an efficient computational method for practical plate analysis.

The material in this book is presented systematically and in detail so the reader can follow without difficulty. A chapter on preliminary mathematical knowledge makes the book self-contained. A special feature of the book is that it connects theoretical treatment and numerical analysis. The comprehensibility of the material has been tested with the author's students for several years. Therefore, it can be used as a textbook. The book contains five chapters:

Chapter 1 gives a brief, elementary description of the basic mathematical tools that will be employed throughout the book in developing the BEM, such as Green's reciprocal identity and Dirac's delta function. This chapter concludes with a section on calculus of variations, which provides the reader with an efficient mathematical tool to derive the governing differential equation together with the associated boundary conditions in complicated structural systems from stationary principles of mechanics. Comprehension of these mathematical concepts helps readers feel confident in their subsequent application.

Chapter 2 presents the direct BEM for the static analysis of thin plates under bending. First, the essential elements of the Kirchhoff plate are discussed. Then, the BEM is formulated in terms of the transverse displacement of the middle surface. The integral representation of the solution and the boundary integral equations are derived in clear, comprehensible steps. Emphasis is on the numerical implementation of the method. A computer program is developed for the complete analysis of plates of arbitrary shape and arbitrary boundary conditions. The program is explained thoroughly and its structure is developed systematically, so the reader can be acquainted with the logic of writing the BEM code in the computer language of preference. The method is illustrated by analyzing several plates.

Chapter 3 presents the BEM for the analysis of more complex plate problems appearing in engineering practice. First, there is a discussion of plate bending under the combined action of membrane forces, which applies to buckling of plates. Then, it follows the analysis of plates resting on any type of elastic foundation, and the large deflections of plates and their postbuckling response. Plates with variable thickness are discussed with application to plate-thickness optimization for maximization of plate stiffness or buckling load. Thick plates are also studied in this chapter, which concludes with the treatment of thin and thick anisotropic plates. As all problems in this chapter are solved by the AEM, its application and numerical implementation are described in detail. Several example problems are solved to demonstrate the efficiency of the solution procedure.

Chapter 4 develops the BEM for linear and nonlinear dynamic analysis of plates, such as free and forced vibrations with or without membrane forces, buckling of plates using the dynamic criterion, and flutter instability of plates under nonconservative loads. Both isotropic and anisotropic plates are analyzed. Plates under aerodynamic loads such as the wings of aircrafts are also discussed. The chapter ends with the application of the BEM to static and dynamic analysis of viscoelastic plates described with differential models of integer and fractional order.

Chapter 5 presents the BEM for the static and dynamic analysis of flat elastic and viscoelastic membranes undergoing large deflections. First, the nonlinear PDEs governing the response of the membrane are derived in terms of the three displacements together with the associated boundary conditions. The resulting boundary and initial boundary value problems are solved by the BEM in conjunction with the principle of the analog equation. Several membranes, elastic and viscoelastic, of various shapes under static and dynamic loads are analyzed.

The book also includes three appendices. **Appendix A** gives useful formulas for the differentiation of the kernel functions and the expressions of tangential derivatives necessary for the treatment of boundary quantities on curvilinear boundaries. **Appendix B** presents the Gauss integration for the numerical evaluation of line and domain integrals. Finally, **Appendix C** describes the time integration method employed for the solution of linear and nonlinear equations of motion.

In closing, the author wishes to express his sincere thanks to his former student and colleague Dr. A.J. Yiotis for carefully reading the manuscript, his suggestions for constructive amendments and for his overall contribution to minimizing the oversights of the text. Warm thanks, also, to Dr. Nikos G. Babouskos, former student and colleague of the author, not only for the careful reading of the manuscript and his suggestions for the improvement of the book, but also for his assistance in developing the computer programs and in producing the numerical results for the examples, most of which are contained in joint publications with the author of the book.

J.T. Katsikadelis
Athens
December 2013

Chapter | one

Preliminary Mathematical Knowledge

CHAPTER OUTLINE

1.1 INTRODUCTION

In this chapter, some mathematical relations required for the development and understanding of the boundary element method (BEM) are presented. Although these relations could have been included in an appendix, they are placed here to show the reader their important role in the theoretical foundation and development of the BEM. They will be used often throughout the book and particularly for the transformation of the differential equations, which govern the response of physical systems within a domain, into integral equations on the boundary. The chapter ends with a section on calculus on variations, which provides the

The Boundary Element Method for Plate Analysis

reader with an efficient mathematical tool to derive the governing differential equation, and especially the associated boundary conditions, in complicated structural systems from stationary principles of mechanics. An understanding of these mathematical concepts will give the reader a feeling of confidence for their subsequent use.

1.2 GAUSS-GREEN THEOREM

The Gauss-Green theorem is a fundamental identity that relates the integral of the derivative of a function over a domain Ω to the integral of that function on its boundary Γ. The domain may be two- or three-dimensional. For simplicity of presentation, this relationship is derived for the two-dimensional case. Consider the plane domain Ω bounded by the curve Γ. We shall work first with the derivative of a function $f = f(x, y)$ with respect to x. The integral over Ω may be written as a double integral, for which the integration is carried out first with respect to x and then with respect to y. Thus, we can write

$$\int_\Omega \frac{\partial f}{\partial x} d\Omega = \int_{y_1}^{y_2} \left(\int_{x_1}^{x_2} \frac{\partial f}{\partial x} dx \right) dy = \int_{y_1}^{y_2} \{f(x_2, y) - f(x_1, y)\} dy \tag{1.1}$$

where

$$x_1 = x_1(y) \quad \text{and} \quad x_2 = x_2(y) \tag{1.2}$$

From the detail of Fig. 1.1, we have

$$\frac{dy}{ds} = \cos\alpha = n_x \quad \Rightarrow \quad dy = n_x ds \tag{1.3a}$$

$$-\frac{dx}{ds} = \sin\alpha = n_y \quad \Rightarrow \quad dx = -n_y ds \tag{1.3b}$$

where n_x and n_y are the components of the unit vector **n**, which is normal to the boundary Γ. The negative sign in Eq. (1.3b) is due to the fact that the dx and the

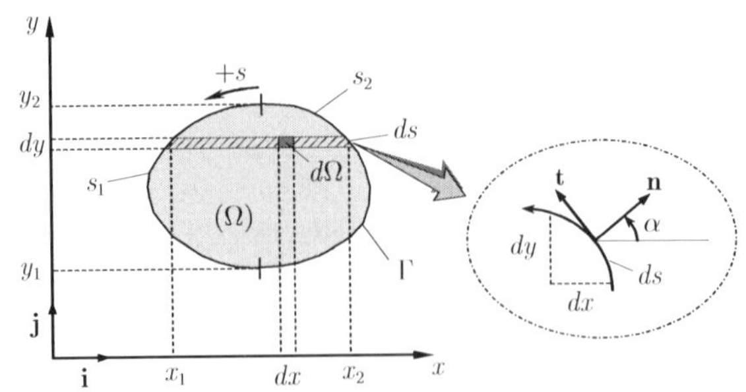

FIGURE 1.1 Integration over a plane domain Ω bounded by a curve Γ.

$\sin\alpha$ have opposite signs when the angle α is measured in the counter-clockwise sense with respect to the positive x–direction (see detail in Fig. 1.1).

Consequently, Eq. (1.1) becomes

$$\int_{y_1}^{y_2} \{f(x_2,y)-f(x_1,y)\}\,dy=\int_{s_2} f(x_2,y)n_x\,ds-\int_{s_1} f(x_1,y)n_x\,ds \tag{1.4}$$

In the previous expression the integration on s_1 is performed in the negative direction (clockwise) when y varies from y_1 to y_2. Using uniform direction for the integration over s, both terms in Eq. (1.4) can be combined in a single expression

$$\int_\Omega \frac{\partial f}{\partial x}\,d\Omega=\int_\Gamma f\,n_x\,ds \tag{1.5}$$

Interchanging x with y in Eq. (1.5), we obtain

$$\int_\Omega \frac{\partial f}{\partial y}\,d\Omega=\int_\Gamma f\,n_y\,ds \tag{1.6}$$

If g is another function of x and y, then Eqs. (1.5) and (1.6) result in

$$\int_\Omega \frac{\partial (fg)}{\partial x}\,d\Omega=\int_\Gamma fg\,n_x\,ds=\int_\Omega g\frac{\partial f}{\partial x}\,d\Omega+\int_\Omega f\frac{\partial g}{\partial x}\,d\Omega \quad\Rightarrow$$

$$\int_\Omega g\frac{\partial f}{\partial x}\,d\Omega=-\int_\Omega f\frac{\partial g}{\partial x}\,d\Omega+\int_\Gamma fg\,n_x\,ds \tag{1.7}$$

$$\int_\Omega \frac{\partial (fg)}{\partial y}\,d\Omega=\int_\Gamma fg\,n_y\,ds=\int_\Omega g\frac{\partial f}{\partial y}\,d\Omega+\int_\Omega f\frac{\partial g}{\partial y}\,d\Omega \quad\Rightarrow$$

$$\int_\Omega g\frac{\partial f}{\partial y}\,d\Omega=-\int_\Omega f\frac{\partial g}{\partial y}\,d\Omega+\int_\Gamma fg\,n_y\,ds \tag{1.8}$$

Equations (1.7) and (1.8) state the integration by parts in two dimensions and are known as the *Gauss-Green theorem.*

1.3 DIVERGENCE THEOREM OF GAUSS

The divergence theorem results readily as an application of the Gauss-Green theorem. Consider the vector field $\mathbf{u}=u\mathbf{i}+v\mathbf{j}$, where $\mathbf{i},\mathbf{j}$ denote the unit vectors along the x and y axes and $u=u(x,y)$, $v=v(x,y)$ its components. Applying Eq. (1.5) for $f=u$ and Eq. (1.6) for $f=v$ and adding, yields

$$\int_\Omega \left(\frac{\partial u}{\partial x}+\frac{\partial v}{\partial y}\right)d\Omega=\int_\Gamma \left(un_x+vn_y\right)ds \tag{1.9}$$

If the coordinates x and y are represented by x_1 and x_2, respectively, then the components of the vector field $\mathbf{u}$ are denoted by $u_i(i=1,2)$ and those of the normal vector $\mathbf{n}$ by n_i. Therefore, Eq. (1.9) can be written as

$$\int_\Omega \left(\frac{\partial u_1}{\partial x_1} + \frac{\partial u_2}{\partial x_2} \right) d\Omega = \int_\Gamma (u_1 n_1 + u_2 n_2)\, ds \tag{1.10}$$

or using the summation convention

$$\int_\Omega \frac{\partial u_i}{\partial x_i} d\Omega = \int_\Gamma u_i n_i \, ds \quad (i=1,2) \tag{1.11}$$

Equations (1.9), (1.10), and (1.11) can also be written using vector notation as

$$\int_\Omega \nabla \cdot \mathbf{u} \; d\Omega = \int_\Gamma \mathbf{u} \cdot \mathbf{n} \; ds \tag{1.12}$$

in which the symbolic vector ∇ is defined as

$$\nabla \equiv \mathbf{i}\frac{\partial}{\partial x} + \mathbf{j}\frac{\partial}{\partial y} = \mathbf{i}_1 \frac{\partial}{\partial x_1} + \mathbf{i}_2 \frac{\partial}{\partial x_2} \tag{1.13}$$

and represents the differential operator that produces the gradient of a scalar field.

The quantity $\nabla \cdot \mathbf{u}$, i.e. the dot product of the vectors ∇ and $\mathbf{u}$, is referred to as the *divergence* of a vector field $\mathbf{u}$ at a point inside the domain Ω, whereas the quantity $\mathbf{u} \cdot \mathbf{n}$ is referred to as the *flux* of the vector field at a point on the boundary Γ. The latter dot product expresses the projection of $\mathbf{u}$ in the direction of $\mathbf{n}$. Eq. (1.12) relates the total divergence to the total flux of a vector field and it is known as the *divergence theorem of Gauss*. It is one of the most important theorems of integral calculus.

1.4 GREEN'S SECOND IDENTITY

Consider the functions $u = u(x, y)$ and $v = v(x, y)$, which are twice continuously differentiable in Ω and once on Γ. Applying Eq. (1.7) for $g = v$, $f = \dfrac{\partial u}{\partial x}$ and Eq. (1.8) for $g = v$, $f = \dfrac{\partial u}{\partial y}$ and adding the resulting equations, we arrive at

$$\begin{aligned} \int_\Omega v \left(\frac{\partial^2 u}{\partial x^2} + \frac{\partial^2 u}{\partial y^2} \right) d\Omega = & -\int_\Omega \left(\frac{\partial u}{\partial x}\frac{\partial v}{\partial x} + \frac{\partial u}{\partial y}\frac{\partial v}{\partial y} \right) d\Omega \\ & + \int_\Gamma v \left(\frac{\partial u}{\partial x} n_x + \frac{\partial u}{\partial y} n_y \right) ds \end{aligned} \tag{1.14}$$

Similarly, applying Eq. (1.7) for $g = u$, $f = \dfrac{\partial v}{\partial x}$ and Eq. (1.8) for $g = u$, $f = \dfrac{\partial v}{\partial y}$ and adding the resulting equations, we obtain

$$\int_\Omega u\left(\frac{\partial^2 v}{\partial x^2}+\frac{\partial^2 v}{\partial y^2}\right)d\Omega = -\int_\Omega\left(\frac{\partial u}{\partial x}\frac{\partial v}{\partial x}+\frac{\partial u}{\partial y}\frac{\partial v}{\partial y}\right)d\Omega + \int_\Gamma u\left(\frac{\partial v}{\partial x}n_x+\frac{\partial v}{\partial y}n_y\right)ds \tag{1.15}$$

Subtracting Eq. (1.15) from Eq. (1.14) yields

$$\int_\Omega (v\nabla^2 u - u\nabla^2 v)\,d\Omega = \int_\Gamma\left(v\frac{\partial u}{\partial n}-u\frac{\partial v}{\partial n}\right)ds \tag{1.16}$$

where ∇^2 is known as the *Laplace operator* or *harmonic operator* and it is defined as

$$\nabla^2 \equiv \nabla\cdot\nabla = \left(\mathbf{i}\frac{\partial}{\partial x}+\mathbf{j}\frac{\partial}{\partial y}\right)\cdot\left(\mathbf{i}\frac{\partial}{\partial x}+\mathbf{j}\frac{\partial}{\partial y}\right) = \frac{\partial^2}{\partial x^2}+\frac{\partial^2}{\partial y^2} \tag{1.17}$$

while

$$\frac{\partial}{\partial n} \equiv \mathbf{n}\cdot\nabla = (n_x\mathbf{i}+n_y\mathbf{j})\cdot\left(\mathbf{i}\frac{\partial}{\partial x}+\mathbf{j}\frac{\partial}{\partial y}\right) = n_x\frac{\partial}{\partial x}+n_y\frac{\partial}{\partial y} \tag{1.18}$$

is the operator that produces the derivative of a scalar function in the direction of $\mathbf{n}$. Equation (1.16) is known as *Green's second identity* for the harmonic operator or *Green's reciprocal identity*.

1.5 ADJOINT OPERATOR

Consider the complete second-order differential equation with variable coefficients

$$L(u) = A\frac{\partial^2 u}{\partial x^2}+2B\frac{\partial^2 u}{\partial x\partial y}+C\frac{\partial^2 u}{\partial y^2}+D\frac{\partial u}{\partial x}+E\frac{\partial u}{\partial y}+Fu=0 \tag{1.19}$$

where $A, B, \ldots, F$ are given functions of x and y in Ω. Multiplying Eq. (1.19) by a function $v = v(x, y)$ and integrating over the domain, we have

$$\int_\Omega vL(u)\;d\Omega = 0 \tag{1.20}$$

Let us assume that v is twice continuously differentiable in Ω and once on Γ. Subsequently, integrating Eq. (1.20) by parts repeatedly until all derivatives of u are eliminated and incorporating Eqs. (1.7) and (1.8), we arrive at

$$\int_\Omega \{vL(u) - uL^*(v)\}\;d\Omega = \int_\Gamma (Xn_x + Yn_y)\;ds \tag{1.21}$$

where

$$L^*(v) = \frac{\partial^2(Av)}{\partial x^2} + 2\frac{\partial^2(Bv)}{\partial x \partial y} + \frac{\partial^2(Cv)}{\partial y^2} - \frac{\partial(Dv)}{\partial x} - \frac{\partial(Ev)}{\partial y} + Fv \tag{1.22}$$

$$X = A\left(v\frac{\partial u}{\partial x} - u\frac{\partial v}{\partial x}\right) + B\left(v\frac{\partial u}{\partial y} - u\frac{\partial v}{\partial y}\right) + \left(D - \frac{\partial A}{\partial x} - \frac{\partial B}{\partial y}\right)uv \tag{1.23}$$

$$Y = B\left(v\frac{\partial u}{\partial x} - u\frac{\partial v}{\partial x}\right) + C\left(v\frac{\partial u}{\partial y} - u\frac{\partial v}{\partial y}\right) + \left(E - \frac{\partial B}{\partial x} - \frac{\partial C}{\partial y}\right)uv \tag{1.24}$$

The differential operator $L^*(\ )$ defined in Eq. (1.22) is referred to as the *adjoint* operator of $L(\ )$. Equation (1.21) is the general form of the Green's second identity (1.16), which results readily as a special case of Eq. (1.21) for $A = C = 1$ and $B = D = E = 0$. The case $F \neq 0$ does not affect Eq. (1.16) as it is equivalent to adding and subtracting the term Fuv in the integral of the left-hand side of this equation. When A, B, C are constants and $D = E = 0$, Eq. (1.22) becomes

$$L^*(v) = A\frac{\partial^2 v}{\partial x^2} + 2B\frac{\partial^2 v}{\partial x \partial y} + C\frac{\partial^2 v}{\partial y^2} + Fv \tag{1.25}$$

Namely, the operator $L^*(\ )$ is identical to $L(\ )$ and in this case $L(\ )$ is called *self-adjoint*.

The character of the solution of Eq. (1.19), as well as the type of the problem to be solved, depends on the quantity $\Delta = B^2 - AC$. We distinguish three types of equations:

(a) Elliptic type, for $\Delta < 0$
(b) Parabolic type, for $\Delta = 0$
(c) Hyperbolic type, for $\Delta > 0$

1.6 DIRAC DELTA FUNCTION

In problems of solid mechanics, we often come across *concentrated loads*, i.e., actions that are applied to a very small region, theoretically at a point, of the space or of the time.

For example, consider the plane elastic body Δ_1 having constant thickness h and occupying the half-plane $-\infty < x < +\infty$, $y \geq 0$ (Fig. 1.2). A circular disc Δ_2

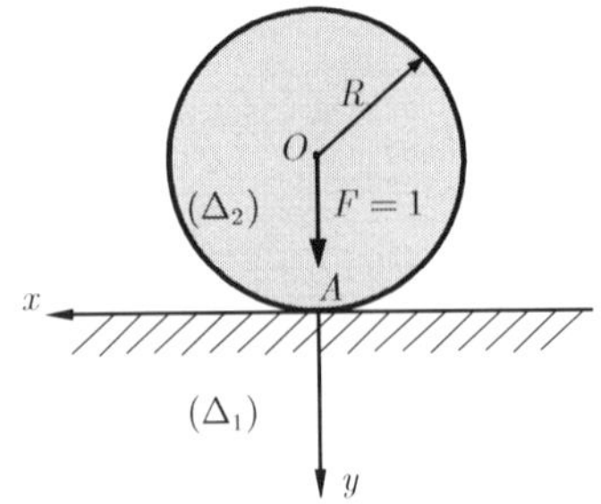

FIGURE 1.2 Circular disc loaded by vertical concentrated unit load F.

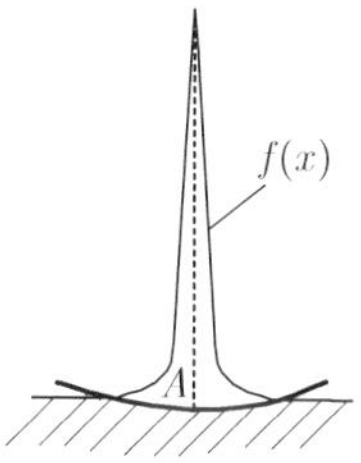

FIGURE 1.3 Force distribution on the boundary of the body.

of the same thickness h and radius R is in contact with the semi-infinite body at point A$(0,0)$ of its free boundary. The circular disc is loaded by a vertical concentrated unit load F as shown in Fig. 1.2. Because of the deformation of the elastic bodies the contact does not take place at a point, but it is extended over a small region. The function $f(x)$ represents the distribution per unit length of the force applied on the boundary of the body Δ_1. This function, which likely has the shape shown in Fig. 1.3, is not *a priori* known. However, we know that it is sufficiently *concentrated* and

$$\int_{-\infty}^{+\infty} f(x)\,dx = 1 \tag{1.26}$$

which expresses that the total force applied on the body Δ_1 is equal to unity. If we skip the problem of determining analytically the function f, we may assume *a priori* a prescribed shape for this function as

$$f_k(x) = \begin{cases} k/2, & |x| < \dfrac{1}{k} \\ 0, & |x| \geq \dfrac{1}{k} \end{cases} \tag{1.27}$$

or

$$f_k(x) = \frac{k}{\pi\left(1 + k^2 x^2\right)} \tag{1.28}$$

where k is a positive number.

Figures 1.4a and 1.4b show that both functions f_k defined in Eqs. (1.27) and (1.28), respectively, become sufficiently *concentrated* for large values of k. Moreover, they satisfy Eq. (1.26), which means that they are statically equivalent to f.

The distribution of $f(x)$ becomes more *"concentrated"* as the bodies become less deformable. This is expressed by increasing the values of k in Eqs. (1.27) and (1.28). The limiting case where the bodies are rigid is expressed by $k \rightarrow \infty$. This produces a fictitious distribution of the unit force per unit length, which is denoted by $\delta(x)$ and is defined as

$$\delta(x) = \lim_{k \rightarrow \infty} f_k(x) \tag{1.29}$$

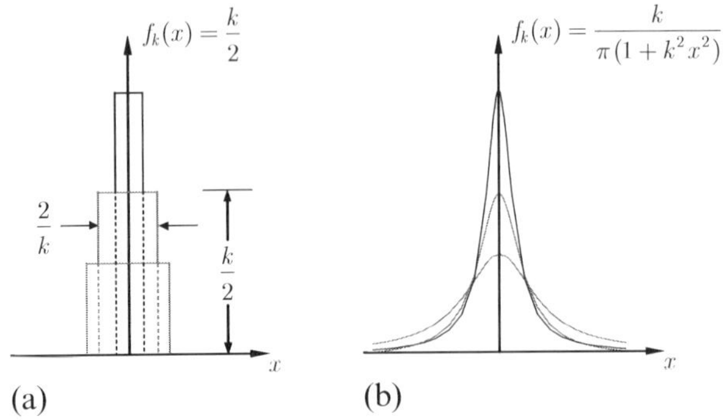

FIGURE 1.4 Functions f_k sufficiently *concentrated* for large values of k.

where $f_k(x)$ is the function defined in Eq. (1.27) or (1.28). The function $\delta(x)$ is known as the *delta function* or *Dirac delta function.*

In mathematics, the delta function is treated in the theory of generalized functions. We give the following definition of the delta function as a generalized function of one and two variables, and we mention some of its properties used in developing the BEM.

The one-dimensional delta function is defined by the relation

$$\int_{-\infty}^{+\infty} \delta(x)\, h(x)\; dx = h(0) \tag{1.30}$$

for a point source applied at the position $x=0$ or by the relation

$$\int_{-\infty}^{+\infty} \delta(x - x_o)\, h(x)\; dx = h(x_o) \tag{1.31}$$

for a point source applied at $x = x_o$. The function $h(x)$ is continuous in a finite interval containing the source point $x=0$ or $x = x_o$. The one-dimensional delta function can also be described by the relations

$$\delta(x) = \begin{cases} 0, & x \neq 0 \\ \infty, & x = 0 \end{cases} \tag{1.32a}$$

$$\int_{-\infty}^{+\infty} \delta(x)\, dx = \int_{-\varepsilon}^{\varepsilon} \delta(x)\, dx = 1 \tag{1.32b}$$

where ε is a positive number. According to this definition, the function $\delta(x)$ has zero value everywhere except at point $x=0$, where it becomes infinite and satisfies Eq. (1.32b).

Equation (1.30) may be obtained from Eq. (1.32b) by applying the mean value theorem of integral calculus. Referring to Fig. 1.4a and choosing $\varepsilon = 1/k$, we obtain

$$\int_{-\infty}^{\infty} h(x)\delta(x)\, dx = \lim_{\varepsilon\to 0}\int_{-\varepsilon}^{\varepsilon} h(x) f_k(x)\, dx = \lim_{\varepsilon\to 0}\left[h(x^*)\frac{1}{2\varepsilon}\right] 2\varepsilon$$
$$= \lim_{\varepsilon\to 0}[h(x^*)] = h(0)$$

The Dirac delta function $\delta(Q-Q_o)$ in two dimensions is defined as

$$\int_{\Omega} \delta(Q-Q_o)h(Q)\, d\Omega_Q = h(Q_o), \quad Q(x,y),\ Q_o(x_o,y_o)\in\Omega \tag{1.33}$$

for an arbitrary function $h(Q)$, which is continuous in the domain Ω containing the point $Q_o(x_o, y_o)$. The two-dimensional delta function may also be described by

$$\delta(Q-Q_o) = \begin{cases} 0, & Q\neq Q_o \\ \infty, & Q = Q_o \end{cases} \tag{1.34a}$$

$$\int_{\Omega} \delta(Q-Q_o)\, d\Omega_Q = \int_{\Omega^*} \delta(Q-Q_o)\, d\Omega_Q = 1, \qquad Q_o(x_o,y_o)\in\Omega^*\subseteq\Omega \tag{1.34b}$$

In accordance to the one-dimensional delta function $\delta(x-x_o)$, the two-dimensional $\delta(Q-Q_o)$ may be defined as the limit of a set of functions. More specifically,

$$\delta(Q-Q_o) = \lim_{k\to\infty} f_k(r), \quad Q(x,y),\ Q_o(x_o,y_o)\in\Omega \tag{1.35}$$

where

$$r = \sqrt{(x-x_o)^2 + (y-x_o)^2}$$

For example, the limit of the function

$$f_k(r) = \begin{cases} k^2/\pi, & r < \dfrac{1}{k} \\ 0, & r \geq \dfrac{1}{k} \end{cases} \tag{1.36a}$$

or

$$f_k(r) = \frac{ke^{-kr^2}}{\pi} \tag{1.36b}$$

for $k\to\infty$ is a two-dimensional delta function. We can also write $\delta(Q-Q_o)$ as the product of two one-dimensional delta functions as

$$\delta(Q-Q_o) = \delta(x-x_o)\ \delta(y-y_o) \tag{1.37}$$

Consider now the transformation

$$x = u(\xi, \eta) \tag{1.38a}$$

$$y = v(\xi, \eta) \tag{1.38b}$$

where u and v are assumed to be single-valued continuous differentiable functions of their arguments. The transformation of the function $\delta(Q - Q_o)$ from the Cartesian coordinates x and y to the curvilinear coordinates ξ and η can be performed in the following way.

It is supposed that under this transformation the point $x = x_o$, $y = y_o$ is mapped onto the point $\xi = \xi_o$, $\eta = \eta_o$. By changing coordinates, the integral of Eq. (1.33)

$$\int_\Omega \delta(x - x_o)\, \delta(y - y_o)\, h(x, y)\, dx\, dy = h(x_o, y_o)$$

becomes

$$\int_\Omega \delta[u(\xi, \eta) - x_o]\, \delta[v(\xi, \eta) - y_o]\, h(u, v)\, |J|\, d\xi\, d\eta = h(x_o, y_o) \tag{1.39}$$

where

$$J = \begin{vmatrix} \dfrac{\partial u}{\partial \xi} & \dfrac{\partial v}{\partial \xi} \\ \dfrac{\partial u}{\partial \eta} & \dfrac{\partial v}{\partial \eta} \end{vmatrix}$$

is the Jacobian of the transformation.

Equation (1.39) states that the symbolic function

$$\delta[u(\xi, \eta) - x_o]\, \delta[v(\xi, \eta) - y_o]\, |J|$$

assigns the value of the function $h(x, y)$ at the point where $u = x_o$, $v = y_o$, i.e., at the point where $\xi = \xi_o, \eta = \eta_o$. Consequently, we may write

$$\delta[u(\xi, \eta) - x_o]\, \delta[v(\xi, \eta) - y_o]\, |J| = \delta(\xi - \xi_o)\, \delta(\eta - \eta_o)$$

which may also be set in the form

$$\delta(x - x_o)\, \delta(y - y_o) = \frac{\delta(\xi - \xi_o)\, \delta(\eta - \eta_o)}{|J|} \tag{1.40}$$

provided that $|J| \neq 0$, a condition which implies that the transformation (1.38) is non-singular (invertible).

We close this section by stating a property of the derivatives of the delta function.

(i) For the mth order derivative of the one-dimensional delta function the following relation is valid:

$$\int_a^b h(x)\frac{d^m\delta(x-x_o)}{dx^m}dx=(-1)^m\frac{d^m h(x_o)}{dx^m}, \quad (a<x_o<b) \tag{1.41}$$

(ii) For the $(m+n)^{th}$ order derivative of the two-dimensional delta function, the following relation is valid:

$$\int_\Omega h(Q)\,\frac{\partial^{m+n}\delta(Q-Q_o)}{\partial x^m\partial y^n}\,d\Omega_Q=(-1)^{m+n}\,\frac{\partial^{m+n}h(Q_o)}{\partial x^m\partial y^n} \tag{1.42}$$

where $Q_o(x_o,y_o)$, $Q(x,y)\in\Omega$.

1.7 CALCULUS OF VARIATIONS; EULER-LAGRANGE EQUATION

In mechanics, one of the applications of the calculus of variations is the derivation of the differential equation and the associated boundary conditions governing the response of complicated structural systems, especially when other approaches give doubtful results. In this context, we present the basics of the calculus of variations. The calculus of variations is a field of mathematical analysis that deals with maximizing or minimizing functionals.

Functionals are mappings from a set of functions to the real numbers. They are often expressed as definite integrals, whose integrand depends on an unknown function or functions and/or certain derivatives. The basic problem of the calculus of variations is to determine the *extremal functions*, i.e., the functions that make the functional attain a stationary value (minimum or maximum) – in other words, those functions where the rate of change of the functional is zero. The elementary part of the theory is concerned with the necessary condition the required function must satisfy. This condition is generally in the form of a differential equation, *the Euler-Lagrange equation*, with boundary conditions, whose solution gives the required function.

A simple example of such a problem is to find the curve $u(x)$ that passes through the points $A(x_1,u_1)$ and $B(x_2,u_2)$ (Fig. 1.5) and produces a surface of revolution about the x axis with a minimum area. Thus, the function $u(x)$ is determined in such a way that the integral representing the area of the surface

$$I=2\pi\int_{x_1}^{x_2} u\left(1+u'^2\right)^{1/2}dx \tag{1.43}$$

is a minimum, and also that $u(x_1)=u_1$ and $u(x_2)=u_2$. Here it is assumed that u_1 and u_2 are nonnegative. We consider all continuous functions with continuous

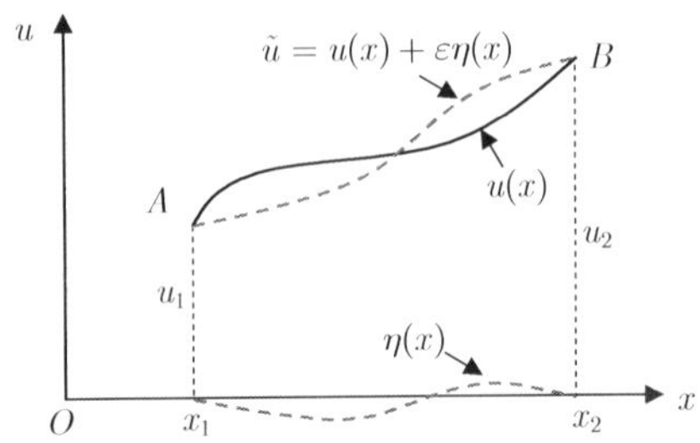

FIGURE 1.5 Function $u(x)$ and a variant $\widetilde{u}(x)$.

derivatives that satisfy the *geometric or essential boundary conditions*, i.e., they pass through the points A and B. These functions constitute the set of *admissible functions*, where we will look for the minimizing function $u(x)$ of the integral (1.43). We suppose that $u(x)$ is the actual minimizing function, and we choose any continuously differentiable function $\eta(x)$ with $\eta(x_1) = \eta(x_2) = 0$ (Fig. 1.5). Then for any constant ε the function $\widetilde{u}(x) = u(x) + \varepsilon\eta(x)$ is an admissible function. The change $\varepsilon\eta(x)$ is called the variation of $u(x)$ and is conventionally denoted by δu,

$$\delta u = \varepsilon\eta(x) \tag{1.44}$$

1.7.1 Euler-Lagrange equation

We consider the integral

$$I(u) = \int_{x_1}^{x_2} F(x, u, u')\, dx \tag{1.45}$$

where we suppose that integrand $F = F(x, u, u')$ has continuous second order derivatives with respect to its three arguments. The integral

$$I(\varepsilon) = \int_{x_1}^{x_2} \widetilde{F}(x, u + \varepsilon\eta, u' + \varepsilon\eta')\, dx \tag{1.46}$$

obtained by replacing $u(x)$ with $u(x) + \varepsilon\eta(x)$ is a function of ε, once the functions $u(x)$ and $\eta(x)$ are specified, and takes on a minimum value when $\varepsilon = 0$. This is possible only if

$$\left.\frac{dI(\varepsilon)}{d\varepsilon}\right|_{\varepsilon=0} = 0 \tag{1.47}$$

Differentiating Eq. (1.46) with respect to ε and noting that the derivative passes under the integral sign we obtain

$$\frac{dI(\varepsilon)}{d\varepsilon} = \int_{x_1}^{x_2} \left(\frac{\partial \widetilde{F}}{\partial u}\eta + \frac{\partial \widetilde{F}}{\partial u'}\eta' \right) dx \tag{1.48}$$

which after integration by parts of the second term becomes

$$\frac{dI(\varepsilon)}{d\varepsilon} = \int_{x_1}^{x_2} \left(\frac{\partial \widetilde{F}}{\partial u} - \frac{d}{dx}\frac{\partial \widetilde{F}}{\partial u'} \right) \eta \, dx + \left[\frac{\partial \widetilde{F}}{\partial u'} \eta(x) \right]_{x_1}^{x_2} \tag{1.49}$$

Noting that $\eta(x_1) = \eta(x_2) = 0$ and $\widetilde{F} \to F$ when $\varepsilon \to 0$, the necessary condition (1.47) becomes

$$\int_{x_1}^{x_2} \left(\frac{\partial F}{\partial u} - \frac{d}{dx}\frac{\partial F}{\partial u'} \right) \eta \, dx = 0 \tag{1.50}$$

Since $\eta(x)$ is arbitrary we conclude that its coefficient in (1.50) must vanish in the interval (x_1, x_2)

$$\frac{\partial F}{\partial u} - \frac{d}{dx}\frac{\partial F}{\partial u'} = 0 \tag{1.51}$$

Thus, if $u(x)$ minimizes (or maximizes) the integral (1.45), it must satisfy the condition (1.51), which is called the *Euler-Lagrange equation* of the functional (1.45). When F is specified, the condition (1.51) yields the differential equation, whose solution in the interval (x_1, x_2) gives the extremal function.

The change in F corresponding to δu is

$$\Delta F = F(x, u + \delta u, u' + \delta u') - F(x, u, u') \tag{1.52}$$

Expanding the right-hand side in Taylor series yields

$$\Delta F = \frac{\partial F}{\partial u}\delta u + \frac{\partial F}{\partial u'}\delta u' + (\text{higher order terms of } \delta u \text{ and } \delta u') \tag{1.53}$$

In analogy with the differential the first two terms in the right-hand side of (1.53) are defined as the variation of F,

$$\delta F = \frac{\partial F}{\partial u}\delta u + \frac{\partial F}{\partial u'}\delta u' \tag{1.54}$$

The operator δ acts as the differential operator d, which for a function $F(x, y)$ of two independent variables reads

$$dF = \frac{\partial F}{\partial x} dx + \frac{\partial F}{\partial y} dy \tag{1.55}$$

There is, however, a significant difference between the two operators: The differential of a function represents a first-order approximation to the change in the function along a particular curve, while the variation of the functional is the first-order approximation to the change in the functional from curve to curve. Note that the complete analogy to the differential would be

$$\delta F = \frac{\partial F}{\partial x}\delta x + \frac{\partial F}{\partial u}\delta u + \frac{\partial F}{\partial u'}\delta u' \tag{1.56}$$

but as x is not varied, it is $\delta x = 0$ and the analogy is complete.

It can be easily verified that the laws of variation of sums, product, ratios, powers, and so forth are completely analogous to differentiation, e.g.,

$$\delta(FG) = G\delta F + F\delta G \tag{1.57a}$$

$$\delta F^n = nF^{n-1}\delta F \tag{1.57b}$$

$$\delta\left(\frac{F}{G}\right) = \frac{G\delta F - F\delta G}{G^2} \tag{1.57c}$$

Moreover, we have

$$\frac{d}{dx}(\delta u) = \varepsilon\frac{d\eta}{dx} = \delta\frac{du}{dx} \tag{1.58}$$

which shows that, if x is an independent variable, the operators d/dx and δ are commutative.

Using this notation the variation of the integral (1.45) is written as

$$\begin{aligned}\delta I &= \int_{x_1}^{x_2} \delta F(x, u, u')\,dx \\ &= \int_{x_1}^{x_2} \left(\frac{\partial F}{\partial u}\delta u + \frac{\partial F}{\partial u'}\delta u'\right) dx\end{aligned} \tag{1.59}$$

From Eq. (1.58) we have $\delta u' = (\delta u)'$. Then integrating by parts the second term in the integrand yields

$$\delta I = \int_{x_1}^{x_2} \left(\frac{\partial F}{\partial u} - \frac{d}{dx}\frac{\partial F}{\partial u'}\right)\delta u\,dx + \left[\frac{\partial F}{\partial u'}\delta u(x)\right]_{x_1}^{x_2} \tag{1.60}$$

or taking into account that $\delta u(x_1) = \delta u(x_2) = 0$ we have

$$\delta I = \int_{x_1}^{x_2} \left(\frac{\partial F}{\partial u} - \frac{d}{dx}\frac{\partial F}{\partial u'}\right)\delta u\,dx \tag{1.61}$$

By virtue of (1.51) we conclude that $\delta I = 0$. Hence, the necessary condition that a functional attains a minimum (or maximum) is the vanishing of its variation.

1.7.2 Natural boundary conditions

When the value of the unknown function $u(x)$ is not prescribed at one or both end points $x=x_1, x_2$, the variation $\delta u(x)$ need not vanish there. However, the right-hand side of Eq. (1.60) must vanish when $u(x)$ is the minimizing (or maximizing) function for all admissible variations $\delta u(x)$. Hence, the second term in Eq. (1.60) must itself vanish:

$$\left[\frac{\partial F}{\partial u'}\delta u(x)\right]_{x_1}^{x_2} = \left[\frac{\partial F}{\partial u'}\delta u(x)\right]_{x=x_2} - \left[\frac{\partial F}{\partial u'}\delta u(x)\right]_{x=x_1} = 0 \tag{1.62}$$

If $u(x)$ is not prescribed at either end point it is $\delta u(x_1) \neq 0$ or $\delta u(x_2) \neq 0$ and consequently Eq. (1.62) holds if

$$\left[\frac{\partial F}{\partial u'}\right]_{x=x_1} = 0 \quad \text{or} \quad \left[\frac{\partial F}{\partial u'}\right]_{x=x_2} = 0 \tag{1.63a,b}$$

The requirements that Eq. (1.63a) hold when $u(x_1)$ is not given, and that (1.63b) hold when $u(x_2)$ is not given are called the *natural boundary conditions*.

On the basis of the above the function $u(x)$ that minimizes (or maximizes) the functional (1.45) results from the solution of the following boundary value problem:

$$\frac{\partial F}{\partial u} - \frac{d}{dx}\frac{\partial F}{\partial u'} = 0 \quad \text{in} \quad x_1 \le x \le x_2 \tag{1.64}$$

$$u(x_1) = u_1 \quad \text{or} \quad \left[\frac{\partial F}{\partial u'}\right]_{x=x_1} = 0 \quad \text{and} \quad u(x_2) = u_2 \quad \text{or} \quad \left[\frac{\partial F}{\partial u'}\right]_{x=x_2} = 0 \tag{1.65a,b}$$

1.7.3 Functional depending on a function of two variables

We consider the functional

$$I = \int_\Omega F(x, y, u, u_{,x}, u_{,y})\, dxdy \tag{1.66}$$

where $u = u(x, y)$, a continuously differentiable function of x and y to be determined so that the functional attains a stationary value, and Ω is a two-dimensional domain in the xy plane with boundary Γ (Fig. 1.6). The procedure to derive the necessary conditions is more complicated here. However, it becomes simple if the required condition is obtained from the requirement that the variation $\delta I = 0$. This yields

$$\begin{aligned}\delta I &= \int_\Omega \delta F(x, y, u, u_{,x}, u_{,y})\, dxdy \\ &= \int_\Omega \left(\frac{\partial F}{\partial u}\delta u + \frac{\partial F}{\partial u_{,x}}\delta u_{,x} + \frac{\partial F}{\partial u_{,y}}\delta u_{,y}\right) dxdy\end{aligned} \tag{1.67}$$

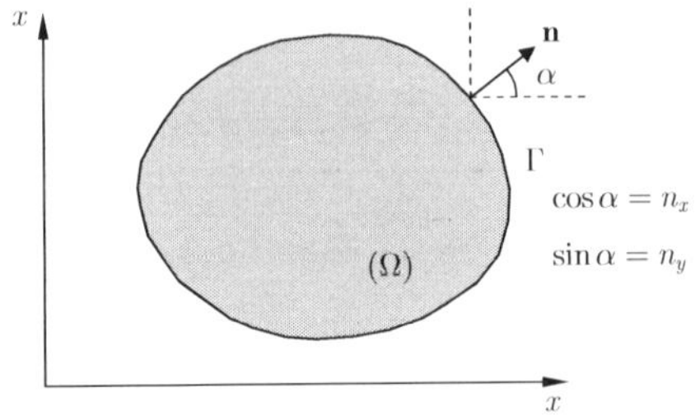

FIGURE 1.6 Domain Ω bounded by a curve Γ.

We next get rid of the $\delta u_{,x}$ and $\delta u_{,y}$ from the two terms in the integrand. This can be achieved using integration by parts. Thus applying the Gauss-Green theorem, Eq. (1.7) and Eq. (1.8), we obtain

$$\begin{aligned}\int_\Omega \frac{\partial F}{\partial u_{,x}}\delta u_{,x}dxdy &= \int_\Omega \frac{\partial F}{\partial u_{,x}}(\delta u)_{,x}dxdy \\ &= -\int_\Omega \frac{\partial}{\partial x}\frac{\partial F}{\partial u_{,x}}\delta u dxdy + \int_\Gamma \frac{\partial F}{\partial u_{,x}}n_x\delta u ds\end{aligned} \tag{1.68a}$$

$$\begin{aligned}\int_\Omega \frac{\partial F}{\partial u_{,y}}\delta u_{,y}dxdy &= \int_\Omega \frac{\partial F}{\partial u_{,y}}(\delta u)_{,y}dxdy \\ &= -\int_\Omega \frac{\partial}{\partial y}\frac{\partial F}{\partial u_{,y}}\delta u dxdy + \int_\Gamma \frac{\partial F}{\partial u_{,y}}n_y\delta u ds\end{aligned} \tag{1.68b}$$

Introducing Eqs. (1.68a,b) in Eq. (1.67) gives

$$\begin{aligned}\delta I &= \int_\Omega \left(\frac{\partial F}{\partial u} - \frac{\partial}{\partial x}\frac{\partial F}{\partial u_{,x}} - \frac{\partial}{\partial y}\frac{\partial F}{\partial u_{,y}}\right)\delta u dxdy \\ &+ \int_\Gamma \left(\frac{\partial F}{\partial u_{,x}}n_x + \frac{\partial F}{\partial u_{,y}}n_y\right)\delta u ds\end{aligned} \tag{1.69}$$

If the function $u(x, y)$ is prescribed on Γ, then $\delta u = 0$ there and the boundary integral in Eq. (1.69) vanishes. Then, since δu is arbitrary in Ω its coefficient in the integrand must vanish. This yields the condition

$$\frac{\partial F}{\partial u} - \frac{\partial}{\partial x}\frac{\partial F}{\partial u_{,x}} - \frac{\partial}{\partial y}\frac{\partial F}{\partial u_{,y}} = 0 \tag{1.70}$$

which is the Euler-Lagrange equation of the functional.

When $u(x, y)$ is not prescribed on the boundary, then $\delta u \neq 0$ on Γ and the boundary integral vanishes only if the natural boundary condition is satisfied:

$$\frac{\partial F}{\partial u_{,x}}n_x + \frac{\partial F}{\partial u_{,y}}n_y = 0 \text{ on } \Gamma \tag{1.71}$$

Thus, the boundary condition on a part $\Gamma_1 \subseteq \Gamma$ reads

$$u = \overline{u} \quad \text{or} \quad \frac{\partial F}{\partial u_{,x}} n_x + \frac{\partial F}{\partial u_{,y}} n_y = 0 \tag{1.72}$$

The overbar designates a prescribed quantity.

1.7.4 Examples

EXAMPLE 1.1

Derive the differential equation for the function that passes through the points $A(x_1, u_1)$ and $B(x_2, u_2)$, (Fig. 1.5), and produces a surface of revolution about the x axis with a minimum area. The required function must minimize the functional (1.43). Here it is:

$$F = u\left(1 + u'^2\right)^{1/2} \tag{a}$$

Hence,

$$\frac{\partial F}{\partial u} = \sqrt{1 + u'^2}, \quad \frac{\partial F}{\partial u'} = \frac{uu'}{\sqrt{1 + u'^2}} \tag{b, c}$$

Inserting (b, c) in Eq. (1.51) gives

$$\sqrt{1 + u'^2} - \frac{d}{dx} \frac{uu'}{\sqrt{1 + u'^2}} = 0 \tag{d}$$

which after reduction gives the differential equation

$$uu'' - u'^2 - 1 = 0 \ \text{ in } \ x_1 < x < x_2 \tag{e}$$

with the boundary conditions

$$u(x_1) = u_1, \quad u(x_2) = u_2 \tag{f}$$

EXAMPLE 1.2

Derive the differential equation of a surface that passes through a space curve and has a minimum area. This is the well-known minimal surface problem or soap bubble problem. In this case, the required minimizing function $u(x, y)$ has prescribed values on the boundary and minimizes the functional

$$I = \int_\Omega \left(1 + u_{,x}^2 + u_{,y}^2\right)^{1/2} dxdy \tag{a}$$

Hence, the functional is of the form (1.66) and the differential equation will result from Eq. (1.70) with

$$F = 1 + u_{,x}^2 + u_{,y}^2 \tag{b}$$

for which we have

$$\frac{\partial F}{\partial u} = 0, \quad \frac{\partial F}{\partial u_{,x}} = \frac{u_{,x}}{\sqrt{1 + u_{,x}^2 + u_{,y}^2}}, \quad \frac{\partial F}{\partial u_{,y}} = \frac{u_{,y}}{\sqrt{1 + u_{,x}^2 + u_{,y}^2}} \tag{c,d,e}$$

Inserting (c, d, e) in Eq. (1.70) gives

$$\frac{\partial}{\partial x} \frac{u_{,x}}{\sqrt{1 + u_{,x}^2 + u_{,y}^2}} + \frac{\partial}{\partial y} \frac{u_{,y}}{\sqrt{1 + u_{,x}^2 + u_{,y}^2}} = 0 \tag{f}$$

and after reduction we obtain the differential equation

$$\left(1 + u_{,y}^2\right) u_{,xx} - u_{,x}\, u_{,y}\, u_{,xy} + \left(1 + u_{,x}^2\right) u_{,yy} = 0 \text{ in } \Omega \tag{g}$$

with the boundary condition

$$u = \overline{u} \text{ on } \Gamma \tag{h}$$

1.8 REFERENCES

This chapter has been designed to give a brief elementary description of the basic mathematical tools that will be employed throughout this book in developing the BEM formulations. The reader may look for more details about the Gauss-Green theorem and the Gauss divergence theorem in most books on calculus as well as in many books on engineering mathematics or mathematical physics, such as Smirnow [1], Hildebrand [2], Kreyszig [3], and Sommerfeld [4]. For a detailed discussion of the Dirac delta function, we refer to the book by Greenberg [5] or to the more advanced books of Roach [6] and Duff and Naylor [7]. For more information on the calculus of variation and the variational methods in mechanics the reader is advised to reference the books of Hildebrand [8] and Washizu [9].

[1] W.I. Smirnow, Lehrgang der höheren Mathematik, Teil II, 6ste Auflage, VEB Deutschder Verlag der Wissenschaften, Berlin, 1964.

[2] F.B. Hildebrand, Advanced Calculus for Applications, Prentice Hall, Englewood Cliffs, New Jersey, 1962.

[3] E. Kreyszig, Advanced Engineering Mathematics, fourth ed., John Wiley & Sons, New York, 1979.

[4] A. Sommerfeld, Partial Differential Equations in Physics, Academic Press, New York and London, 1967.

[5] M. Greenberg, Application of Green's Functions in Science and Engineering, Prentice Hall, Englewood Cliff, New Jersey, 1971.

[6] G.F. Roach, Green's Functions, Van Nostrand Reinhold Company, London, 1970.

[7] G.F.D. Duff, D. Naylor, Differential Equations of Applied Mathematics, John Wiley & Sons, New York, 1966.

[8] F.B. Hildebrand, Methods of Applied Mathematics, Dover Publications, Inc., New York, 1965
[9] K. Washizu, Variational Methods in Elasticity and Plasticity, Pergamon Press, Oxford, 1975.

PROBLEMS

1.1 Convert the domain integral

$$\int_{\Omega} f \, d\Omega$$

to boundary line integral on Γ, when the function f is

(i) $f = x$
(ii) $f = y$
(iii) $f = xy$
(iv) $f = x^2$
(v) $f = y^2$
(vi) $f = x^2 + y^2$
(vii) $f = \cos x$

1.2 Derive the operator ∇^2 in polar coordinates (r, θ) and convert the following domain integral to a boundary line integral on Γ:

$$\int_{\Omega} \ell nr \; d\Omega, \quad r = \sqrt{x^2 + y^2}$$

1.3 Evaluate the integrals

(i) $\int_a^b \delta(x - x_o) \; dx, \;\; a < x_o < b$

(ii) $\int_a^b \delta(kx) f(x) \; dx, \;\; a < 0 < b$

(iii) $\int_a^b \delta(-x) \; dx, \;\; a < 0 < b$

(iv) $\int_a^b \delta^{(n)}(x) \phi(x) \; dx, \;\; a < 0 < b$

1.4 Show that

(i) $\delta(-x) = \delta(x), \;\; a < 0 < b$

(ii) $\delta(ax + by)\, \delta(cx + dy) = \dfrac{\delta(x)\delta(y)}{|ad - bc|}$

1.5 Transform the delta function $\delta(P - P_o)$ for the points $P(x, y)$ and $P_o(x_o, y_o)$ into polar coordinates r and θ.

1.6 Derive Eq. (1.21) along with Eqs. (1.22) through (1.24).

1.7 Show that $\int_{\Omega} \nabla^2 u \, d\Omega = \int_{\Gamma} \dfrac{\partial u}{\partial n} ds.$

1.8 Given $L(u) = \nabla^2 u + \mathbf{a} \cdot \nabla u + cu$, where $\mathbf{a} = a_x \mathbf{i} + a_y \mathbf{j}$ is an arbitrary vector, show that

(i) $L^*(u) = \nabla^2 u - \nabla \cdot (\mathbf{a}u) + cu$,

(ii) $vL(u) - uL^*(v) = \nabla \cdot (v\nabla u - u\nabla v + \mathbf{a}uv)$, and

(iii) derive Green's identity for the operator $L(u)$.

1.9 The total potential energy of a string supported at its ends and subjected to a uniform tension S and a transverse load $f(x)$ is given as

$$\Pi = \int_0^l \left(\frac{S}{2} u'^2 - fu \right) dx$$

Derive the deferential equation for the transverse displacement $u(x)$ by applying the principle of minimum potential energy.

1.10 The total potential energy of a membrane supported on its boundary and subjected to a uniform tension S and a transverse load $f(x)$ is given as

$$\Pi = \int_0^l \left[\frac{S}{2} \left(u_{,x}^2 + u_{,y}^2 \right) - fu \right] dx$$

Derive the deferential equation for the transverse displacement $u(x, y)$ by applying the principle of minimum potential energy.

Chapter | two

BEM for Plate Bending Analysis

CHAPTER OUTLINE

The Boundary Element Method for Plate Analysis

2.1 INTRODUCTION

This chapter presents the direct BEM for the static analysis of thin plates under bending. The problem is formulated in terms of the transverse displacement of the middle surface. The development of the BEM for the plate problem is analogous to that for the plane potential problem (see [1], Chapter 3). There is, however, an essential difference. Here, the governing partial differential equation is of the fourth order. Therefore, the derivation of the boundary integral equations is more complicated and their numerical solution requires special care.

In the development of the BEM for plates several formulations have been proposed, direct and indirect as well as formulations for solving special plate problems. A description of these methods is presented by Katsikadelis [2]. The direct BEM, however, is the most widely used, because it is not restricted to specific plate shapes or boundary conditions and it is formulated in terms of boundary quantities, which have direct physical meaning in the thin plate theory, namely the transverse displacement, the normal slope, the bending moment, and the equivalent shear force. The integral representation of the solution of the plate equation in terms of these boundary quantities is based on the generalized Rayleigh-Green identity and can be found in earlier mathematical books, e.g., Bergmann and Schiffer [3]. Forbes and Robinson [4] were the first to apply the direct BEM. They derived the generalized Rayleigh-Green identity from Betti's reciprocal theorem. But their formulation was good only for plate with smooth boundaries because they ignored the fictitious corner forces. Next, Bezine [5], following the procedure presented by Katsikadelis et al. [6], derived a second boundary integral equation for the normal derivative and applied the direct BEM to polygonal plates taking into account the corner terms without placing points on them. Finally, Stern [7] was the first to give a complete description of the basic integral equations for plates with non-smooth boundaries and evaluate the free term coefficients at corner points.

In the previous papers the numerical results were limited as they were used to test the effectiveness of the derived equations. In a paper published by Hartmann and Zotemantel [8] the authors focused their interest on the numerical implementation of the direct BEM and discussed computational problems arising from the application of the method to problems of engineering praxis. The direct BEM for static analysis of plates has also been used by many other authors. Most of them aimed at reformulating the direct BEM for a better numerical solution of the boundary integral equations or at developing numerical schemes to obtain improved numerical results or to solve other special plate bending problems.

The relevant literature on plate analysis with the BEM is very large, either as published articles or as concise chapters in books [9,10]. Extensive literature is presented in [11] as well as in some review articles [12,13]. This literature, however, is by no means complete. The *Journal of Engineering Analysis with Boundary Elements* and the proceedings of BEM conferences constitute a rich source for the subject. The basic ingredients of the direct BEM for plates are the Rayleigh-Green identity, the fundamental solution of the biharmonic equation, the integral representation of the deflection, the boundary integral equations and their numerical solution. These are the main topics that will be discussed in the following sections. However, the essential elements of the thin plate theory and the pertinent equations, which will be used in the subsequent sections, should be first presented. The presentation is concise but complete. Detailed treatment is given in books on plate theory and analysis [14–18].

2.2 THIN PLATE THEORY

2.2.1 Definition, model, and loading of the plate

A plate is an elastic cylindrical or prismatic body whose thickness h (height) is small compared to the dimensions of its base. The plane that bisects the plate thickness is the middle surface of the plate. The intersection of the middle surface of the plate with the lateral surface of the prismatic body defines the boundary of the plate. Thus, the boundary of the plate may be a closed plane curve, a polygon, or a mixed line composed of rectilinear or curvilinear segments (Fig. 2.1).

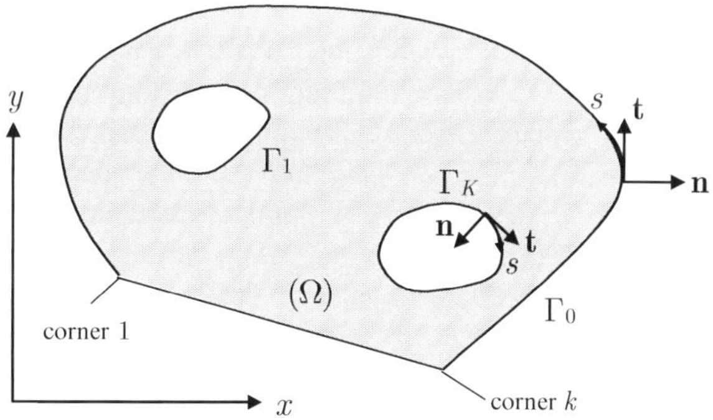

FIGURE 2.1 Multiply connected domain Ω occupied by the plate in the xy plane bounded by the boundary $\Gamma = \cup_{k=0}^{K} \Gamma_k$.

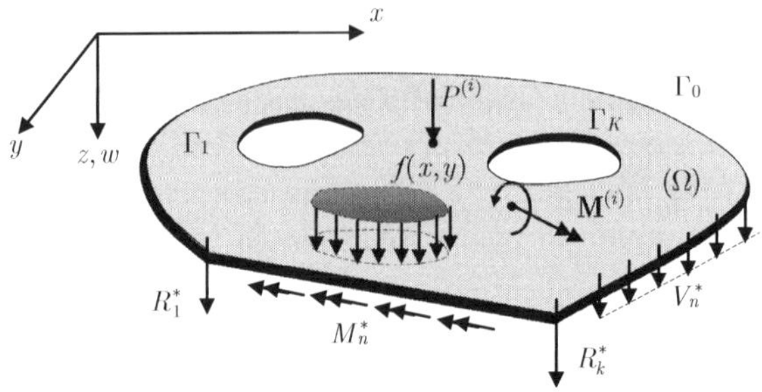

FIGURE 2.2 Loads acting on the plate.

The plate is defined by its middle surface, thickness, and boundary. In the plate analysis, the middle surface coincides with the xy plane. The plate may have holes, i.e., the domain it occupies may be multiply connected. The arc length variable (Fig. 2.1) is measured counter-clockwise along the outer boundary, whereas clockwise along the inner boundaries.

From the structural point of view a plate is a straight flat surface structure whose thickness is slight compared to the other dimensions and which is supported at the edge in various ways (clamped, simply supported, guided, free (not supported), or elastically supported edge) and is loaded by transverse loads, i.e., loads normal to the middle surface (Fig. 2.2). We may distinguish the following loads:

Loads acting in the interior of the plate:

(a) Load densities $f(x, y)$ acting in the direction of the z axis, distributed on the whole middle surface or on a part of it.

(b) Concentrated loads $P^{(i)}$ at a point (x_i, y_i).

(c) Concentrated moments $\boldsymbol{M}^{(i)}\left(M_x^{(i)} M_y^{(i)}\right)$ at a point (x_i, y_i).

Loads acting on the boundary of the plate:

(a) Distributed transverse forces $V_n^*(s)$ and distributed bending moments $M_n^*(s)$.

(b) Concentrated forces R_k^* at corner points k.

2.2.2 Thin plates

With regard to their thickness we distinguish between thin and thick plates. The theory of thin plates is based on the following assumptions:

1. The material of the plate is linearly elastic, homogeneous, and isotropic.
2. The plate thickness is small compared to the other dimensions.
3. The deformations are such that straight lines, initially normal to the middle surface, remain straight, inextensional ($\varepsilon_z = 0$), and normal to the deformed middle surface. This is a consequence of neglecting the transverse shear deformation $\left(\gamma_{xz}, \gamma_{yz} = 0\right)$.
4. The stress normal to the middle surface is negligibly small ($\sigma_z = 0$).
5. The strains are small compared to the unity.
6. The transverse displacement $w(x, y)$ is small compared to the plate thickness (usually $w_{\max}/h < 0.1$).

The last assumption is a consequence of neglecting the squares and the products of the slopes of the middle surface in the strain-displacement relations and it is valid only in the linear theory of thin plates. This implies that the membrane strain and the membrane forces produced by bending are negligibly small.

The above assumptions are attributed to Kirchhoff and Love and the linear plate theory is known as the Kirchhoff or Kirchhoff-Love theory of plates. As the thickness of the plate increases the shear deformation in the transverse direction cannot be disregarded and, consequently, assumption (3) is no more valid. Thick plates are described by other equations derived on the basis of various theories [19–21] (see also Section 3.7).

2.2.3 Strain-displacement and stress-strain relations for thin plates

The strain-displacement relations for the plate can be derived by considering the length of a line element before and after deformation. Thus, if $d\mathbf{r}(dx, dy, 0)$ is a line element, which prior to deformation is parallel to the xy plane, and $d\mathbf{r}'(dx', dy', dz')$ is the same element after deformation, we have

$$dx' = dx + d\bar{u} = \left[1 + \frac{\partial \bar{u}}{\partial x}\right] dx + \frac{\partial \bar{u}}{\partial y} dy \tag{2.1a}$$

$$dy' = dy + d\bar{v} = \frac{\partial \bar{v}}{\partial x} dx + \left[1 + \frac{\partial \bar{v}}{\partial y}\right] dy \tag{2.1b}$$

$$dz' = 0 + d\bar{w} = \frac{\partial \bar{w}}{\partial x} dx + \frac{\partial \bar{w}}{\partial y} dy \tag{2.1c}$$

where

$$\bar{u} = \bar{u}(x, y, z) \tag{2.2a}$$

$$\bar{v} = \bar{v}(x, y, z) \tag{2.2b}$$

$$\bar{w} = \bar{w}(x, y) \tag{2.2c}$$

are the components of the displacement vector at a point (x, y, z) through the thickness of the plate. Note that the transverse displacement $\bar{w}$ is independent of z due to assumption 3 for thin plates.

The symbols $\bar{u}$, $\bar{v}$, $\bar{w}$,... provided with an overbar denote quantities at any point through the plate thickness as distinguished from u, v, w,... which refer to the corresponding quantities at the middle plane ($z = 0$). Hence,

$$\bar{u}(x, y, 0) = u(x, y) \tag{2.3a}$$

$$\bar{v}(x, y, 0) = v(x, y) \tag{2.3b}$$

$$\bar{w}(x, y) = w(x, y) \tag{2.3c}$$

The length of the deformed element is

$$ds' = |d\mathbf{r}'| = \sqrt{(1+2\bar{\varepsilon}_x)dx^2 + (1+2\bar{\varepsilon}_y)dy^2 + 2\bar{\gamma}_{xy}dxdy} \tag{2.4}$$

where $\bar{\varepsilon}_x$, $\bar{\varepsilon}_y$, $\bar{\gamma}_{xy}$ are the extensional and shearing strain components at any point through the plate thickness and are given as

$$\bar{\varepsilon}_x = \frac{\partial \bar{u}}{\partial x} + \frac{1}{2}\left[\left(\frac{\partial \bar{u}}{\partial x}\right)^2 + \left(\frac{\partial \bar{v}}{\partial x}\right)^2 + \left(\frac{\partial w}{\partial x}\right)^2\right] \tag{2.5a}$$

$$\bar{\varepsilon}_y = \frac{\partial \bar{v}}{\partial y} + \frac{1}{2}\left[\left(\frac{\partial \bar{u}}{\partial y}\right)^2 + \left(\frac{\partial \bar{v}}{\partial y}\right)^2 + \left(\frac{\partial w}{\partial y}\right)^2\right] \tag{2.5b}$$

$$\bar{\gamma}_{xy} = \left(\frac{\partial \bar{u}}{\partial y} + \frac{\partial \bar{v}}{\partial x}\right) + \left[\frac{\partial \bar{u}}{\partial x}\frac{\partial \bar{u}}{\partial y} + \frac{\partial \bar{v}}{\partial x}\frac{\partial \bar{v}}{\partial y} + \frac{\partial w}{\partial x}\frac{\partial w}{\partial y}\right] \tag{2.5c}$$

The intermediate class of deformations is defined by the limitations that the strains are small compared with the unity and the rotations about the x and y axes are moderately large, while the rotations about the z axis are negligibly small [22,23]. This implies that the first two terms in the square brackets of Eqs. (2.5) are substantially smaller than the last term in the same bracket and it enables us to write

$$\bar{\varepsilon}_x = \frac{\partial \bar{u}}{\partial x} + \frac{1}{2}\left(\frac{\partial w}{\partial x}\right)^2 \tag{2.6a}$$

$$\bar{\varepsilon}_y = \frac{\partial \bar{v}}{\partial y} + \frac{1}{2}\left(\frac{\partial w}{\partial y}\right)^2 \tag{2.6b}$$

$$\bar{\gamma}_{xy} = \frac{\partial \bar{u}}{\partial y} + \frac{\partial \bar{v}}{\partial x} + \frac{\partial w}{\partial x}\frac{\partial w}{\partial y} \tag{2.6c}$$

Equations (2.6) can be further simplified by omitting the squares and the product of the rotations in them. This assumption leads to the following expressions

$$\bar{\varepsilon}_x = \frac{\partial \bar{u}}{\partial x} \tag{2.7a}$$

$$\bar{\varepsilon}_y = \frac{\partial \bar{v}}{\partial y} \tag{2.7b}$$

$$\bar{\gamma}_{xy} = \frac{\partial \bar{u}}{\partial y} + \frac{\partial \bar{v}}{\partial x} \tag{2.7c}$$

Equations (2.6) form the base for the nonlinear theory of thin plates with moderately large deflections known as the von Kármán plate theory, whereas Eqs. (2.7) form the base for the classical linear plate theory.

On the basis of the assumptions for the thin plate theory, the displacement components $\bar{u}$, $\bar{v}$, $\bar{w}$ at any point in the plate may be expressed in terms of the corresponding quantities u, v, w at points of the middle surface. In Fig. (2.3) the section of the plate by the plane $y=$ constant is shown before and after deformation. The point Q on the normal AB, with coordinates x, y, z before deformation, is displaced at the position Q', with coordinates $x+\bar{u}$, $y+\bar{v}$, $z+\bar{w}$, after deformation.

Noting that $\tan\theta \approx \theta \approx \dfrac{\partial w}{\partial x}$ and referring to Fig. (2.3), we obtain

$$\bar{u} = u - z\frac{\partial w}{\partial x} \tag{2.8a}$$

Similarly, by considering a section of the plate by the plane $x=$ constant, we obtain

$$\bar{v} = v - z\frac{\partial w}{\partial y} \tag{2.8b}$$

Introducing Eqs. (2.8a,b) into Eqs. (2.7) gives

$$\bar{\varepsilon}_x = \varepsilon_x - z\frac{\partial^2 w}{\partial x^2} \tag{2.9a}$$

$$\bar{\varepsilon}_y = \varepsilon_y - z\frac{\partial^2 w}{\partial y^2} \tag{2.9b}$$

$$\bar{\gamma}_{xy} = \gamma_{xy} - 2z\frac{\partial^2 w}{\partial x \partial y} \tag{2.9c}$$

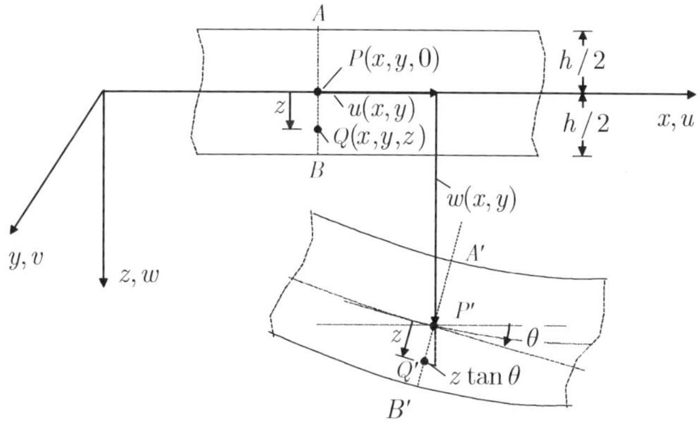

FIGURE 2.3 Section of the plate by the xz plane before and after deformation.

where ε_x, ε_y and γ_{xy} are the components of strain at the middle surface and are given as

$$\varepsilon_x = \frac{\partial u}{\partial x} \tag{2.10a}$$

$$\varepsilon_y = \frac{\partial v}{\partial y} \tag{2.10b}$$

$$\gamma_{xy} = \frac{\partial u}{\partial y} + \frac{\partial v}{\partial x} \tag{2.10c}$$

Equations (2.10) are valid for thin plate theory. In absence of inplane forces it is $u = v = 0$, which implies $\varepsilon_x = \varepsilon_y = \gamma_{xy} = 0$, i.e., there is no stretching of the middle surface.

For small slopes the quantities $-\partial^2 w/\partial x^2, -\partial^2 w/\partial y^2, \partial^2 w/\partial x \partial y$ represent the curvatures along the x and y axes and the twist of the deflected middle surface. Hence, using the notation

$$\kappa_x = -\frac{\partial^2 w}{\partial x^2} \tag{2.11a}$$

$$\kappa_y = -\frac{\partial^2 w}{\partial y^2} \tag{2.11b}$$

$$\kappa_{xy} = \frac{\partial^2 w}{\partial x \partial y} \tag{2.11c}$$

and taking into account that there is no stretching of the middle surface, Eqs. (2.9) are written as

$$\bar{\varepsilon}_x = z k_x \tag{2.12a}$$

$$\bar{\varepsilon}_y = z k_y \tag{2.12b}$$

$$\bar{\gamma}_{xy} = -2 z k_{xy} \tag{2.12c}$$

The strain-stress relations for thin plates are derived from the generalized Hook's law for a three-dimensional isotropic medium with $\bar{\sigma}_z = 0$. Thus, we have

$$\bar{\varepsilon}_x = \frac{1}{E}(\bar{\sigma}_x - \nu \bar{\sigma}_y) \tag{2.13a}$$

$$\bar{\varepsilon}_y = \frac{1}{E}(\bar{\sigma}_y - \nu \bar{\sigma}_x) \tag{2.13b}$$

$$\bar{\gamma}_{xy} = \frac{2(1+\nu)}{E} \bar{\tau}_{xy} \tag{2.13c}$$

$\bar{\sigma}_x$, $\bar{\sigma}_y$, $\bar{\tau}_{xy}$ are the stress components at any point through the thickness with E being the modulus of elasticity and ν Poisson's ratio.

Equations (2.13) are inverted to yield

$$\bar{\sigma}_x = \frac{E}{1-\nu^2}\left(\bar{\varepsilon}_x + \nu\bar{\varepsilon}_y\right) \tag{2.14a}$$

$$\bar{\sigma}_y = \frac{E}{1-\nu^2}\left(\bar{\varepsilon}_y + \nu\bar{\varepsilon}_x\right) \tag{2.14b}$$

$$\bar{\tau}_{xy} = G\bar{\gamma}_{xy} \tag{2.14c}$$

where $G = E/2(1+\nu)$ is the shear modulus.

2.2.4 Stress resultants and constitutive equations for thin plates

The stress resultants on a cross-section are expressed by forces and moments per unit length along it. Their positive directions are shown in Fig. (2.4). The stress resultants are related to the internal stresses by the equations

$$M_x = \int_{-h/2}^{h/2} \bar{\sigma}_x z dz \tag{2.15a}$$

$$M_y = \int_{-h/2}^{h/2} \bar{\sigma}_y z dz \tag{2.15b}$$

$$M_{xy} = -\int_{-h/2}^{h/2} \bar{\tau}_{xy} z dz \tag{2.15c}$$

$$M_{yx} = \int_{-h/2}^{h/2} \bar{\tau}_{yx} z dz \tag{2.15d}$$

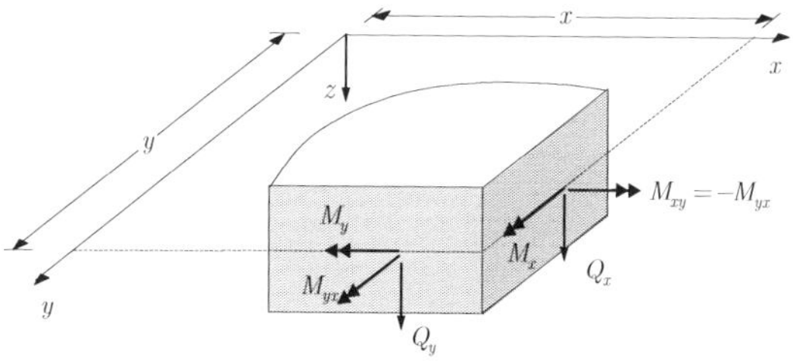

FIGURE 2.4 Stress resultants.

$$Q_x = \int_{-h/2}^{h/2} \overline{\tau}_{xz}\, dz \tag{2.15e}$$

$$Q_y = \int_{-h/2}^{h/2} \overline{\tau}_{yz}\, dz \tag{2.15f}$$

where M_x, M_y are the bending moments, M_{xy}, M_{yx} are the twisting moments, and Q_x, Q_y are the shear forces. Since $\overline{\tau}_{xy} = \overline{\tau}_{yx}$, it follows from Eqs. (2.15c,d) that $M_{xy} = -M_{xy}$.

It must be noted here that the twisting moment on a cross-section is defined as positive if the vector representing it is directed outwards. This explains the appearance of the minus sign in the expression for M_{xy}.

On the base of Eqs. (2.12), Eqs. (2.14) imply that the stresses $\overline{\sigma}_x$, $\overline{\sigma}_y$ and $\overline{\tau}_{xy}$ vary linearly through the thickness of the plate. Their distribution is shown in Fig. (2.5).

The shear stresses $\overline{\tau}_{xz}$ and $\overline{\tau}_{yz}$ are related to the corresponding shear strains by Hook's law:

$$\overline{\gamma}_{xz} = \frac{1}{G}\overline{\tau}_{xz} \tag{2.16a}$$

$$\overline{\gamma}_{yz} = \frac{1}{G}\overline{\tau}_{yz} \tag{2.16b}$$

The equilibrium of a plate element in the z direction requires that the shear stresses $\overline{\tau}_{xz}$ and $\overline{\tau}_{yz}$ do not vanish. This also yields no-vanishing shear strains on the basis of Eqs. (2.16). This, however, is inconsistent with the assumption that $\overline{\gamma}_{xz}$ and $\overline{\gamma}_{yz}$ vanish. Nevertheless, we can match vanishing shear strains with nonvanishing shear stresses by assuming that the thin plate reacts in shear deformation with a fictitious infinitely large shear modulus. The variation of shear stresses along the thickness follows the parabolic law (see Eqs. 2.22d,e).

Introducing Eqs. (2.14) into Eqs. (2.15) and integrating yields the equations

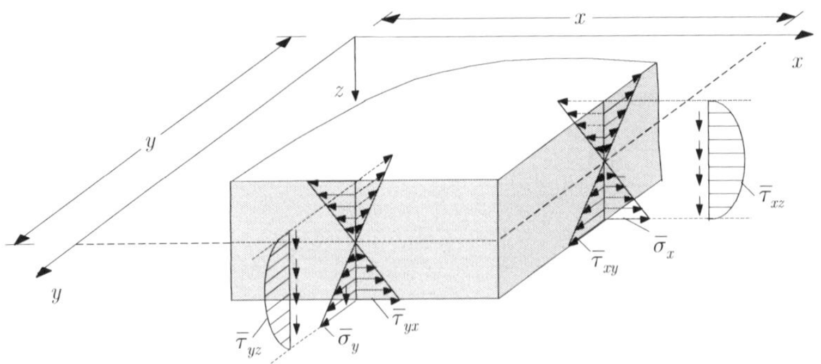

FIGURE 2.5 Distribution of stresses through the plate thickness.

$$M_x = D(\kappa_x + \nu\kappa_y) \tag{2.17a}$$

$$M_y = D(\kappa_y + \nu\kappa_x) \tag{2.17b}$$

$$M_{xy} = -M_{yx} = D(1-\nu)\kappa_{xy} \tag{2.17c}$$

where

$$D = \frac{Eh^3}{12(1-\nu^2)} \tag{2.18}$$

is the bending or flexural stiffness of the plate. Equations (2.17) are the constitutive equations of the plate.

The strain-displacement relations enable us to express the stress resultants in terms of the displacement components. Thus, substituting Eqs. (2.11) in Eqs. (2.17) we obtain

$$M_x = -D\left(\frac{\partial^2 w}{\partial x^2} + \nu\frac{\partial^2 w}{\partial y^2}\right) \tag{2.19a}$$

$$M_y = -D\left(\frac{\partial^2 w}{\partial y^2} + \nu\frac{\partial^2 w}{\partial x^2}\right) \tag{2.19b}$$

$$M_{xy} = -M_{yx} = D(1-\nu)\frac{\partial^2 w}{\partial x \partial y} \tag{2.19c}$$

The corresponding expressions for the shear forces Q_x and Q_y are obtained from the equilibrium of the moments acting on a plate element about the x and y axes. Thus, referring to Fig. 2.6 we obtain

$$Q_x = \frac{\partial M_x}{\partial x} + \frac{\partial M_{yx}}{\partial y} \tag{2.20a}$$

$$Q_y = \frac{\partial M_y}{\partial y} - \frac{\partial M_{xy}}{\partial x} \tag{2.20b}$$

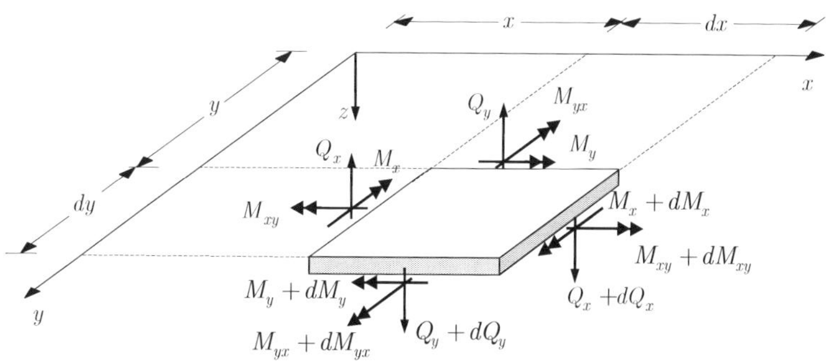

FIGURE 2.6 Plate element with stress resultants.

or using Eqs. (2.19)

$$Q_x = -D\frac{\partial}{\partial x}\nabla^2 w \tag{2.21a}$$

$$Q_y = -D\frac{\partial}{\partial y}\nabla^2 w \tag{2.21b}$$

The stresses through the thickness of the plate are given in terms of the stress resultants by the following expressions:

$$\bar{\sigma}_x = \frac{M_x}{I}z \tag{2.22a}$$

$$\bar{\sigma}_y = \frac{M_y}{I}z \tag{2.22b}$$

$$\bar{\tau}_{xy} = -\frac{M_{xy}}{I}z \tag{2.22c}$$

$$\bar{\tau}_{xz} = \frac{3}{2}\frac{Q_x}{h}\left[1-4\left(\frac{z}{h}\right)^2\right] \tag{2.22d}$$

$$\bar{\tau}_{yz} = \frac{3}{2}\frac{Q_y}{h}\left[1-4\left(\frac{z}{h}\right)^2\right] \tag{2.22e}$$

where $I = h^3/12$ is the moment of inertia of the cross-section of the plate per unit length.

The stress resultants with respect to an orthogonal system of axes n, t, rotated by an angle α with respect to the x axis, can be readily obtained from the respective transformation of stresses. Thus, we take

$$M_n = M_x \cos^2\alpha + M_y \sin^2\alpha - 2M_{xy} \sin\alpha \cos\alpha \tag{2.23a}$$

$$M_t = M_x \sin^2\alpha + M_y \cos^2\alpha + 2M_{xy} \sin\alpha \cos\alpha \tag{2.23b}$$

$$M_{nt} = M_{xy}\left(\cos^2\alpha - \sin^2\alpha\right) + \left(M_x - M_y\right)\sin\alpha \cos\alpha \tag{2.23c}$$

$$Q_n = Q_x \cos\alpha + Q_y \sin\alpha \tag{2.23d}$$

$$Q_t = -Q_x \sin\alpha + Q_y \cos\alpha \tag{2.23e}$$

Introducing Eqs. (2.19) and (2.21) into (2.23) we obtain the stress resultants on the inclined cross-section as

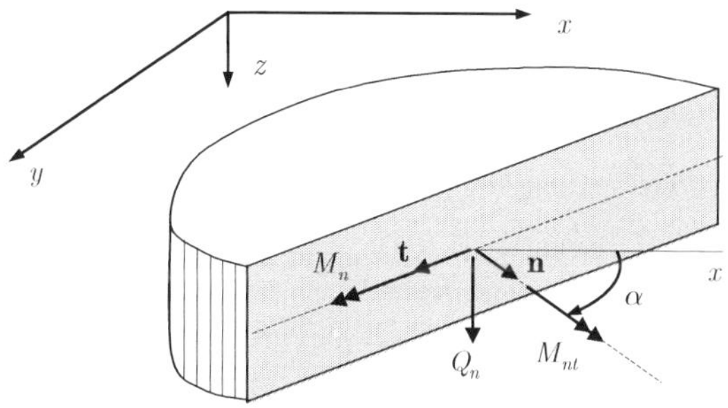

FIGURE 2.7 Stress resultants on an inclined cross-section, $\measuredangle x, \mathbf{n} = \alpha$.

$$M_n = -D\left(\frac{\partial^2 w}{\partial n^2} + \nu\frac{\partial^2 w}{\partial t^2}\right) \tag{2.24a}$$

$$M_t = -D\left(\frac{\partial^2 w}{\partial t^2} + \nu\frac{\partial^2 w}{\partial n^2}\right) \tag{2.24b}$$

$$M_{nt} = D(1-\nu)\frac{\partial^2 w}{\partial n \partial t} \tag{2.24c}$$

$$Q_n = -D\frac{\partial}{\partial n}\nabla^2 w \tag{2.24d}$$

$$Q_t = -D\frac{\partial}{\partial t}\nabla^2 w \tag{2.24e}$$

The sign convention is shown in Fig. 2.7.

2.2.5 Differential equation of the plate

The governing equation of the plate can be readily derived by taking the equilibrium of a plate element in the z-direction. However, in this section the equation is derived by applying the principle of the minimum total potential energy and employing the mathematical tool of the calculus of variations (Chapter 1 and [24,25]). This method generates not only the governing differential equation but also the associated boundary conditions in a straightforward and impeccable manner.

A plate with a boundary of arbitrary shape and having corners will be considered (Fig. 2.1). The plate is subjected to transverse load density $f(x, y)$ and edge load densities M_n^*, V_n^*. Moreover, a concentrated transverse load R_k^* $(k = 1,2,\ldots,K)$ may be applied at a corner k (Fig. 2.2).

The boundary of the plate may be elastically supported in the transverse direction on springs with translational stiffness $k_T(s)$ as well as elastically restrained by springs having rotational stiffness $k_R(s)$. The corner points may be elastically supported in the transverse direction on isolated springs with stiffness $k_c^{(k)}$.

The total potential energy Π of a plate is the sum of the strain energy U_b due to bending, the elastic energy of the yielding support U_s and the potential V of the external applied loads. Thus, we have

$$\Pi = U_b + U_s + V \tag{2.25}$$

Taking into account the assumptions of the thin plate theory, the strain energy due to bending is given by the volume integral

$$U_b = \frac{1}{2}\int_V \left(\bar{\sigma}_x \bar{\varepsilon}_x + \bar{\sigma}_y \bar{\varepsilon}_y + \bar{\tau}_{xy}\bar{\gamma}_{xy}\right) dV \tag{2.26}$$

Using Eqs. (2.14) and (2.12) and integrating through the plate thickness, Eq. (2.26) gives

$$U_b = \frac{D}{2}\int_\Omega \left[\kappa_x^2 + \kappa_y^2 + 2\nu\kappa_x\kappa_y + 2(1-\nu)\kappa_{xy}^2\right] d\Omega \tag{2.27}$$

which by virtue of the strain-displacement relations (2.11) becomes

$$U_b = \frac{D}{2}\int_\Omega \left[w_{,xx}^2 + w_{,yy}^2 + 2\nu w_{,xx}\, w_{,yy} + 2(1-\nu) w_{,xy}^2\right] d\Omega \tag{2.28}$$

For the sake of convenience, the derivatives are denoted here by subscripts preceded by a comma.

The elastic energy due to the yielding of the support is expressed as

$$U_s = \frac{1}{2}\int_\Gamma \left(k_T w^2 + k_R w_{,n}^2\right) ds + \frac{1}{2}\sum_k k_c^{(k)} w_k^2 \tag{2.29}$$

while the potential of the external loads is given as

$$V = -\int_\Omega f w d\Omega - \int_\Gamma \left(V_n^* w - M_n^* w_{,n}\right) ds - \sum_k R_k^* w_k \tag{2.30}$$

Therefore, Eq. (2.25) is written as

$$\begin{aligned} \Pi = &\int_\Omega F\left(w, w_{,xx}, w_{,yy}, w_{,xy}\right) d\Omega \\ &+ \frac{1}{2}\int_\Gamma \left(k_T w^2 + k_R w_{,n}^2\right) ds + \frac{1}{2}\sum_k k_c^{(k)} w_k^2 \\ &- \int_\Gamma \left(V_n^* w - M_n^* w_{,n}\right) ds - \sum_k R_k^* w_k \end{aligned} \tag{2.31}$$

where

$$F = \frac{D}{2}\left[w_{,xx}^2 + w_{,yy}^2 + 2\nu w_{,xx}\, w_{,yy} + 2(1-\nu) w_{,xy}^2\right] - fw$$

The equilibrium of the plate requires that the first variation of the total potential, $\delta\Pi$, vanishes. Thus, using the properties of the operator δ described in Section 1.7 we obtain

$$\begin{aligned}\delta\Pi &= \int_\Omega \left(\frac{\partial F}{\partial w}\delta w + \frac{\partial F}{\partial w_{,xx}}\delta w_{,xx} + \frac{\partial F}{\partial w_{,yy}}\delta w_{,yy} + \frac{\partial F}{\partial w_{,xy}}\delta w_{,xy}\right) d\Omega \\ &+ \int_\Gamma (k_T w\delta w + k_R w_{,n}\, \delta w_{,n}) ds \\ &- \int_\Gamma \left(V_n^* \delta w - M_n^* \delta w_{,n}\right) ds + \sum_k \left[k_c^{(k)} w_k - R_k^*\right]\delta w_k = 0\end{aligned} \tag{2.32}$$

Taking into account that

$$\frac{\partial F}{\partial w} = -f \tag{2.33a}$$

$$\frac{\partial F}{\partial w_{,xx}} = D\left(w_{,xx} + \nu w_{,yy}\right) = -M_x \tag{2.33b}$$

$$\frac{\partial F}{\partial w_{,xy}} = 2D(1-\nu) w_{,xy} = 2M_{xy} \tag{2.33c}$$

$$\frac{\partial F}{\partial w_{,yy}} = D\left(w_{,yy} + \nu w_{,xx}\right) = -M_y \tag{2.33d}$$

Eq. (2.32) can be written as

$$\begin{aligned}\delta\Pi &= \int_\Omega \left(-M_x \delta w_{,xx} + 2M_{xy}\delta w_{,xy} - M_y \delta w_{,yy}\right) d\Omega \\ &+ \int_\Gamma (k_T w\delta w + k_R w_{,n}\, \delta w_{,n}) ds - \int_\Omega f\delta w d\Omega \\ &- \int_\Gamma \left(V_n^* \delta w - M_n^* \delta w_{,n}\right) ds + \sum_k \left[k_c^{(k)} w_k - R_k^*\right]\delta w_k = 0\end{aligned} \tag{2.34}$$

Eq. (2.34) expresses the principle of virtual work for a plate in bending [24]. The domain integrals in the above equation can be integrated twice by parts using the *Gauss-Green theorem* (see Chapter 1) to eliminate the derivatives from the variations $\delta w_{,xx}$, $\delta w_{,xy}$, $\delta w_{,yy}$. Then using Eqs. (2.23d,e) and taking into account that

$$w_{,x} = w_{,n}\cos\alpha - w_{,t}\sin\alpha \tag{2.35a}$$

$$w_{,y} = w_{,n}\sin\alpha + w_{,t}\cos\alpha \tag{2.35b}$$

Eq. (2.34) becomes

$$\begin{aligned}\delta\Pi = &\int_\Omega \left(-M_{x,xx} + 2M_{xy,xy} - M_{y,yy} - f\right)\delta w d\Omega \\ &+ \int_\Gamma \left[\left(-M_n + k_R w_{,n} + M_n^*\right)\delta w_{,n} + M_{nt}\delta w_{,t}\right] ds \\ &+ \int_\Gamma \left(Q_n + k_\Gamma w - V_n^*\right)\delta w ds + \sum_k \left[k_c^{(k)} w_k - R_k^*\right]\delta w_k = 0\end{aligned} \tag{2.36}$$

Moreover, noting that $w_{,t} = w_{,s}$ and M_{nt} is discontinuous at the corner points, the integration by parts of the last term in the first line integral of Eq. (2.36) gives

$$\int_\Gamma M_{nt}\delta w_{,s} ds = -\int_\Gamma M_{nt,s}\delta w ds - \sum_k [\![M_{nt}]\!]_k \delta w \tag{2.37}$$

where $[\![M_{nt}]\!]_k = M_{nt}^{(+)} - M_{nt}^{(-)}$ designates the jump of discontinuity of the twisting moment at the corner point k. Thus, Eq. (2.36) becomes

$$\begin{aligned}\delta\Pi = &\int_\Omega \left(-M_{x,xx} - 2M_{xy,xy} + M_{y,yy} - f\right)\delta w d\Omega \\ &+ \int_\Gamma \left[\left(-M_n + k_R w_{,n} + M_n^*\right)\delta w_{,n} + \left(Q_n - M_{nt,s} + k_\Gamma w - V_n^*\right)\delta w\right] ds \\ &+ \sum_k \left[k_c^{(k)} w_k - [\![M_{nt}]\!]_k - R_k^*\right]\delta w_k = 0\end{aligned} \tag{2.38}$$

Inasmuch as the quantities δw, $\delta w_{,n}$ and δw_k are arbitrary, Eq. (2.38) yields the following Euler-Lagrange differential equation:

$$M_{x,xx} - 2M_{xy,xy} + M_{y,yy} = -f \quad \text{in } \Omega \tag{2.39}$$

and the boundary conditions

$$Q_n - M_{nt,s} + k_T w = V_n^* \quad \text{or} \quad w = w^* \quad \text{on } \Gamma \tag{2.40a}$$

$$M_n - k_R w_{,n} = M_n^* \quad \text{or} \quad w_{,n} = w_{,n}^* \quad \text{on } \Gamma \tag{2.40b}$$

$$k_c^{(k)} w_k - [\![M_{nt}]\!]_k = R_k^* \quad \text{or} \quad w_k = w_k^* \quad \text{at corner } k \tag{2.40c}$$

It should be noted that the star superscript denotes a prescribed quantity. Physically, the quantity $[\![M_{nt}]\!]_k$ represents a fictitious concentrated corner force, as it can be readily deduced from Eq. (2.40c). On the basis of Fig. 2.8, Kirchhoff has given a heuristic interpretation of this fictitious force. Note that the direction of

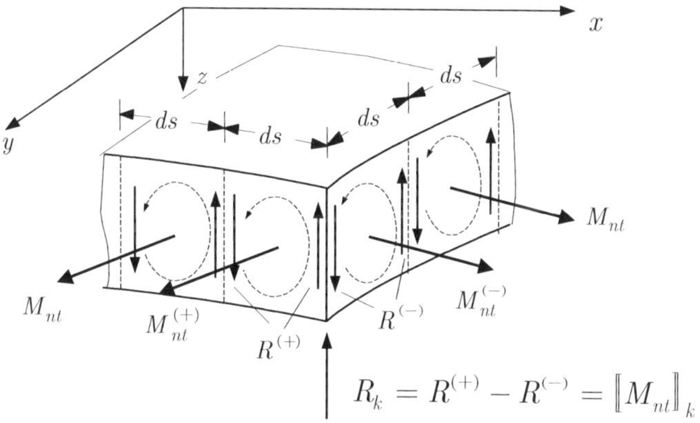

FIGURE 2.8 Kirchhoff's device for interpreting the fictitious corner force.

the positive fictitious force is opposite to the z axes. Concentrated forces may also arise at points where generally the twisting moment along the boundary is discontinuous, e.g., points of change of the boundary conditions.

The quantity

$$V_n = Q_n - M_{nt,s} \tag{2.41}$$

which appears in Eq. (2.40a) is termed the equivalent or the effective shear force on the boundary. It represents the total reaction of the support along the boundary of the plate. It also explains the reason why Q_n and M_{nt} cannot be independently specified in the thin plate theory. This reduces the number of the boundary conditions to two as required from the well-posedness of the boundary value problem for the fourth-order elliptic partial differential equation.

Eq. (2.39) is the plate equation, which by virtue of Eqs. (2.19) takes the form in terms of the displacements

$$D\nabla^4 w = f \quad \text{in}\,\Omega \tag{2.42}$$

Where

$$\nabla^4 = \left(\nabla^2\right)^2 = \frac{\partial^4}{\partial x^4} + 2\frac{\partial^4}{\partial x^2 y^2} + \frac{\partial^4}{\partial y^4} \tag{2.43}$$

represents the biharmonic operator.

2.2.6 Boundary conditions

The boundary conditions (2.40a,b) for the thin plate theory can be written in the form

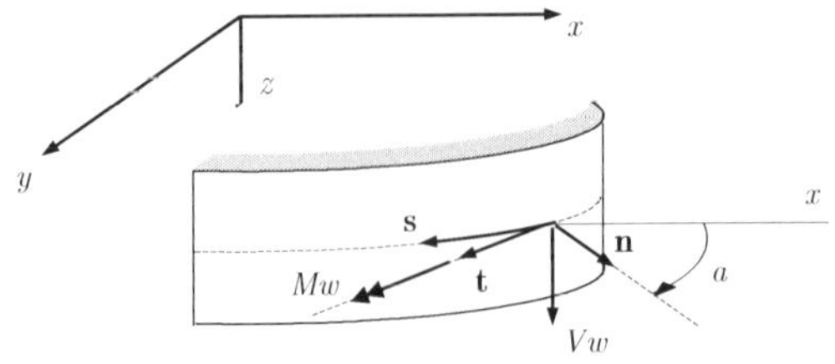

FIGURE 2.9 Boundary forces.

$$\alpha_1 w + \alpha_2 Vw = \alpha_3 \quad \text{on } \Gamma \tag{2.44a}$$

$$\beta_1 \frac{\partial w}{\partial n} + \beta_2 Mw = \beta_3 \quad \text{on } \Gamma \tag{2.44b}$$

where $\alpha_i = \alpha_i(s)$, $\beta_i = \beta_i(s)$ $(i = 1, 2, 3)$ are functions specified on the boundary Γ and M, V are differential operators defined as

$$M = -D\left[\nabla^2 + (\nu - 1)\frac{\partial^2}{\partial t^2}\right] \tag{2.45}$$

$$V = -D\left[\frac{\partial}{\partial n}\nabla^2 - (\nu - 1)\frac{\partial}{\partial s}\left(\frac{\partial^2}{\partial n \partial t}\right)\right] \tag{2.46}$$

The quantities Mw and Vw represent the densities of the bending moment M_n and of the equivalent shear force V_n on the boundary, respectively (Fig. 2.9).

The boundary conditions (2.44) are the most general linear boundary conditions for the plate. The conventional boundary conditions are obtained from them by specifying appropriately the functions α_i and β_i. Thus, an edge is:
(a) Clamped $(w = 0, \partial w/\partial n = 0)$ if

$$\alpha_1 = 1, \alpha_2 = 0, \alpha_3 = 0, \beta_1 = 1, \beta_2 = 0, \beta_3 = 0$$

(b) Simply supported $(w = 0, M_n = 0)$ if

$$\alpha_1 = 1, \alpha_2 = 0, \alpha_3 = 0, \beta_1 = 0, \beta_2 = 1, \beta_3 = 0$$

(c) Free $(M_n = 0, V_n = 0)$ if

$$\alpha_1 = 0, \alpha_2 = 1, \alpha_3 = 0, \beta_1 = 0, \beta_2 = 1, \beta_3 = 0$$

(d) Guided $(\partial w/\partial n = 0, V_n = 0)$ if

$$\alpha_1 = 0, \alpha_2 = 1, \alpha_3 = 0, \beta_1 = 1, \beta_2 = 0, \beta_3 = 0$$

Notice that at points where the boundary conditions change, the functions α_i, β_i are discontinuous.

For the existence and uniqueness of the solution of the plate equation the interested reader is advised to reference [3].

For plates with a nonsmooth boundary, the boundary conditions (2.44) should be supplemented by the corner condition

$$c_{1k}w + c_{2k}[\![Tw]\!]_k = c_{3k} \tag{2.47}$$

where c_{ik} $(i=1,2,3)$ are constants specified at the corner point k and T is a differential operator defined as

$$T = D(1-\nu)\frac{\partial^2}{\partial n \partial t} \tag{2.48}$$

Hence, Tw is the twisting moment M_{nt} along the boundary and $[\![Tw]\!]_k$ represents its jump of discontinuity at the corner point k. The positive directions of the boundary forces are shown in Fig. (2.8) and Fig. (2.9).

For curved boundaries it is convenient to use intrinsic coordinates s and n, i.e., the arc length s and the coordinate n along the outward normal to the boundary (see Fig. 2.10). These are curvilinear coordinates and the derivatives on the boundary are expressed by the following relations [26]:

$$\frac{\partial w}{\partial t} = \frac{\partial w}{\partial s} \tag{2.49a}$$

$$\frac{\partial^2 w}{\partial t^2} = \frac{\partial^2 w}{\partial s^2} + \kappa \frac{\partial w}{\partial n} \tag{2.49b}$$

$$\frac{\partial^2 w}{\partial n \partial t} = \frac{\partial^2 w}{\partial s \partial n} - \kappa \frac{\partial w}{\partial s} \tag{2.49c}$$

where $\kappa = \kappa(s)$ is the curvature of the boundary at point s, which is evaluated from the relation

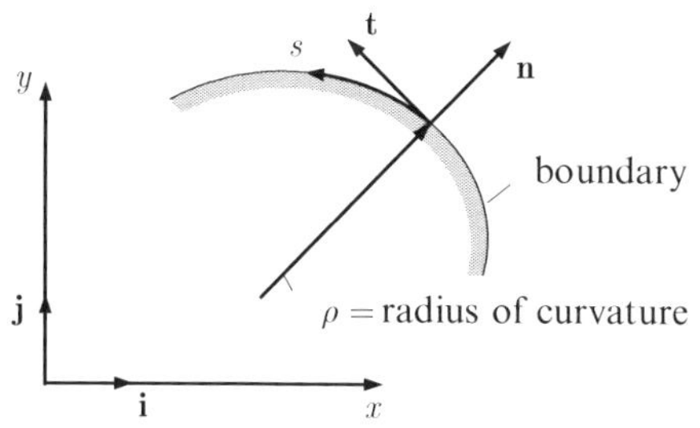

FIGURE 2.10 Intrinsic coordinates.

$$\kappa = \frac{d\phi}{ds} = \frac{d\phi/dp}{ds/dp} = \frac{\dot{x}\ddot{y} - \dot{y}\ddot{x}}{\left(\dot{x}^2 + \dot{y}^2\right)^{1/2}} \tag{2.50}$$

where $x(p)$, $y(p)$ are the parametric equations of the boundary. Here the dot designates differentiation with respect to the parameter p. If the boundary is specified by the equation $y = y(x)$, then Eq. (2.50) yields

$$\kappa = -\frac{y''}{\left(1 + y'^2\right)^{3/2}} \tag{2.51a}$$

The minus sign results because the vector $\mathbf{n}$ is taken opposite to that defined by the principal normal to the curve. Moreover, if the boundary is defined by polar coordinates, Eq. (2.50) becomes

$$\kappa = \frac{r^2 + 2r_{,\theta}^2 - rr_{,\theta\theta}}{\left(r^2 + r_{,\theta}^2\right)^{3/2}} \tag{2.51b}$$

For a rectilinear edge it is $\kappa = 0$ and the variable t is identical to s.

Using relations (2.49) the stress resultants on the boundary are written as

$$M_n = -D\left[\nabla^2 w + (\nu - 1)\left(\frac{\partial^2 w}{\partial s^2} + \kappa \frac{\partial w}{\partial n}\right)\right] \tag{2.52a}$$

$$V_n = -D\left[\frac{\partial}{\partial n}\nabla^2 w - (\nu - 1)\frac{\partial}{\partial s}\left(\frac{\partial^2 w}{\partial s \partial n} - \kappa \frac{\partial w}{\partial s}\right)\right] \tag{2.52b}$$

$$M_{nt} = -D(\nu - 1)\left(\frac{\partial^2 w}{\partial s \partial n} + \kappa \frac{\partial w}{\partial s}\right) \tag{2.52c}$$

$$M_t = -D\left[\nu \nabla^2 w - (\nu - 1)\left(\frac{\partial^2 w}{\partial s^2} + \kappa \frac{\partial w}{\partial n}\right)\right] \tag{2.52d}$$

2.3 DIRECT BEM FOR THE PLATE EQUATION

2.3.1 Rayleigh-Green identity

Consider the two-dimensional multiply connected domain Ω of the xy plane occupied by the plate. The domain is bounded by the $K + 1$ non-intersecting curves $\Gamma_0, \Gamma_1, \ldots, \Gamma_K$, forming the boundary $\Gamma = \cup_{k=0}^{K} \Gamma_k$ of the plate, which may be piecewise smooth, i.e., it may have a finite number of corners (Fig. 2.11).

The derivation of the integral representation of the solution for the plate equation requires the establishment of a reciprocal identity for the biharmonic operator (2.43). We can get such an identity starting from Green's reciprocal

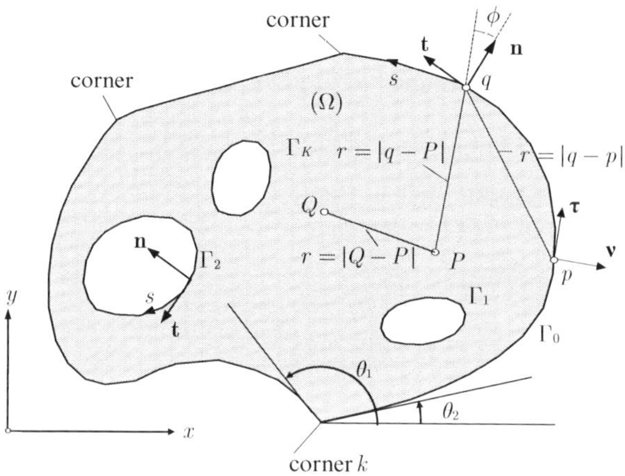

FIGURE 2.11 Plate geometry and notation.

identity for the Laplace operator, (Eq. 1.16), valid also for multiply connected domains ([1], Section 4.7). Thus for any two functions $u = u(x, y)$ and $\tilde{u} = \tilde{u}(x, y)$ we have

$$\int_\Omega (\tilde{u}\nabla^2 u - u\nabla^2 \tilde{u})\, d\Omega = \int_\Gamma \left(\tilde{u}\frac{\partial u}{\partial n} - u\frac{\partial \tilde{u}}{\partial n} \right) ds \tag{2.53}$$

Applying Eq. (2.53) for $u = \nabla^2 w$ and $\tilde{u} = v$ yields

$$\int_\Omega (v\nabla^4 w - \nabla^2 w \nabla^2 v)\, d\Omega = \int_\Gamma \left(v\frac{\partial}{\partial n}\nabla^2 w - \nabla^2 w \frac{\partial v}{\partial n} \right) ds \tag{2.54}$$

Similarly, the same equation yields for $u = w$ and $\tilde{u} = \nabla^2 v$

$$\int_\Omega (\nabla^2 v \nabla^2 w - w\nabla^4 v)\, d\Omega = \int_\Gamma \left(\nabla^2 v \frac{\partial w}{\partial n} - w\frac{\partial}{\partial n}\nabla^2 u \right) ds \tag{2.55}$$

Adding Eqs. (2.54) and (2.55) gives

$$\begin{aligned} &\int_\Omega (v\nabla^4 w - w\nabla^4 v)\, d\Omega \\ &= \int_\Gamma \left(v\frac{\partial}{\partial n}\nabla^2 w - w\frac{\partial}{\partial n}\nabla^2 v - \frac{\partial v}{\partial n}\nabla^2 w + \frac{\partial w}{\partial n}\nabla^2 v \right) ds \end{aligned} \tag{2.56}$$

The above relation is the Green's reciprocal identity for the biharmonic operator. It is known as the *Rayleigh-Green identity*. It can also be derived by repeated integration by parts of the integral $\int_\Omega v\nabla^4 w d\Omega$ using the Gauss-Green theorem (see Section 1.4).

Eq. (2.56) holds for any two functions v and w, which are four times continuously differentiable inside Ω and three times on Γ. The integral in the right-hand side involves the quantities w, $\partial w/\partial n$, $\nabla^2 w$ and $\partial \nabla^2 w/\partial n$. When the function w represents the deflection of the middle surface of the plate, $\partial w/\partial n$ is its slope in the direction normal to the boundary. Except for special cases, the other two quantities do not have a direct physical meaning. For example, when the edge of the plate is clamped we have $w=0$, $\partial w/\partial n=0$, which imply also $\partial^2 w/\partial s^2=0$ and $\partial^2 w/\partial s\partial n=0$. Thus, Eqs. (2.52a,b) reduce to

$$M_n=-D\nabla^2 w \tag{2.57a}$$

$$V_n=-D\frac{\partial}{\partial n}\nabla^2 w \tag{2.57b}$$

i.e., the quantities $\nabla^2 w$ and $\partial \nabla^2 w/\partial n$ represent the bending moment and the effective shear force at the boundary, respectively, multiplied by $-1/D$. The reciprocal identity (2.56) has been employed by many investigators to develop BEM formulations for the plate bending problem with specific boundary conditions, especially in the early stages of the method, or indirect BEMs [2].

From the above observation it becomes clear that Eq. (2.56) cannot be used to derive boundary integral equations for a direct BEM that can treat all types of boundary conditions, unless it is converted into an equivalent form involving boundary quantities having direct physical meaning. Though other methods have been reported to derive such an identity, the procedure presented herein is that developed by Katsikadelis [26], which, we think, is easier to be followed by the reader. For this purpose we write the boundary integral in Eq. (2.56) into the following form after adding and subtracting suited terms to create the boundary quantities Mw, Mv and Vw, Vv in the boundary integral

$$\begin{aligned}\int_\Gamma\left(v\frac{\partial}{\partial n}\nabla^2 w-w\frac{\partial}{\partial n}\nabla^2 w-\frac{\partial v}{\partial n}\nabla^2 w+\frac{\partial w}{\partial n}\nabla^2 v\right)ds=\\ -\frac{1}{D}\int_\Gamma\left(v\,Vw-w\,Vv-\frac{\partial v}{\partial n}Mw+\frac{\partial w}{\partial n}Mv\right)ds+I_\Gamma\end{aligned} \tag{2.58}$$

Where

$$I_\Gamma=(\nu-1)\int_\Gamma\left(v\frac{\partial}{\partial s}\frac{\partial^2 w}{\partial n\partial t}-w\frac{\partial}{\partial s}\frac{\partial^2 v}{\partial n\partial t}+\frac{\partial v}{\partial n}\frac{\partial^2 w}{\partial t^2}-\frac{\partial w}{\partial n}\frac{\partial^2 v}{\partial t^2}\right)ds \tag{2.59}$$

Using Eqs. (2.49), the above equation becomes

$$I_\Gamma=(\nu-1)\int_\Gamma\left[v\frac{\partial}{\partial s}\left(\frac{\partial^2 w}{\partial n\partial t}\right)-w\frac{\partial}{\partial s}\left(\frac{\partial^2 v}{\partial n\partial t}\right)+\frac{\partial v}{\partial n}\frac{\partial^2 w}{\partial s^2}-\frac{\partial w}{\partial n}\frac{\partial^2 v}{\partial s^2}\right]ds \tag{2.60}$$

Subsequently, integrating Eq. (2.60) by parts, using Eq. (2.48), and noting that $\frac{\partial w}{\partial n}, \frac{\partial w^2}{\partial n \partial t}, \frac{\partial v}{\partial n}, \frac{\partial^2 v}{\partial n \partial t}$ are discontinuous at the corner points, we obtain the following expression for I_Γ:

$$I_\Gamma = \frac{1}{D} \sum_k (v[\![Tw]\!]_k - w[\![Tv]\!]_k) + (\nu - 1) \sum_k \left[\!\!\left[\frac{\partial v}{\partial t}\frac{\partial w}{\partial n} - \frac{\partial w}{\partial t}\frac{\partial v}{\partial n} \right]\!\!\right]_k \tag{2.61}$$

Eq. (2.61) can be further simplified by showing that the terms in the second sum vanish identically. Thus, referring to Fig. (2.12) we obtain

$$\frac{\partial w}{\partial n^{(+)}} = \frac{\partial w}{\partial n^{(-)}} \cos\alpha + \frac{\partial w}{\partial t^{(-)}} \sin\alpha \tag{2.62a}$$

$$\frac{\partial w}{\partial t^{(+)}} = -\frac{\partial w}{\partial n^{(-)}} \sin\alpha + \frac{\partial w}{\partial t^{(-)}} \cos\alpha \tag{2.62b}$$

$$\frac{\partial v}{\partial n^{(+)}} = \frac{\partial v}{\partial n^{(-)}} \cos\alpha + \frac{\partial v}{\partial t^{(-)}} \sin\alpha \tag{2.62c}$$

$$\frac{\partial v}{\partial t^{(+)}} = -\frac{\partial v}{\partial n^{(-)}} \sin\alpha + \frac{\partial v}{\partial t^{(-)}} \cos\alpha \tag{2.62d}$$

The typical term in the second sum in the right-hand side of Eq. (2.61) is written in expanded form as

$$\left[\!\!\left[\frac{\partial v}{\partial n}\frac{\partial w}{\partial t} - \frac{\partial w}{\partial n}\frac{\partial v}{\partial t} \right]\!\!\right]_k =$$
$$\left(\frac{\partial v}{\partial n^{(+)}}\frac{\partial w}{\partial t^{(+)}} - \frac{\partial v}{\partial t^{(+)}}\frac{\partial w}{\partial n^{(+)}} \right)_k - \left(\frac{\partial v}{\partial n^{(-)}}\frac{\partial w}{\partial t^{(-)}} - \frac{\partial w}{\partial t^{(-)}}\frac{\partial w}{\partial n^{(-)}} \right)_k$$

Substituting now Eqs. (2.62) in the previous equation yields

$$\left[\!\!\left[\frac{\partial v}{\partial n}\frac{\partial w}{\partial t} - \frac{\partial w}{\partial n}\frac{\partial v}{\partial t} \right]\!\!\right]_k = 0 \tag{2.63}$$

On the basis of Eqs. (2.61) and (2.63), Eq. (2.58) becomes

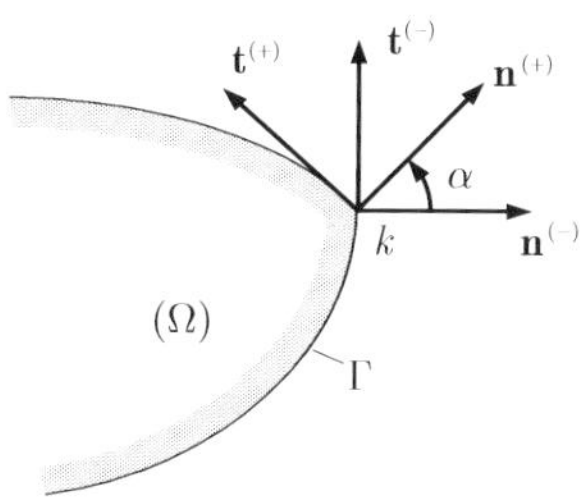

FIGURE 2.12 Corner point k on Γ.

$$D\int_{\Omega}\left(v\nabla^4 w - w\nabla^4 v\right)d\Omega = -\int_{\Gamma}\left(vVw - wVv - \frac{\partial v}{\partial n}Mw + \frac{\partial w}{\partial n}Mv\right)ds + \sum_k \left(v[\![Tw]\!]_k - w[\![Tv]\!]_k\right) \tag{2.64}$$

The reciprocal identity (2.64) is known as the *generalized Rayleigh-Green identity*. The boundary integral now involves quantities that have direct geometrical and physical meaning, namely, the deflection w, its normal slope $\partial w/\partial n$, the bending moment $Mw = M_n$ and the effective shear force $Vw = V_n$ on the boundary. After rearranging the terms, Eq. (2.64) can be also written as

$$\begin{aligned}\int_{\Omega} v\left(D\nabla^4 w\right)d\Omega + \int_{\Gamma}\left(vVw - \frac{\partial v}{\partial n}Mw\right)ds - \sum_k v[\![Tw]\!]_k = \\ \int_{\Omega} w\left(D\nabla^4 v\right)d\Omega + \int_{\Gamma}\left(wVv - \frac{\partial w}{\partial n}Mv\right)ds - \sum_k w[\![Tv]\!]_k\end{aligned} \tag{2.65}$$

The above relation states Betti's reciprocal theorem for the plate. Note that the corner terms express the work of the fictitious corner forces. This proves that a consistent application of Betti's theorem for the plate bending problem should include the work done by the fictitious corner forces. The fictitious forces are directed in the negative z direction, when produced by positive twisting moments (see Fig. 2.8). It is apparent that Eq. (2.64) can be derived by direct application of Betti's reciprocal theorem by considering the following two states of stress due to two different sets of loads:

State I:

$$w, f = D\nabla^4 w \quad \text{in } \Omega \text{ and } w,\ w_{,n},\ w_k,\ Mw,\ Vw,\ [\![Tw]\!]_k \quad \text{on } \Gamma \tag{2.66a}$$

State II:

$$v, \tilde{f} = D\nabla^4 v \quad \text{in } \Omega \text{ and } v,\ v_{,n},\ v_k,\ Mv,\ Vv,\ [\![Tv]\!]_k \quad \text{on } \Gamma \tag{2.66b}$$

Betti's theorem without considering the corner forces has been employed in [4] to derive the Rayleigh-Green identity for the plate flexure problem. Another procedure to derive the Rayleigh-Green identity is that based on a bilinear form resulting from the strain energy of the plate [3]. This procedure has been used in [5,7]. The Rayleigh-Green identity obtained by the latter procedure differs in the sign of the corner term, because the twisting moment M_{xy} is defined with an opposite sign.

2.3.2 The fundamental solution

The fundamental solution of the biharmonic equation is a singular particular solution to equation

$$D\nabla^4 v = \delta(Q - P) \tag{2.67}$$

where $\delta(Q-P)$ is the Dirac delta function representing the load density at a point $Q:\{\xi, n\}$ due to a concentrated transverse unit force at point P:{x, y}. Physically, the solution of Eq. (2.67) is a two-point function $v = v(Q, P)$ that represents the deflection at point Q (field point) due to a unit concentrated force at point P (source point) in an infinite plate having flexural stiffness D. The fundamental solution $v(Q, P)$ is axisymmetric with respect to the source point P and can be established employing a procedure analogous to that employed for the Laplace equation (see [1], Section 3.2).

We consider the homogeneous equation of (2.67), which in polar coordinates with origin the point P is written as

$$\nabla^2(\nabla^2 v) = \frac{1}{r}\frac{d}{dr}\left(r\frac{d}{dr}\right)\left[\frac{1}{r}\frac{d}{dr}\left(r\frac{dv}{dr}\right)\right] = 0, \quad r \neq 0 \tag{2.68}$$

where $r = |Q - P| = \left[(\xi - x)^2 + (\eta - y)\right]^{1/2}$ is the distance between points P and Q.

Repeated integration of Eq. (2.68) yields

$$v(r) = C_1 r^2 \ln r + C_2 r^2 + C_3 \ln r + C_4 \tag{2.69}$$

where $C_i\,(i = 1, 2, 3, 4)$ are arbitrary integration constants that can be determined from the following physical considerations:

(i) $v(r)$ must be finite at $r = 0$.

(ii) $v(r)$ must be bounded at the infinity.

(iii) The transverse forces on any infinitely small portion of the plate including the unit concentrated force must be in equilibrium.

Since we seek a particular solution it can be set as $C_4 = 0$. Moreover, conditions (i) and (ii) require that C_2 and C_3 vanish. Thus, the solution (2.69) becomes

$$v(r) = C_1 r^2 \ln r \tag{2.70}$$

If Γ_0 denotes the boundary of a small domain Ω_0 including point Q then condition (iii) requires that

$$\int_{\Gamma_0} V v ds - \sum_k [\![Tv]\!]_k + 1 = 0 \tag{2.71}$$

Note that Eq. (2.71) is derived from Eq. (2.64) if $w = 1$ and v satisfies Eq. (2.67).

On the basis of Eqs. (2.46) and (2.48), we can write

$$Vv = -D\frac{\partial}{\partial n}\nabla^2 v - \frac{\partial}{\partial s}Tv \tag{2.72}$$

and Eq. (2.71) is now written as

$$-D\int_{\Gamma_o}\frac{\partial}{\partial n}\nabla^2 v ds-\int_{\Gamma_o}\frac{\partial}{\partial s}Tv ds-\sum_k [\![Tv]\!]_k+1=0$$

Taking into account that

$$\int_{\Gamma_o}\frac{\partial}{\partial s}Tv ds=-\sum_k [\![Tv]\!]_k$$

we obtain

$$-D\int_{\Gamma_o}\frac{\partial}{\partial n}\nabla^2 v ds+1=0 \tag{2.73}$$

Hence, Eq. (2.73) holds for any boundary Γ_o, smooth or nonsmooth. From Eq. (2.70), we obtain

$$\nabla^2 v=\frac{1}{r}\frac{d}{dr}\left(r\frac{dv}{dr}\right)=4C_1(1+\ln r)$$

and

$$\frac{\partial}{\partial n}\nabla^2 v=4C_1\frac{1}{r}r_n$$

where for convenience r_n denotes the derivative $r_{,n}$ The normal derivative is taken with respect to the integration point (ξ,η). Thus, using Eq. (A.7) of Appendix A we have

$$\frac{\partial}{\partial n}\nabla^2 v= C_1\frac{4}{r}\cos\phi,\quad \phi=\text{angle}(r,\mathbf{n})$$

Without limiting the generality, we choose for Ω_o a small circle centered at P $(r=0)$ having radius ρ (see Fig. 2.13). Then $\phi=0$ and Eq. (2.73) is written as

$$-DC_1\int_{\theta=0}^{\theta=2\pi}\frac{4}{\rho}\rho d\theta+1=0$$

which yields

$$C_1=\frac{1}{8\pi D} \tag{2.74}$$

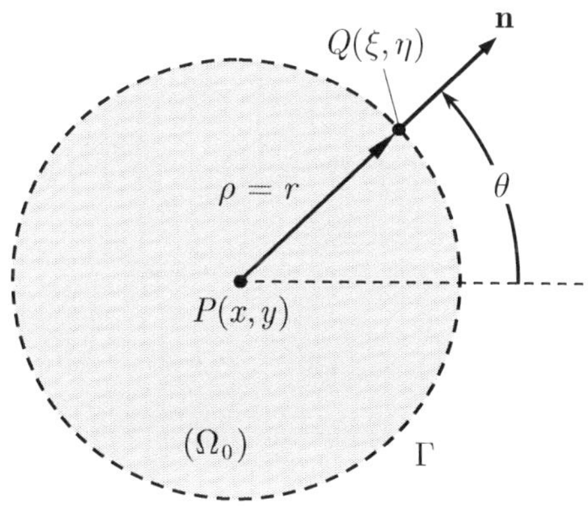

FIGURE 2.13 Circular domain Ω_0 of radius ρ with a source P at its center.

With this value of C_1 the fundamental solution (2.70) becomes

$$v = \frac{1}{8\pi D} r^2 \ln r \tag{2.75}$$

Obviously, the fundamental solution is symmetric with respect to the points P and Q, i.e., $v(P, Q) = v(Q, P)$. Therefore, the role of the source and field point can be interchanged.

2.3.3 Integral representation of the solution

The integral representation of the solution for the plate equation (2.42) results from the Rayleigh-Green identity (2.64). We formulate this relation for the pair of functions w and v satisfying Eqs. (2.42) and (2.67), respectively. In the following, the derivatives $r_{,x}$, $r_{,y}$, $r_{,n}$, $r_{,t}$ will be also designated for convenience by r_x, r_y, r_n, r_t.

On the basis of Eqs. (2.45), (2.46), and (2.48), the kernel functions in Eq. (2.64) are obtained as

$$v = \frac{1}{8\pi D} r^2 \ln r \tag{2.76a}$$

$$\frac{\partial v}{\partial n} = \frac{1}{8\pi D} r r_n (1 + 2\ln r) \tag{2.76b}$$

$$Mv = -\frac{1}{8\pi}\left[2(1+\nu)\ln\ r + (3+\nu)r_n^2 + (1+3\nu)r_t^2\right] \tag{2.76c}$$

$$Vv = -\frac{1}{4\pi r}\left[2r_n + (1-\nu)(r_n - \kappa r)\left(r_n^2 - r_t^2\right)\right] \tag{2.76d}$$

$$Tv = \frac{1}{4\pi}(1-\nu) r_n r_t \tag{2.76e}$$

where

$$r_n = \frac{\partial r}{\partial n_q} = \cos\phi \tag{2.77a}$$

$$r_t = \frac{\partial r}{\partial t_q} = -\sin\phi \tag{2.77b}$$

$$r = |q - P|, \quad q:\{\xi, \eta\} \in \Gamma, \quad P:\{x, y\} \in \Omega \tag{2.77c}$$

and $\kappa = \kappa(s)$ is the curvature of the boundary.

We consider a point P in the interior of the domain Ω. The kernel functions are continuous everywhere except possibly at P, where $r = 0$. To this end we form the domain Ω^* which results from Ω by deleting a small circle Ω_ρ with radius ρ centered at P. The radius is chosen so that $\Omega_\rho \subset \Omega$ (Fig. 2.14). Applying Eq. (2.64) for the domain $\Omega^* = \Omega - \Omega_\rho$ and noting that $\delta(q - P) = 0$ when P is outside Ω^* we obtain

$$\begin{aligned} 0 = \int_{\Omega^*} vf d\Omega + \int_\Gamma \left(vVw - wVv - \frac{\partial v}{\partial n} Mw + \frac{\partial w}{\partial n} Mv \right) ds \\ - \sum_k \left(v[\![Tw]\!]_k - w[\![Tv]\!]_k \right) + I_{\Gamma_\rho} \end{aligned} \tag{2.78}$$

where

$$I_{\Gamma_\rho} = \int_{\Gamma_\rho} \left(vVw - wVv - \frac{\partial v}{\partial n} Mw + \frac{\partial w}{\partial n} Mv \right) ds \tag{2.79}$$

For points q on Γ_ρ it is $r = \rho$, $\phi = \pi$, $r_n = -1$, $r_t = 0$, $\kappa = -1/\rho$. Hence, Eqs. (2.76) become

$$v = \frac{1}{8\pi D} \rho^2 \ln \rho \tag{2.80a}$$

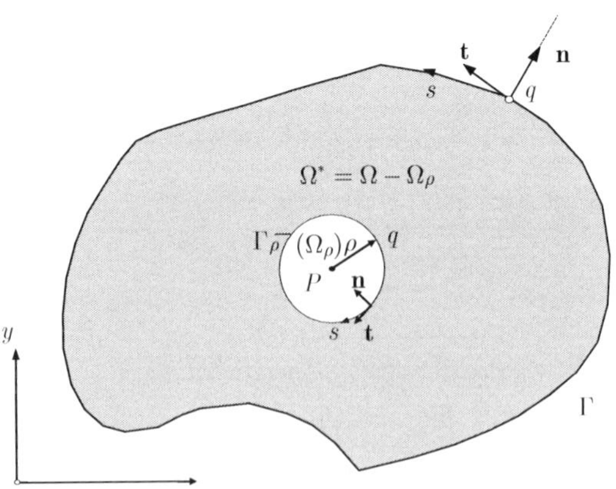

FIGURE 2.14 Domains Ω^*,Ω_ρ and notation.

$$\frac{\partial v}{\partial n} = -\frac{1}{8\pi D}\rho(1+2\ln\rho) \tag{2.80b}$$

$$Mv = -\frac{1}{8\pi}[2(1+\nu)\ln\rho + (3+\nu)] \tag{2.80c}$$

$$Vv = \frac{1}{2\pi}\frac{1}{\rho} \tag{2.80d}$$

$$Tv = 0 \tag{2.80e}$$

We next examine the limit in Eq. (2.79) when $\rho \to 0$. In this case, Ω_ρ shrinks to a point and $\Omega^* \to \Omega$. Before taking the limit in Eq. (2.79), the terms in the integral are evaluated using the mean value theorem of the integral calculus. Thus, denoting by $w_0, \left(\frac{\partial w}{\partial n}\right)_0$, $(Mw)_0$, $(Vw)_0$ the respective values at a point $q_0 \in \Gamma_\rho$ and noting that $\lim_{\rho\to 0} q_0 = P$ and $ds = \rho d\theta$, $0 \le \theta \le 2\pi$, we obtain

$$\lim_{\rho\to 0}\int_{\Gamma_\rho} vVwds = \lim_{\rho\to 0}\left[(Vw)_0\int_0^{2\pi}\frac{1}{8\pi D}\rho^2\ln\rho\rho d\theta\right] = 0 \tag{2.81a}$$

$$\lim_{\rho\to 0}\int_{\Gamma_\rho} wVvds = \lim_{\rho\to 0}\left[w_0\int_0^{2\pi}\frac{1}{2\pi}\frac{1}{\rho}\rho d\theta\right] = \lim_{\rho\to 0} w(q_0) = w(P) \tag{2.81b}$$

$$\lim_{\rho\to 0}\int_{\Gamma_\rho}\frac{\partial v}{\partial n}Mwds = \lim_{\rho\to 0}\left[(Mw)_0\int_0^{2\pi}\frac{-1}{8\pi D}\rho(1+2\ln\rho)\rho d\theta\right] = 0 \tag{2.81c}$$

$$\lim_{\rho\to 0}\int_{\Gamma_\rho}\frac{\partial w}{\partial n}Mvds = \lim_{\rho\to 0}\left[\left(\frac{\partial w}{\partial n}\right)_0\int_0^{2\pi}\frac{-1}{8\pi}[(2+\nu)\ln\rho + (3+\nu)]\rho d\theta\right] = 0 \tag{2.81d}$$

Substituting the above limits into Eq. (2.79) yields

$$\lim_{\rho\to 0} I_{\Gamma_\rho} = -w(P) \tag{2.82}$$

and in the limit Eq. (2.78) gives the integral representation of the solution in Ω in the form

$$\begin{aligned} w(P) &= \int_\Omega v(P,Q)f(Q)\,d\Omega_Q \\ &+ \int_\Gamma\left[v(P,q)\,Vw(q) - w(q)\,V_q v(P,q) - \frac{\partial v(P,q)}{\partial n_q}Mw(q) + \frac{\partial w(q)}{\partial n_q}M_q v(P,q)\right]ds_q \\ &- \sum_k \left(v(P,q_k)[\![T_q w(q_k)]\!] - w(q_k)[\![T_q v(P,q_k)]\!]\right) \end{aligned} \tag{2.83}$$

where $P, Q \in \Omega, q \in \Gamma$. The subscript in the differentials $d\Omega$, ds and in the operators $\frac{\partial}{\partial n}$, M, V, T indicates the point with respect to which integration or differentiation is performed. Except where it is required for clarification, the arguments of the functions and subscripts in Eq. (2.83) will be omitted. Hence, this equation is written as

$$w(P) = \iint_{\Omega} vf d\Omega + \int_{\Gamma} \left[vVw - wVv - \frac{\partial v}{\partial n} Mw + \frac{\partial w}{\partial n} Mv \right] ds \\ - \sum_{k} \left(v[\![Tw]\!]_k - w[\![Tv]\!]_k \right) \tag{2.84}$$

Using Eq. (2.56) and applying the same procedure we readily obtain the integral representation also in the form

$$w(P) = \int_{\Omega} vf d\Omega \\ - D \int_{\Gamma} \left[v \frac{\partial}{\partial n} \nabla^2 w - w \frac{\partial}{\partial n} \nabla^2 v - \frac{\partial v}{\partial n} \nabla^2 w + \frac{\partial w}{\partial n} \nabla^2 v \right] ds \tag{2.85}$$

We leave the proof to the reader.

2.3.4 Boundary integral equations

Eq. (2.84) permits the evaluation of the deflection $w(P)$ at a point P in the interior of the plate, if the values of the deflection w, its normal slope $\partial w/\partial n$, the bending moment Mw and effective shear force Vw on the boundary Γ, along with the corner forces $[\![Tw]\!]_k$, are known. However, in a well-posed plate problem only two of the four boundary quantities can be prescribed (Section 2.2.6). Consequently, it becomes necessary to determine the two unknown (not specified) quantities on the boundary. This is achieved by formulating two coupled boundary integral equations whose solution gives the unknown quantities.

2.3.4.1 FIRST BOUNDARY INTEGRAL EQUATION

The first boundary integral equation is obtained by formulating Eq. (2.64) in terms of the same functions w and v and by placing point P at a point p on the boundary Γ. For generality, we suppose that p is a corner point with internal angle $\alpha\pi$; for points p where the boundary is smooth it is $\alpha = 1$.

When point p is on Γ, the distance r vanishes when q coincides with p. From Eqs. (2.76a,b) we conclude that the kernels v and $\partial v/\partial n$ vanish for $r = 0$. Therefore, the corresponding boundary integrals are regular. Moreover, it is $\phi \rightarrow \pm\pi/2$ as $r \rightarrow 0$, hence Eq. (2.76e) gives $Tv = 0$. Furthermore, we see from Eqs. (2.76c) and (2.76d) that the kernels Mv and Vv behave like $\ln r$ and $1/r$ for $r \rightarrow 0$, i.e., they exhibit a logarithmic and a Cauchy type singularity, respectively. Working as previously, we consider a domain Ω^* resulting from Ω by deleting a small domain Ω_ρ, which includes all points of Ω within a distance $d < \rho$ (Fig. 2.15).

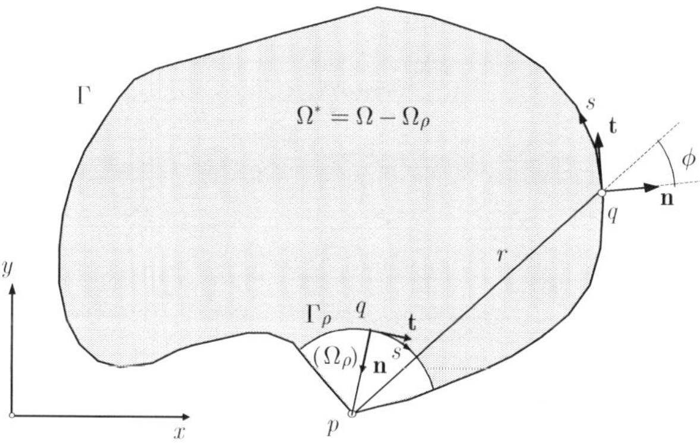

FIGURE 2.15 Domains Ω^*,Ω_ρ for a boundary point p and notation.

We apply the reciprocal identity (2.64) for the domain Ω^*. Point p is outside Ω^*, hence we have

$$\begin{aligned} 0 = &\int_{\Omega^*} vf d\Omega \\ &+ \int_{\Gamma-\Gamma_\rho}\left(vVw - wVv - \frac{\partial v}{\partial n}Mw + \frac{\partial w}{\partial n}Mv\right)ds + I_{\Gamma_\rho} \\ &-\sum_{k^*}\left(v[\![Tw]\!]_{k^*} - w[\![Tv]\!]_{k^*}\right) - [\![\ ]\!]_p \end{aligned} \tag{2.86}$$

where

$$I_{\Gamma_\rho} = \int_{\Gamma_\rho}\left(vVw - wVv - \frac{\partial v}{\partial n}Mw + \frac{\partial w}{\partial n}Mv\right)ds \tag{2.87}$$

$$[\![\ ]\!]_p = \lim_{\rho\to 0}\left([\![\ ]\!]_{k^{(-)}} + [\![\ ]\!]_{k^{(+)}}\right) \tag{2.88}$$

$[\![\ ]\!]_{k^{(\pm)}}$ denotes the additional corner terms arising on the new corners $k^{(-)}$ and $k^{(+)}$ (Fig. 2.16) and $\sum_{k^*}$ denotes the sum of the corner terms without the contribution of corner point p.

In the limit as $\rho \to 0$ it is

$$\lim_{\rho\to 0}\int_{\Gamma-\Gamma_\rho}(.)ds = \int_\Gamma(.)ds \tag{2.89a}$$

$$\lim_{\rho\to 0}\int_{\Omega^*}(.)ds = \int_\Omega(.)ds \tag{2.89b}$$

$$\lim_{\rho\to 0}(\theta_1 - \theta_2) = \alpha\pi \tag{2.89c}$$

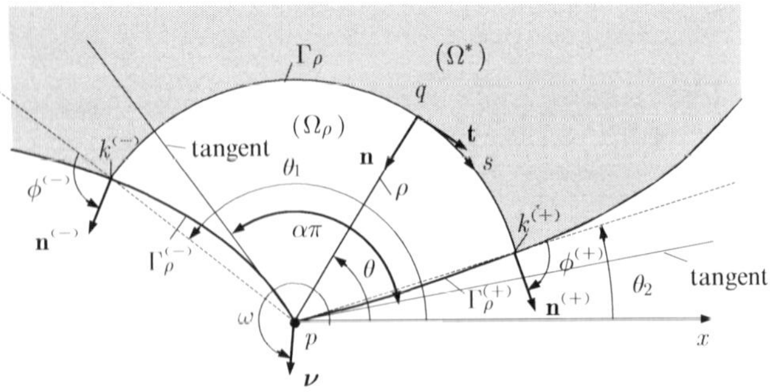

FIGURE 2.16 Domain Ω_ρ for a corner point and notation.

On the basis of Eqs. (2.80), we obtain

$$\lim_{\rho\to 0}\int_{\Gamma_\rho} vVwds = \lim_{\rho\to 0}\left[(Vw)_0\int_{\theta_1}^{\theta_2}\frac{1}{8\pi D}\rho^2\ln\rho\rho d(-\theta)\right]=0 \quad (2.90a)$$

$$\begin{aligned}\lim_{\rho\to 0}\int_{\Gamma_\rho} wVvds &= \lim_{\rho\to 0}\left[w_0\int_{\theta_1}^{\theta_2}\frac{1}{2\pi}\frac{1}{\rho}\rho d(-\theta)\right]\\ &= \frac{1}{2\pi}\lim_{\rho\to 0}[(\theta_1-\theta_2)w(q_0)]=\frac{\alpha}{2}w(p)\end{aligned} \quad (2.90b)$$

$$\lim_{\rho\to 0}\int_{\Gamma_\rho}\frac{\partial v}{\partial n}Mwds = \lim_{\rho\to 0}\left[(Mw)_0\int_{\theta_1}^{\theta_2}\frac{-1}{8\pi D}\rho(1+2\ln\rho)\rho d(-\theta)\right]=0 \quad (2.90c)$$

$$\lim_{\rho\to 0}\int_{\Gamma_\rho}\frac{\partial w}{\partial n}Mvds = \lim_{\rho\to 0}\left[\left(\frac{\partial w}{\partial n}\right)_0\int_{\theta_1}^{\theta_2}\frac{-1}{8\pi}[(2+\nu)\ln\rho+(3+\nu)]\rho d(-\theta)\right]=0 \quad (2.90d)$$

Substituting Eqs. (2.90) in Eq. (2.87) yields

$$I_{\Gamma_\rho} = -\frac{\alpha}{2}w(p) \quad (2.91a)$$

Further, from Eqs. (2.80a,e) we see that $\ell im_{\rho\to 0}\, v=0$ and $\ell im_{\rho\to 0}\, Tv=0$. Thus, Eq. (2.88) yields

$$[\![\ ]\!]_p = 0 \quad (2.91b)$$

Thus, taking into account Eqs. (2.91a,b), Eq. (2.86) becomes

$$\begin{aligned}\frac{\alpha}{2}w(p) = \int_\Omega vfd\Omega + \int_\Gamma\left(vVw - wVv - \frac{\partial v}{\partial n}Mw + \frac{\partial w}{\partial n}Mv\right)ds\\ -\sum_k\left(v[\![Tw]\!]_k - w[\![Tv]\!]_k\right)\end{aligned} \quad (2.92)$$

It should be noted that, if p is a corner point, the relevant corner term must be omitted from the sum in Eq. (2.92).

Equations (2.84) and (2.92) can be combined in

$$\varepsilon(P)w(P)=\int_{\Omega} vfd\Omega+\int_{\Gamma}\left(vVw-wVv-\frac{\partial v}{\partial n}Mw+\frac{\partial w}{\partial n}Mv\right)ds \\ -\sum_{k}\left(v[\![Tw]\!]_k-w[\![Tv]\!]_k\right) \tag{2.93a}$$

where $\varepsilon(P)$ is the free term coefficient defined as

$$\varepsilon(P)=\begin{cases} 1 & P \quad \text{inside} \quad \Omega \\ \dfrac{\alpha}{2} & P=p \quad \text{on} \quad \Gamma \\ 0 & P \quad \text{outside} \quad \Omega \end{cases} \tag{2.93b}$$

Apparently, for points P where the boundary is smooth it is $\alpha=1$, hence $\varepsilon(P)=1/2$.

2.3.4.2 SECOND BOUNDARY INTEGRAL EQUATION

The derivation of the second independent boundary integral equation requires more attention. A general method for deriving it systematically for smooth boundaries has been presented by Katsikadelis et al. [6] who formulated the boundary integral equations for the biharmonic equation. In this method the directional derivative of the deflection given by Eq. (2.84) is taken with respect to point P in a fixed direction $\mathbf{m}(m_x, m_y)$. Subsequently, the point P approaches a point p on Γ and the direction $\mathbf{m}$ is chosen to coincide with the outward normal to the boundary $\boldsymbol{\nu}(\nu_x, \nu_y)$ at that point (Fig. 2.11). This method has been employed also to derive the second boundary integral equation for the plate bending problem in [5]. However, in the following we will derive the second boundary integral equation following a procedure similar to that presented in [7]. This method permits the integration of the hypersingular integrals whose kernels behave like $1/r^2$ when $r\rightarrow 0$.

We formulate the reciprocal identity (2.64) in terms of the functions $\hat{w}=w-w(p)$ and v. Due to the singular behavior of the derivatives of the kernel functions on the boundary, we consider the domain Ω^* as defined previously (Fig. 2.15). Thus, we have

$$0=\int_{\Omega^*} vfd\Omega+\int_{\Gamma^*}\left\{vVw-[w-w(p)]Vv-\frac{\partial v}{\partial n}Mw+\frac{\partial w}{\partial n}Mv\right\}ds+I_{\Gamma\rho} \\ -\sum_{k^*}\left(v[\![\mathrm{T}w]\!]_{k^*}-[w-w(p)][\![Tv]\!]_{k^*}\right)-[\![\ ]\!]_p \tag{2.94}$$

in which the symbols $\sum_{k^*}$, $I_{\Gamma\rho}$, $[\![\ ]\!]_p$ and $\int_{\Gamma^*}(.)ds$ have the same meaning as in Eq. (2.86). Note that in deriving the above relation it has been taken into account that

$$\frac{\partial}{\partial n}\hat{w}=\frac{\partial}{\partial n}[w-w(p)]=\frac{\partial w}{\partial n} \tag{2.95a}$$

$$M\hat{w} = M[w - w(p)] = Mw \tag{2.95b}$$

$$V\hat{w} = V[w - w(p)] = Vw \tag{2.95c}$$

because the derivatives in the boundary integral are taken with respect to point q and consequently, $w(p)$ is constant during differentiation.

Now differentiating Eq. (2.94) with respect to the direction $\boldsymbol{\nu}$ yields

$$0 = \int_{\Omega^*} v_1 f d\Omega + \int_{\Gamma^*} \left\{ v_1\, Vw - [w - w(p)]\, Vv_1 - \frac{\partial v_1}{\partial n} Mw + \frac{\partial w}{\partial n} Mv_1 \right\} ds$$
$$- \sum_{k^*} \left(v_1 [\![Tw]\!]_{k^*} - [w - w(p)] [\![Tv_1]\!]_{k^*} \right) + I_{\Gamma\rho} - [\![\;]\!]_p \tag{2.96}$$

The kernels in Eq. (2.96) are obtained from Eqs. (2.76) by differentiating with respect to the direction $\boldsymbol{\nu}$. Thus, we take

$$v_1 = \frac{1}{8\pi D} r r_\nu (1 + 2\ell n r) \tag{2.97a}$$

$$\frac{\partial v_1}{\partial n} = \frac{1}{8\pi D} [2(r_n r_\nu + r_t r_\tau)\ell n r + 3 r_n r_\nu + r_t r_\tau] \tag{2.97b}$$

$$M v_1 = -\frac{1}{4\pi r} [(1+\nu) r_\nu + 2(1-\nu) r_n r_t r_\tau] \tag{2.97c}$$

$$V v_1 = -\frac{1}{4\pi r^2} \{ [(-3+\nu) + 2(1-\nu) r_t^2](r_n r_\nu - r_t r_\tau)$$
$$+ 4(1-\nu)(r_n - \kappa r) r_n r_t t_\tau \} \tag{2.97d}$$

$$T v_1 = \frac{1-\nu}{4\pi r} (r_t^2 - r_n^2) r_\tau \tag{2.97e}$$

Next, we evaluate the limits in Eq. (2.96) as ρ tends to 0. Without restricting the generality, we assume that p is a corner point, Fig. 2.16. We recall that it is $\phi = \pi$, $r_n = -1$, $r_t = 0$, $\kappa = -1/\rho$ for points q on $\Gamma\rho$. Hence, Eqs. (2.97) become

$$v_1 = \frac{1}{8\pi D} \rho (1 + 2\ell n \rho) r_\nu \tag{2.98a}$$

$$\frac{\partial v_1}{\partial n} = -\frac{1}{8\pi D} (3 + 2\ell n \rho) r_\nu \tag{2.98b}$$

$$M v_1 = -\frac{1+\nu}{4\pi} \frac{r_\nu}{\rho} \tag{2.98c}$$

$$V v_1 = \frac{-3+\nu}{4\pi} \frac{r_\nu}{\rho^2} \tag{2.98d}$$

$$T v_1 = -\frac{1-\nu}{4\pi} \frac{r_\tau}{\rho} \tag{2.98e}$$

Referring to Fig. (2.16) we have

$$\frac{\partial w}{\partial n_q} = -\left[w_{,x}(q)\cos\theta + w_{,y}(q)\sin\theta\right] \tag{2.99}$$

Expanding now $\hat{w}$ in Taylor series at point p we obtain

$$\hat{w} = w - w(p) = \left[w_{,x}(p)\cos\theta + w_{,y}(p)\sin\theta\right]\rho + O\left(\rho^2\right) \tag{2.100}$$

Let

$$\nu_x = \cos\omega \quad \text{and} \quad \nu_y = \sin\omega \tag{2.101}$$

be the direction cosines of a vector $\boldsymbol{\nu}$ at point p, then

$$r_\nu = r_x \nu_x + r_y \nu_y = -\left(\nu_x\,\cos\theta + \nu_y\,\sin\theta\right) \tag{2.102}$$

We choose first $\omega = 0$, hence $r_\nu = r_x = -\cos\theta$ and we evaluate the following limits:

$$\begin{aligned}
A_x &= -\lim_{\rho\to 0}\int_{\Gamma\rho}[w - w(p)]\,V v_1\,ds \\
&= -\lim_{\rho\to 0}\int_{\theta_1}^{\theta_2}\left[\left(w_{,x}(p)\cos\theta + w_{,y}(p)\sin\theta\right)\rho + O(\rho^2)\right]\frac{-3+\nu}{4\pi}\frac{1}{\rho^2}(-\cos\theta)\rho d(-\theta) \\
&= \frac{3-\nu}{4\pi}\int_{\theta_1}^{\theta_2}\left(w_{,x}(p)\cos^2\theta + w_{,y}(p)\sin\theta\,\cos\theta\right)d\theta \\
&= \frac{3-\nu}{4\pi}\left\{w_{,x}(p)\left(\frac{1}{2}[\theta]_1^2 + \frac{1}{4}[\sin 2\theta]_1^2\right) + w_{,y}(p)\left(-\frac{1}{4}[\cos 2\theta]_1^2\right)\right\}
\end{aligned} \tag{2.103}$$

and

$$\begin{aligned}
B_x &= \lim_{\rho\to 0}\int_{\Gamma\rho}\frac{\partial w}{\partial n}M v_1\,ds \\
&= \lim_{\rho\to 0}\int_{\theta_1}^{\theta_2}\left[-\left(w_{,x}(q)\cos\theta + w_{,y}(q)\sin\theta\right)\right]\frac{-(1+\nu)}{4\pi}\frac{1}{\rho}(-\cos\theta)\rho d(-\theta) \\
&= \frac{1+\nu}{4\pi}\int_{\theta_1}^{\theta_2}\left(w_{,x}(q)\cos^2\theta + w_{,y}(q)\sin\theta\,\cos\theta\right)d\theta \\
&= \frac{1+\nu}{4\pi}\left\{w_{,x}(p)\left(\frac{1}{2}[\theta]_1^2 + \frac{1}{4}[\sin 2\theta]_1^2\right) + w_{,y}(p)\left(-\frac{1}{4}[\cos 2\theta]_1^2\right)\right\}
\end{aligned} \tag{2.104}$$

In obtaining the latter expression it was set as

$$\lim_{\rho\to 0} w_{,x}(q) = w_{,x}(p) \quad \text{and} \quad \lim_{\rho\to 0} w_{,y}(q) = w_{,y}(p)$$

Adding Eqs. (2.103) and (2.104) yields

$$\begin{aligned}(I_{\Gamma\rho})_x &= A_x + B_x \\ &= \frac{1}{\pi}\left\{ w_{,x}(p)\left(\frac{1}{2}[\theta]_1^2 + \frac{1}{4}[\sin 2\theta]_1^2\right) + w_{,y}(p)\left(-\frac{1}{4}[\cos 2\theta]_1^2\right)\right\}\end{aligned} \tag{2.105}$$

Choosing now $\omega = \pi/2$, then $r_\nu = -\sin\theta$ and working as previously, we obtain

$$(I_{\Gamma\rho})_y = \frac{1}{\pi}\left\{ w_{,x}(p)\left(-\frac{1}{4}[\cos 2\theta]_1^2\right) + w_{,y}(p)\left(\frac{1}{2}[\theta]_1^2 - \frac{1}{4}[\sin 2\theta]_1^2\right)\right\} \tag{2.106}$$

If the vector $\boldsymbol{\nu}$ is directed in an arbitrary direction ($\nu_x, \ \nu_y$), we have

$$\begin{aligned} I_{\Gamma\rho} &= (I_{\Gamma\rho})_x \nu_x + (I_{\Gamma\rho})_y \nu_y \\ &= \frac{1}{2\pi} w_{,x}(p)\left\{\left([\theta]_1^2 + \frac{1}{2}[\sin 2\theta]_1^2\right)\nu_x - \frac{1}{2}[\cos 2\theta]_1^2 \nu_y\right\} \\ &+ \frac{1}{2\pi} w_{,y}(p)\left\{-\frac{1}{2}[\cos 2\theta]_1^2 \nu_x + \left([\theta]_1^2\right) - \frac{1}{2}[\sin 2\theta]_1^2 \nu_y\right\}\end{aligned} \tag{2.107}$$

We next evaluate the nonzero contribution from the corner terms at $k^{(-)}$ and $k^{(+)}$

$$\begin{aligned}(w - w_p)[\![Tv_1]\!]_{k^{(-)}} &= [w_{,x}(p)\cos\theta_1 + w_{,y}(p)\sin\theta_1]\rho\left[-\frac{1-\nu}{4\pi}\frac{r_\tau}{\rho} - \frac{1-\nu}{4\pi\rho}(r_t^2 - r_n^2)r_\tau\right] \\ &= -\frac{1-\nu}{4\pi}[w_{,x}(p)\cos\theta_1 + w_{,y}(p)\sin\theta_1][1 + (r_t^2 - r_n^2)]r_\tau\end{aligned} \tag{2.108}$$

$$\begin{aligned}(w - w_p)[\![Tv_1]\!]_{k^{(+)}} &= [w_{,x}(p)\cos\theta_2 + w_{,y}(p)\sin\theta_2]\rho\left[\frac{1-\nu}{4\pi\rho}(r_t^2 - r_n^2)r_\tau + \frac{1-\nu}{4\pi}\frac{r_\tau}{\rho}\right] \\ &= \frac{1-\nu}{4\pi}[w_{,x}(p)\cos\theta_2 + w_{,y}(p)\sin\theta_2][(r_t^2 - r_n^2) + 1)]r_\tau\end{aligned} \tag{2.109}$$

Noting that (Fig. 2.16)

$$\lim_{\rho\to 0}\left[r_t^2 - r_n^2\right] = -\lim_{\rho\to 0}\cos 2\phi = 1 \tag{2.110}$$

and

$$r_\tau = r_x(-\sin\omega) + r_y\cos\omega = \nu_y\cos\theta - \nu_x\sin\theta$$

we obtain from Eqs. (2.108) and (2.109)

$$\lim_{\rho\to 0}[\![\ ]\!]_{k^{(-)}} = -\frac{1-\nu}{2\pi}\big[w_{,x}(p)\big(\cos^2\theta_1\nu_y - \cos\theta_1\sin\theta_1\nu_x\big) + w_{,y}(p)\big(\sin\theta_1\cos\theta_1\nu_y - \sin^2\theta_1\nu_x\big)\big]$$

$$\lim_{\rho\to 0}[\![\ ]\!]_{k^{(+)}} = \frac{1-\nu}{2\pi}\big[w_{,x}(p)\big(\cos^2\theta_2\nu_y - \cos\theta_2\sin\theta_2\nu_x\big) + w_{,y}(p)\big(\sin\theta_2\cos\theta_2\nu_y - \sin^2\theta_2\nu_x\big)\big]$$

Therefore,

$$\begin{aligned}[\![\ ]\!]_p &= \lim_{\rho\to 0}\left([\![\ ]\!]_{k^{(-)}} + [\![\ ]\!]_{k^{(+)}}\right) \\ &= \frac{1-\nu}{2\pi}\left[w_{,x}(p)\left(\left[\cos^2\theta\right]_1^2\nu_y - \frac{1}{2}[\sin 2\theta]_1^2\nu_x\right)\right. \\ &\left. + w_{,y}(p)\left(\frac{1}{2}[\sin 2\theta]_1^2\nu_y - \left[\sin^2\theta\right]_1^2\nu_x\right)\right]\end{aligned} \tag{2.111}$$

Taking now into account that

$$[\theta]_1^2 = \theta_2 - \theta_1 = -\alpha\pi \tag{2.112a}$$

$$[\cos 2\theta]_1^2 = \left[1 - 2\sin^2\theta\right]_1^2 = -2\left[\sin^2\theta\right]_1^2 \tag{2.112b}$$

$$\left[\cos^2\theta\right]_1^2 = -\left[\sin^2\theta\right]_1^2 \tag{2.112c}$$

we obtain from Eqs. (2.107) and (2.111)

$$I_{\Gamma\rho} - [\![\ ]\!]_p = -\left[\alpha_x w_{,x}(p) + \alpha_y w_{,y}(p)\right] \tag{2.113}$$

where

$$\alpha_x = \frac{\alpha}{2}\nu_x - \frac{\nu}{2\pi}\left[\frac{1}{2}\sin 2\theta\nu_x + \sin^2\theta\nu_y\right]_{\theta_1}^{\theta_2} \tag{2.114}$$

$$\alpha_y = \frac{\alpha}{2}\nu_y - \frac{\nu}{2\pi}\left[\sin^2\theta\nu_x - \frac{1}{2}\sin 2\theta\nu_y\right]_{\theta_1}^{\theta_2} \tag{2.115}$$

Finally, taking into account Eq. (2.89a) and using Eq. (2.113), Eq. (2.96) becomes

$$\begin{aligned}\alpha_x w_{,x}(p) + \alpha_y w_{,y}(p) &= \int_\Omega v_1 f d\Omega \\ &+ \int_\Gamma\left(v_1\, Vw - [w - w(p)]\, V v_1 - \frac{\partial v_1}{\partial n}Mw + \frac{\partial w}{\partial n}M v_1\right) ds \\ &- \sum_k\left(v_1[\![Tw]\!]_k - [w - w(p)][\![T v_1]\!]_k\right)\end{aligned} \tag{2.116}$$

For points p where the boundary is smooth the normal vector $\boldsymbol{\nu}$ to the boundary is uniquely determined, $\alpha = 1, \theta_2 = \theta_1 + \pi$ and the left-hand side of Eq. (2.116) reduces to $\dfrac{\partial w(p)}{\partial n}$.

Differentiating Eq. (2.84) with respect to the direction $\boldsymbol{\nu}$ and applying the resulting expression for $w = 1$ we obtain

$$-\int_\Gamma V v_1 \, ds + \sum_k (\llbracket T v_1 \rrbracket_k) = 0 \tag{2.117}$$

Thus, on the base of Eq. (2.117), Eq. (2.116) gives the second boundary integral equation in the form

$$\begin{aligned} \alpha_x w_x(p) + \alpha_y w_y(p) &= \int_\Omega v_1 f d\Omega \\ &+ \int_\Gamma \left(v_1 Vw - w V v_1 - \frac{\partial v_1}{\partial n} Mw + \frac{\partial w}{\partial n} M v_1 \right) ds \\ &- \sum_k \left(v_1 \llbracket Tw \rrbracket_k - w \llbracket T v_1 \rrbracket_k \right) \end{aligned} \tag{2.118}$$

It should be noted that, if p is a corner point, the relevant corner term must be omitted from the sum in Eqs. (2.116), (2.117) and (2.118).

Equations (2.92) and (2.116) or its equivalent (2.118) constitute a pair of compatibility relations for the four boundary functions $w, \partial w/\partial n, Mw$ and Vw along with the corner forces. Of all the four boundary functions only two are determined by the support and loading conditions on the boundary. The two unknown boundary functions can be established from the solution of Eqs. (2.92) and (2.118) if they are treated as coupled boundary integral equations. At the corner points on the boundary, due to the abrupt change in the orientation of the normal, the normal slope, the bending moment, and the equivalent shear force may have discontinuity and a concentrated force may exist. Therefore, there are eight boundary values associated with a corner; two values for the normal slope $w_{,n}^{-}, w_{,n}^{+}$, two values of the bending moment Mw^{-}, Mw^{+}, two values of the effective shear force Vw^{-}, Vw^{+}, one value of the displacement w and one value of the concentrated force $\llbracket Tw \rrbracket$. Therefore, eight independent relations are required at each corner point. One of these is obtained from Eq. (2.92) and two others from Eq. (2.118). The five additional independent relations are furnished by the boundary conditions and the asymptotic smoothness requirement of the solution in the neighborhood of the corner [27,28]. Some of the more common cases are given in Table 2.1 [13].

2.3.5 Stress resultants

Once the unspecified boundary quantities are established from the solution of the boundary integral equations, the deflection at any point $P(x, y)$ is evaluated from Eq. (2.84). The bending moments M_x, M_y, M_{xy} and the shear forces Q_x, Q_y at a point P in the interior of the plate are evaluated from Eqs. (2.19)

TABLE 2.1 Corner boundary conditions (C=clamped, SS=simply supported, F=free)

Corner support		Relations
− side	+ side	
C	C	$w=0,\ Mw^-=Mw^+=0,\ w_{,n}^-=w_{,n}^+=0$
SS	SS	$w=0,\ Mw^-=Mw^+=0,\ w_{,n}^-=w_{,n}^+=0$
F	F	$[\![Tw]\!]=0,\ Mw^-=Mw^+=0,\ Vw^-=Vw^+=0$
C	SS	$w=0,\ w_{,n}^-=w_{,n}^+=0,\ Mw^-=Mw^+=0$
C	F	$w=0,\ w_{,n}^-=0,\ w_{,n}^+=0,\ Mw^+=0, Vw^+=0$
SS	F	$w=0,\ Mw^-=Mw^+, Vw^+=0,\ w_{,n}^+=-w_{,n}^-\cos\alpha\pi$

and (2.21), respectively. Since they are combinations of the second and third derivatives, it suffices to evaluate only $w_{,xx}$, $w_{,yy}$, $w_{,xy}$ and $\nabla^2 w_{,x}$, $\nabla^2 w_{,y}$. These quantities may be evaluated from the computed values of the deflections using numerical differentiation. However, the accuracy of the results improves and the computer time is considerably reduced if they are evaluated from their integral representations. The expressions of the required derivatives can be established by direct differentiation of the deflection, Eq. (2.84). Inasmuch as the kernels in the integrand are continuous functions, the derivative passes to the kernels of the integral. Therefore, it is necessary to find the expressions of the derivatives of the kernels of Eq. (2.84). The resulting thus expressions are too lengthy and complicated. Nevertheless, for the interested to use them, they are given in Appendix A. Another way would be the use of a symbolic language, e.g., MAPLE, to find the expressions of the derivatives of the kernels and insert them directly in the computer program to avoid possible mistakes. In the following, we present another procedure, which uses simpler expressions for the kernel derivatives.

This procedure uses the integral representation (2.85), which we rewrite here for convenience

$$w(P)=\int_{\Omega} v f d\Omega - D\int_{\Gamma}\left(v\nabla^2 w_{,n} - w\nabla^2 v_{,n} - v_{,n}\nabla^2 w + w_{,n}\nabla^2 v\right) ds \qquad (2.119)$$

In Eq. (2.119), the boundary quantities w and $w_{,n}$ are known from the solution of the boundary integral equations, while $\nabla^2 w$, $\nabla^2 w_{,n}$ are unknown. These quantities, however, can be readily evaluated as follows.

Taking the Laplacian of Eq. (2.119) and noting that $\nabla^4 v=0$, $\nabla^4 v_{,n}=0$ we obtain

$$\nabla^2 w(P)=\int_{\Omega}\nabla^2 v f d\Omega - D\int_{\Gamma}\left(\nabla^2 v\nabla^2 w_{,n} - \nabla^2 v_{,n}\nabla^2 w\right) ds \qquad (2.120)$$

Equations (2.119) and (2.120) for $P\rightarrow p\in\Gamma$ yield [2,29]

$$\frac{1}{2}w(P)=\int_{\Omega}vfd\Omega-D\int_{\Gamma}\left(v\nabla^2 w_{,n}-w\nabla^2 v_{,n}-v_{,n}\nabla^2 w+w_{,n}\nabla^2 v\right)ds \quad (2.121)$$

$$\frac{1}{2}\nabla^2 w(P)=\int_{\Omega}\nabla^2 vfd\Omega-D\int_{\Gamma}\left(\nabla^2 v\nabla^2 w_{,n}-\nabla^2 v_{,n}\nabla^2 w\right)ds \quad (2.122)$$

where v, $v_{,n}$ are given by Eqs. (2.76a,b) and

$$\nabla^2 v=\frac{1}{2\pi D}(\ln r+1) \quad (2.123)$$

$$\nabla^2 v_{,n}=\frac{1}{2\pi D}\frac{r_n}{r} \quad (2.124)$$

Equations (2.121) and (2.122) are two coupled boundary integral equations, which are solved to yield the boundary quantities $\nabla^2 w$ and $\nabla^2 w_{,n}$. Subsequently, the required second and third derivatives are obtained by direct differentiation of Eq. (2.119). Thus, we have

$$\begin{aligned} w_{,xx}(P) &= \int_{\Omega} v_{,xx} fd\Omega \\ &\quad -D\int_{\Gamma}\left(v_{,xx}\nabla^2 w_{,n}-w\nabla^2 v_{,nxx}-v_{,nxx}\nabla^2 w+w_{,n}\nabla^2 v_{,xx}\right)ds \end{aligned} \quad (2.125a)$$

$$\begin{aligned} w_{,yy}(P) &= \int_{\Omega} v_{,yy} fd\Omega \\ &\quad -D\int_{\Gamma}\left(v_{,yy}\nabla^2 w_{,n}-w\nabla^2 v_{,nyy}-v_{,nyy}\nabla^2 w+w_{,n}\nabla^2 v_{,yy}\right)ds \end{aligned} \quad (2.125b)$$

$$\begin{aligned} w_{,xy}(P) &= \int_{\Omega} v_{,xy} fd\Omega \\ &\quad -D\int_{\Gamma}\left(v_{,xy}\nabla^2 w_{,n}-w\nabla^2 v_{,nxy}-v_{,nxy}\nabla^2 w+w_{,n}\nabla^2 v_{,xy}\right)ds \end{aligned} \quad (2.125c)$$

$$\nabla^2 w_{,x}(P)=\int_{\Omega}\nabla^2 v_{,x} fd\Omega-D\int_{\Gamma}\left(\nabla^2 v_{,x}\nabla^2 w_{,n}-\nabla^2 v_{,nx}\nabla^2 w\right)ds \quad (2.125d)$$

$$\nabla^2 w_{,y}(P)=\int_{\Omega}\nabla^2 v_{,y} fd\Omega-D\int_{\Gamma}\left(\nabla^2 v_{,y}\nabla^2 w_{,n}-\nabla^2 v_{,ny}\nabla^2 w\right)ds \quad (2.125e)$$

where the kernels of the above integral representations are given as

$$v_{,xx}=\frac{1}{8\pi D}\left(2\ln r+2r_x^2+1\right) \quad (2.126a)$$

$$v_{,yy}=\frac{1}{8\pi D}\left(2\ln r+2r_y^2+1\right) \quad (2.126b)$$

$$v_{,xy}=\frac{1}{4\pi D}r_x r_y \quad (2.126c)$$

$$v_{,nxx}=\frac{1}{4\pi D}\frac{1}{r}\left(r_n+2r_xr_yr_t\right) \tag{2.126d}$$

$$v_{,nyy}=\frac{1}{4\pi D}\frac{1}{r}\left(r_n-2r_xr_yr_t\right) \tag{2.126e}$$

$$v_{,nxy}=\frac{1}{4\pi D}\frac{1}{r}\left(r_y^2-r_x^2\right)r_t \tag{2.126f}$$

$$\nabla^2 v_{,x}=\frac{1}{2\pi D}\frac{r_x}{r} \tag{2.126g}$$

$$\nabla^2 v_{,y}=\frac{1}{2\pi D}\frac{r_y}{r} \tag{2.126h}$$

$$\nabla^2 v_{,xx}=\frac{1}{2\pi D}\frac{r_y^2-r_x^2}{r^2} \tag{2.126i}$$

$$\nabla^2 v_{,yy}=\frac{1}{2\pi D}\frac{r_x^2-r_y^2}{r^2} \tag{2.126j}$$

$$\nabla^2 v_{,xy}=-\frac{1}{\pi D}\frac{r_xr_y}{r^2} \tag{2.126k}$$

$$\nabla^2 v_{,nx}=\frac{r_tr_y-r_nr_x}{2\pi Dr^2} \tag{2.126l}$$

$$\nabla^2 v_{,ny}=-\frac{r_tr_x+r_nr_y}{2\pi Dr^2} \tag{2.126m}$$

$$\nabla^2 v_{,nxx}=\frac{1}{\pi Dr^3}\left[r_n\left(r_x^2-r_y^2\right)-2r_xr_yr_t\right] \tag{2.126n}$$

$$\nabla^2 v_{,nyy}=\frac{1}{\pi Dr^3}\left[r_n\left(r_y^2-r_x^2\right)+2r_xr_yr_t\right] \tag{2.126o}$$

$$\nabla^2 v_{,nxy}=\frac{1}{\pi Dr^3}\left[r_t\left(r_x^2-r_y^2\right)+2r_xr_yr_n\right] \tag{2.126p}$$

The kernels (2.126) are regular, because point P is in the interior of the domain, while the integration point lies on the boundary, hence $r\neq 0$. Apparently, these expressions of the kernels are much simpler than those given in Appendix A.

2.4 NUMERICAL SOLUTION OF THE BOUNDARY INTEGRAL EQUATIONS

2.4.1 BEM with constant elements

The boundary integral equations (2.92) and (2.118) are solved using the boundary element technique (see [1]). Thus the boundary is discretized into

TABLE 2.2 Critical angles for moments and shear forces (C = clamped, SS = simply supported, F = free) [31]

Boundary conditions		Moment	Shear force
− side	+ side		
C	C	180°	126.283696°
C	SS	128.726699°	90°
C	F	95.349059°	52.054347°
SS	SS	90°	60°
SS	F	90°	51.123345°
F	F	180°	77.753311°

N boundary elements, which may be constant, linear, or parabolic. Unlike the integral equations arising in the potential and the elastostatic problems (see [1]), where the kernels exhibit a logarithmic or a Cauchy type singularity, some kernels in the integral equations for the plate bending problem may be hypersingular. Moreover, the need to include the corner terms in the integral equations and to treat the singularities of the bending moment and the effective shear force at these points, when the internal corner angle exceeds the values given in Table 2.2, requires special care in solving the integral equations [8].

Several techniques have been reported in the literature, some of them using higher-order boundary elements, which may differently approximate the boundary quantities [8,30]. Nevertheless, all these techniques fail to encounter the oscillatory distribution of the computed effective shear force near the corner or at points of change of the boundary conditions. Besides, the higher-order elements, e.g., parabolic or cubic, rather complicate the numerical solution, at least its programming. On the other hand, the use of constant boundary elements simplifies the numerical evaluation of the singular kernels, especially those with hypersingular behavior, and facilitates the programming of the numerical procedure, while the desired accuracy can be achieved by increasing the number of the boundary elements. Therefore, the BEM with constant elements is utilized here to solve the boundary integral equations for the plate problem with some amendments to evaluate the corner forces. A detailed presentation of the numerical procedure is given in Chapter 4 of Ref. [1].

The boundary is divided into N constant elements. Thus, the distribution of the boundary quantities w, $w_{,n}$ Mw, Vw are taken constant on each element and equal to their value at the nodal point, which lies at the midpoint of the element. Designating the values of these quantities at the $i-th$ node with w^i, w^i_n, M^i_n, V^i_n ($i = 1,2,\ldots,N$) and with w^k_c, R^k ($k = 1,2,\ldots,N_c$) the displacements and concentrated forces at the corners, we obtain

(i) N equations from Eq. (2.92) when applied at the N element nodal points

$$\frac{1}{2}w^i+\sum_{j=1}^{N}[Vv]^{ij}w^j-\sum_{j=1}^{N}[Mv]^{ij}w_n^j-\sum_{k=1}^{N_c}[\![Tv]\!]^{ik}w_c^k=\sum_{j=1}^{N}[v]^{ij}Vw^j-\sum_{j=1}^{N}[v_n]^{ij}Mw^j-\sum_{k=1}^{N_c}[v]^{ik}R^k+F_1{}^i \tag{2.127}$$

$i=1,2,\ldots,N$

(ii) N_c equations from Eq. (2.92) when applied at the N_c corner points

$$\frac{\alpha}{2\pi}w_c^l+\sum_{j=1}^{N}[Vv]^{lj}w^j-\sum_{j=1}^{N}[Mv]^{lj}w_n^j-\sum_{\substack{k=1\\k\neq l}}^{N_c}[\![Tv]\!]^{lk}w_c^k=\sum_{j=1}^{N}[v]^{lj}Vw^j-\sum_{j=1}^{N}[v_n]^{lj}Mw^j-\sum_{\substack{k=1\\k\neq l}}^{N_c}[v]^{lk}R^k+F_2{}^l \tag{2.128}$$

$l=1,2,\ldots,N_c$

(iii) N equations from Eq. (2.118) when applied at the N element nodal points

$$\frac{1}{2}w_n^i+\sum_{j=1}^{N}[Vv_1]^{ij}w^j-\sum_{j=1}^{N}[Mv_1]^{ij}w_n^j-\sum_{k=1}^{N_c}[\![Tv_1]\!]^{ik}w_c^k=\sum_{j=1}^{N}[v_1]^{ij}Vw^j-\sum_{j=1}^{N}[v_{1n}]^{ij}Mw^j-\sum_{k=1}^{N_c}[v_1]^{ik}R^k+F_3{}^i \tag{2.129}$$

$i=1,2,\ldots,N$

In previous equations it has been set

$$[v]^{ij}=\int_{\Gamma_j}v(p_i,q)\,ds_q,\qquad [v_n]^{ij}=\int_{\Gamma_j}\frac{\partial v(p_i,q)}{\partial n_q}\,ds_q \tag{2.130a,b}$$

$$[Mv]^{ij}=\int_{\Gamma_j}Mv(p_i,q)\,ds_q,\qquad [Vv]^{ij}=\int_{\Gamma_j}Vv(p_i,q)\,ds_q \tag{2.130c,d}$$

$$[v_1]^{ij}=\int_{\Gamma_j}v_1(p_i,q)\,ds_q,\qquad [v_{1n}]^{ij}=\int_{\Gamma_j}\frac{\partial v_1(p_i,q)}{\partial n_q}\,ds_q \tag{2.130e,f}$$

$$[Mv_1]^{ij}=\int_{\Gamma_j}Mv_1(p_i,q)\,ds_q,\qquad [Vv_1]^{ij}=\int_{\Gamma_j}Vv_1(p_i,q)\,ds_q \tag{2.130g,h}$$

$$[\![Tv]\!]^{ik}=[\![Tv(p_i,q_k)]\!],\qquad [\![Tv_1]\!]^{ik}=[\![Tv_1(p_i,q_k)]\!] \tag{2.130i,j}$$

$$F_1^i=\int_{\Omega}fv(p_i,Q)\,d\Omega_Q \tag{2.130k}$$

$$F_2^l=\int_{\Omega}fv(p_l,Q)\,d\Omega_Q \tag{2.130l}$$

$$F_3^i = \int_\Omega f v_1(p_i, Q)\, d\Omega_Q \tag{2.130m}$$

where Γ_j is the segment (straight line) on which the j –*th* node is located and over which integration is carried out, and p_i is the nodal point of the i –*th* element; p_l, q_k are corner points.

Equations (2.127), (2.128), and (2.129) can be combined as

$$\mathbf{H}\begin{Bmatrix} \mathbf{w} \\ \mathbf{w}_c \\ \mathbf{w}_n \end{Bmatrix} = \mathbf{G}\begin{Bmatrix} \mathbf{V} \\ \mathbf{R} \\ \mathbf{M} \end{Bmatrix} + \mathbf{F} \tag{2.131}$$

where

$$\mathbf{H} = \begin{bmatrix} [Vv]^{ij} + \frac{1}{2}\delta_{ij} & -[\![Tv]\!]^{ik} & -[Mv]^{ij} \\ [Vv]^{lj} + \frac{\alpha}{2\pi}\delta_{lj} & -[\![Tv]\!]^{lk} & -[Mv]^{lj} \\ [Vv_1]^{ij} & -[\![Tv_1]\!]^{ik} & -[Mv_1]^{ij} + \frac{1}{2}\delta_{ij} \end{bmatrix}_{(2N+N_c)\times(2N+N_c)} \tag{2.132}$$

$$\mathbf{G} = \begin{bmatrix} [v]^{ij} & -[v]^{ik} & -[v_n]^{ij} \\ [v]^{lj} & -[v]^{lk} & -[v_n]^{lj} \\ [v_1]^{ij} & -[v_1]^{ik} & -[v_{1n}]^{ij} \end{bmatrix}_{(2N+N_c)\times(2N+N_c)} \tag{2.133}$$

$$\mathbf{F} = \begin{Bmatrix} \mathbf{F}_1 \\ \mathbf{F}_2 \\ \mathbf{F}_3 \end{Bmatrix} \tag{2.134}$$

$$\mathbf{F}_1 = \{F_1{}^i\}_{N\times 1},\quad \mathbf{F}_2 = \{F_2{}^l\}_{N_c\times 1},\quad \mathbf{F}_3 = \{F_3{}^i\}_{N\times 1} \tag{2.135a,b,c}$$

$$\mathbf{w} = \{w^i\}_{N\times 1},\quad \mathbf{w}_c = \{w_c^l\}_{N_c\times 1},\quad \mathbf{w}_n = \{w_n^i\}_{N\times 1} \tag{2.136a,b,c}$$

$$\mathbf{V} = \{Vw^i\}_{N\times 1},\quad \mathbf{R} = \{R^k\}_{N_c\times 1},\quad \mathbf{M} = \{Mw^i\}_{N\times 1} \tag{2.137a,b,c}$$

Equation (2.131) constitutes a system of $2N + N_c$ linear equations for the $4N + 2N_c$ unknown boundary quantities. The additional required equations result from the specified boundary conditions. Thus, applying Eqs. (2.44) at the N element nodal points and Eq. (2.47) at the N_c corner points yields

$$\overline{\mathbf{H}}\begin{Bmatrix} \mathbf{w} \\ \mathbf{w}_c \\ \mathbf{w}_n \end{Bmatrix} = \overline{\mathbf{G}}\begin{Bmatrix} \mathbf{V} \\ \mathbf{R} \\ \mathbf{M} \end{Bmatrix} + \overline{\mathbf{F}} \tag{2.138}$$

where

$$\overline{\mathbf{H}} = \begin{bmatrix} \boldsymbol{\alpha}_1 & \mathbf{0} & \mathbf{0} \\ \mathbf{0} & \mathbf{c}_1 & \mathbf{0} \\ \mathbf{0} & \mathbf{0} & \boldsymbol{\beta}_1 \end{bmatrix},\quad \overline{\mathbf{G}} = \begin{bmatrix} \boldsymbol{\alpha}_2 & \mathbf{0} & \mathbf{0} \\ \mathbf{0} & \mathbf{c}_2 & \mathbf{0} \\ \mathbf{0} & \mathbf{0} & \boldsymbol{\beta}_2 \end{bmatrix},\quad \overline{\mathbf{F}} = \begin{Bmatrix} \boldsymbol{\alpha}_3 \\ \mathbf{c}_3 \\ \boldsymbol{\beta}_3 \end{Bmatrix} \tag{2.139a,b,c}$$

$\boldsymbol{\alpha}_i$, $\boldsymbol{\beta}_i$ $(i=1,2)$ are $N\times N$ diagonal matrices, $\mathbf{c}_i$ are $N_c\times N_c$ $(i=1,2)$ diagonal matrices, and $\boldsymbol{\alpha}_3$, $\boldsymbol{\beta}_3$, $\mathbf{c}_3$ are vectors with respective dimensions.

Equations (2.131) and (2.138) are combined and solved to yield the boundary quantities $\mathbf{w}$, $\mathbf{w}_c$, $\mathbf{w}_n$, $\mathbf{V}$, $\mathbf{R}$, $\mathbf{M}$, which are used in Eq. (2.84) to obtain the deflection at any point $P(x, y)$ inside Ω. Using the same boundary discretization, Eq. (2.84) becomes

$$w(P)=F_i+\sum_{j=1}^{N}\left([v]_P^j Vw^j-[Vv]_P^j w^j-[v_n]_P^j M w^j+[Mv]_P^j w_n^j\right) \\ -\sum_{k=1}^{N_c}\left(v_P^k R^k-[\![Tv]\!]_P^k w^k\right) \tag{2.140}$$

In this case the influence matrices $[v]_P^j, [Vv]_P^j, [v_n]_P^j, [Mv]_P^j$ and $v_P^k, [\![Tv]\!]_P^k$ are evaluated with $r=\left|q_j-P\right|$ and $r=|q_k-P|$.

2.4.2 Evaluation of the line integrals

The formulation of the matrices **G** and **H** defined in Eqs. (2.132) and (2.133) requires the evaluation of the line integrals (2.130a)-(2.130 h). Two cases are distinguished:

(i) *Off-diagonal elements*, $i\neq j$
In this case the point $p_i(x_i, y_i)$ lies outside the j–element (see Fig. 2.17), which means that the distance $r=|q-p_i|$ does not vanish and, consequently, the integral is regular. Thus, the standard Gaussian integration described in Appendix B is employed for its evaluation.

(ii) *Diagonal elements*, $i=j$
In this case the node p_i coincides with node p_j and r lies on the element. Consequently, the distance r takes the value zero and some of the line integrals may be singular or hypersingular. Hence, their evaluation requires special care.

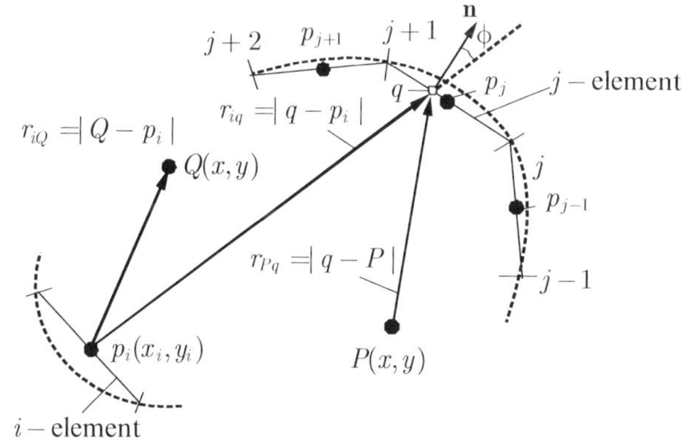

FIGURE 2.17 Nodal-point location and notation in constant element discretization.

We observe that the angle $\phi = \measuredangle r, \mathbf{n}$ is either $\phi = \pi/2$ or $\phi = 3\pi/2$, which yields $r_n = \cos\phi = 0$, $r_t = -\sin\phi = \mp 1$. Moreover, it is $\kappa = 0$ for constant elements. On the base of Eqs. (2.76) we have for $r \to 0$

$$\lim_{r\to 0} v = 0 \tag{2.141a}$$

$$\lim_{r\to 0}\frac{\partial v}{\partial n} = 0 \tag{2.141b}$$

$$\lim_{r\to 0} Mv = -\frac{1}{8\pi}\lim_{r\to 0}[2(1+\nu)\ln r + (1+3\nu)] \tag{2.141c}$$

$$\lim_{r\to 0} Vv = -\lim_{r\to 0}\left\{\frac{1}{4\pi r}[2r_n + (1-\nu)(r_n - \kappa r)(r_n^2 - r_t^2]\right\} \tag{2.141d}$$

Apparently, the integrals $[v]^{ii}$, $[v_n]^{ii}$ are regular, while the integrals $[Mv]^{ii}$ $[Vv]^{ii}$ are singular exhibiting a logarithmic ($\ln r$) and a Cauchy type ($1/r$) singularity, respectively. They can be evaluated analytically (see [1], Section 4.3). Thus, we obtain

$$[Mv]^{ii} = -\frac{(1+\nu)}{2\pi}\frac{l_i}{2}[\ln(l_i/2) - 1] + (1+3\nu)l_i \tag{2.142a}$$

$$[Vv]^{ii} = 0. \tag{2.142b}$$

Similarly, taking into account that $r_\nu = -\cos\phi = 0$, $r_\tau = \sin\phi = \pm 1$ Eqs. (2.97) yield

$$\lim_{r\to 0} v_1 = 0 \tag{2.143a}$$

$$\lim_{r\to 0}\frac{\partial v_1}{\partial \nu} = \frac{1}{8\pi D}\lim_{r\to 0}[2(r_n r_\nu + 1)\ln r + 1] \tag{2.143b}$$

$$\lim_{r\to 0} Mv_1 = -\lim_{r\to 0}\left\{\frac{1}{4\pi r}[(1+\nu)r_\nu + 2(1-\nu)r_n\right\} \tag{2.143c}$$

$$\lim_{r\to 0} Vv_1 = \lim_{r\to 0}\left[-(1+\nu)\frac{r_n r_\nu}{4\pi r^2} - (1+\nu)\frac{1}{4\pi r^2} - (1-\nu)\frac{r_n^2}{\pi r^2} + (1-\nu)\frac{\kappa r_n}{\pi r}\right] \tag{2.143d}$$

Therefore, the integrals $[v_1]^{ii}$ are regular, while the integrals $[v_{1n}]^{ii}$,$[Mv_1]^{ii}$ and $[Vv_1]^{ii}$ exhibit logarithmic ($\ln r$), Cauchy type ($1/r$), and $1/r^2$, singularities. They can be evaluated analytically. The first two are obtained as (see [1], Section 4.3).

$$[v_{1n}]^{ii} = \frac{l_i}{8\pi D}[2\ln(l_i/2) - 1] \tag{2.144a}$$

$$[Mv_1]^{ii} = 0 \tag{2.144b}$$

Finally, the integrals $[Vv_1]^{ii}$ require special treatment. It includes terms of the form

$$I(\xi) = \int_a^b \frac{1}{(\xi - \xi_i)^2} d\xi \tag{2.145}$$

This integral does not exist, either as an improper or a singular integral. But it appears in Vv_1 i.e., finite on the boundary. Therefore, from the physical point of view, this integral must exist on Γ. The following definition can be assigned to this integral if $a < \xi_i < b$ [31]:

$$I(\xi) = \int_a^b \frac{1}{(\xi - \xi_i)^2} d\xi \overset{def}{=} \lim_{\varepsilon \to 0} \left[\int_a^{\xi_i - \varepsilon} \frac{1}{(\xi - \xi_i)^2} d\xi + \int_{\xi_i + \varepsilon}^b \frac{1}{(\xi - \xi_i)^2} d\xi - \frac{2}{\varepsilon} \right] \tag{2.146}$$

This definition coincides with Hadamard's definition of the hypersingular integral [32]. Integrating by parts the integrals in the right-hand side of Eq. (2.146) and taking the limit, we obtain

$$I(\xi) = \int_a^b \frac{1}{(\xi - \xi_i)^2} d\xi = \frac{1}{a - \xi_i} - \frac{1}{b - \xi_i} \tag{2.147}$$

which for $a = -l_i/2$, $b = l_i/2$ and $\xi_i = 0$ gives

$$I(\xi) = -\frac{4}{l_i} \tag{2.148}$$

Using Eq. (2.148) and integrating analytically the Cauchy singularity, we obtain

$$[Vv_1]^{ii} = -\frac{(1+\nu)}{\pi l_i} \tag{2.149}$$

The previous procedure can be avoided if the indirect method (see [1], Section 5.4.3) is employed for the evaluation of $[Vv_1]^{ii}$ as follows.

The function $w = 1$ is a solution of the plate equation, Eq. (2.42), for f = 0. Thus, applying Eq. (2.129) for $w = 1$ and taking into account that $w_{,n} = Mw = Vw = 0$ on the boundary and ${F_3}^i = 0$, we obtain

$$\sum_{j=1}^{N} [Vv_1]^{ij} - \sum_{k=1}^{N_c} [\![Tv_1]\!]^{ik} = 0 \tag{2.150}$$

which yields

$$[Vv_1]^{ii} = -\sum_{\substack{j=1 \\ j \neq i}}^{N} [Vv_1]^{ij} + \sum_{k=1}^{N_c} [\![Tv_1]\!]^{ik} \tag{2.151}$$

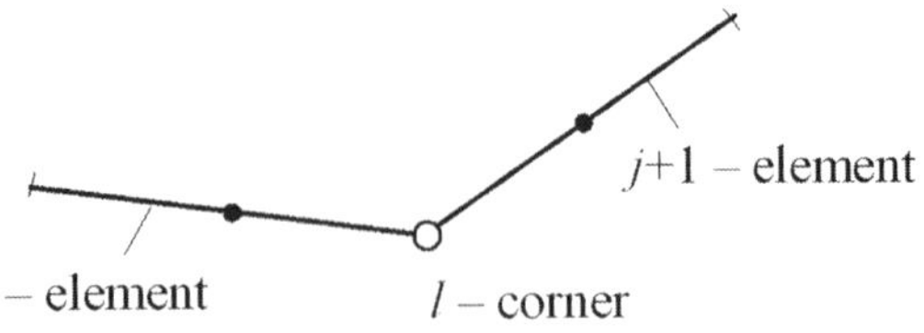

FIGURE 2.18 Corner point and adjacent elements.

Singular boundary integrals appear also in Eq. (2.128) when the integration is performed on the elements adjacent to corner l (Fig. 2.18). In this case the point where $r=0$ is one of the element ends. The integrals $[v]^{lj}$ and $[v_n]^{lj}$ are regular and they are evaluated using Gaussian integration. The integrals, $[Mv]^{lj}$ and $[Vv]^{lj}$ however, exhibit logarithmic and Cauchy type singularity, respectively, and they are evaluated analytically (see [1], Section 4.3). Thus, we have

$$\begin{aligned}[Mv]^{lj} &= -\frac{1}{8\pi}\left[2(1+\nu)\int_0^{l_j}\ln r ds + (1+3\nu)\int_0^{l_j} ds\right] \\ &= -\frac{l_j}{8\pi}\left[2(1+\nu)\left[\left(\ln l_j - 1\right) + (1+3\nu)\right]\right.\end{aligned} \tag{2.152a}$$

Similarly, we obtain

$$[Mv]^{l,j+1} = -\frac{l_{j+1}}{8\pi}\left[2(1+\nu)\left[\left(\ln l_{j+1} - 1\right) + (1+3\nu)\right]\right. \tag{2.152b}$$

Further, taking into account that for the constant element it is $\int_{\Gamma_j}\frac{\cos\phi}{r}ds=0$ (see [1], Section 4.3) and performing analytical integration we obtain

$$[Vv]^{lj} = [Vv]^{l,j+1} = 0 \tag{2.153}$$

Finally, it is $Tv=0$ for $r=0$. Hence,

$$[\![Tv]\!]^{ll} = 0 \tag{2.154}$$

It should be noted that it is convenient to treat points where the boundary conditions change as corner points.

2.4.3 Evaluation of the domain integrals

In evaluating the domain integrals (2.130 k, l, m) we may distinguish the following loading cases:

(a) The plate is subjected to a concentrated load F at a point Q_0. In this case the loading can be represented as

$$f(Q) = F\delta(Q - Q_0) \tag{2.155}$$

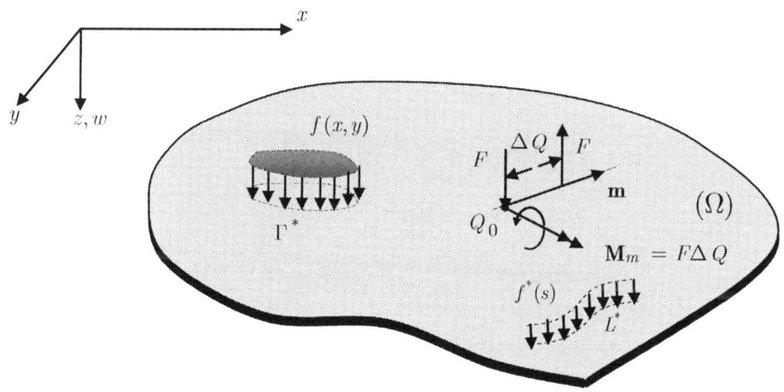

FIGURE 2.19 Concentrated moment, patch and line load.

Using relation (1.33) the domain integrals (2.130 k,l,m) are evaluated as

$$F_1^i = Fv(p_i - Q_0), \quad F_2^l = Fv(p_l - Q_0), \quad F_3^i = Fv_1(p_i - Q_0) \tag{2.156}$$

(b) The plate is subjected to a concentrated moment M_m at a point Q_0 acting about the vector normal to $\mathbf{m}(m_x, m_y)$. Referring to Fig. 2.19, the loading can be represented as

$$\begin{aligned} f(Q) &= \lim_{\Delta Q \to 0} \left\{ F\Delta Q \frac{\delta(Q-Q_0) - \delta[Q-(Q_0+\Delta Q)]}{\Delta Q} \right\} \\ &= M_m \frac{\partial \delta(Q-Q_0)}{\partial m} \end{aligned} \tag{2.157}$$

Using relation (1.42) we obtain

$$F_1^i = -M_m \frac{\partial v(p_i - Q_0)}{\partial m} \tag{2.158a}$$

$$F_2^l = -M_m \frac{\partial v(p_l - Q_0)}{\partial m} \tag{2.158b}$$

$$F_3^i = -M_m \frac{\partial v_1(p_i - Q_0)}{\partial m} \tag{2.158c}$$

(c) The plate is subjected to line load $f^*(s)$ acting along a line L^*. In this case the domain integrals become line integrals on L^*, i.e.,

$$\int_\Omega v f d\Omega = \int_{L*} v f^*(s) ds \tag{2.159a}$$

$$\int_\Omega v_1 f d\Omega = \int_{L*} v_1 f^*(s) ds \tag{2.159b}$$

The above integrals can be approximated using constant elements. Thus, dividing the curve L^* into L elements, the integrals (2.130 k,l,m) can be evaluated from the expressions

$$F_1^i = \sum_{k=1}^{L} f_k^* \int_{L_k} v(p_i, s)\, ds \tag{2.160a}$$

$$F_2^l = \sum_{k=1}^{L} f_k^* \int_{L_k} v(p_l, s)\, ds \tag{2.160b}$$

$$F_3^i = \sum_{k=1}^{L} f_k^* \int_{L_k} v_1(p_i, s)\, ds \tag{2.160c}$$

(d) The plate is subjected to a uniform or a linear varying loading distributed over an area $\Omega^* \subseteq \Omega$ of the plate bounded by the curve Γ^*. In this case, which is very usual in engineering praxis, it is $\nabla^2 f = 0$ and the domain integrals can be readily converted to boundary line integrals on Γ^* using Green's identity. Thus, applying Eq. (1.6) of Chapter 1 for the functions $v = f$ and $u = \hat{u}$, where $\hat{u}$ is a particular solution of the equation

$$\nabla^2 \hat{u} = v, \quad v = r^2 \ln r / 8\pi D \tag{2.161}$$

we obtain

$$\int_{\Omega*} v f d\Omega = \int_{\Gamma*} (f \hat{u}_{,n} - f_{,n} \hat{u})\, ds \tag{2.162}$$

The function $\hat{u}$ results from the integration of Eq. (2.161). Thus, we have

$$\hat{u} = r^4 (2 \ln r - 1)/256\pi D \tag{2.163}$$

and

$$\hat{u}_{,n} = r^3 (2 \ln r - 0.5) r_{n} / 64\pi D \tag{2.164}$$

Differentiating Eq. (2.162) with respect to $\boldsymbol{\nu}$ we obtain

$$\int_{\Omega^*} v_1 f d\Omega = \int_{\Gamma^*} (f \hat{u}_{1,n} - f_{,n} \hat{u}_1)\, ds \tag{2.165}$$

where

$$\hat{u}_1 = r^3 (2 \ln r - 0.5) r_{\nu} / 64\pi D \tag{2.166a}$$

$$\hat{u}_{1,n} = \frac{1}{64\pi D} r^2 [r_n r_{\nu} (6 \ln r + 0.5) + r_{\tau} r_t (2 \ln r - 0.5)] \tag{2.166b}$$

Hence, the integrals (2.130 k, l, m) can be evaluated from the expressions

$$F_1^i = \sum_{j=1}^{N} \int_{\Gamma_j} \left[f\hat{u}(r_{ij})_{,n} - f_{,n}\hat{u}(r_{ij})\right] ds \tag{2.167a}$$

$$F_2^l = \sum_{j=1}^{N} \int_{\Gamma_j} \left[f\hat{u}(r_{lj})_{,n} - f_{,n}\hat{u}(r_{lj})\right] ds \tag{2.167b}$$

$$F_3^i = \sum_{j=1}^{N} \int_{\Gamma_j} \left[f\hat{u}_1(r_{ij})_{,n} - f_{,n}\hat{u}_1(r_{ij})\right] ds \tag{2.167c}$$

(e) The plate is subjected to an arbitrarily distributed loading over an area $\Omega^* \subseteq \Omega$ of the plate bounded by the curve Γ^*. In this case, the domain integrals can be converted to boundary line integrals using the Rayleigh-Green identity. Thus, applying Eq. (2.56) for the functions $v = r^2 \ln r / 8\pi D$ and $w = \hat{w}$, where $\hat{w}$ is a particular solution of the equation

$$\nabla^4 \hat{w} = f \tag{2.168}$$

we obtain

$$\int_{\Omega^*} v f d\Omega = \varepsilon \hat{w}(P) + \int_{\Gamma^*} \left(v\nabla^2 \hat{w}_{,n} - \hat{w}\frac{\partial}{\partial n}\nabla^2 v - v_{,n}\nabla^2 \hat{w} + \hat{w}_{,n}\nabla^2 v \right) ds \tag{2.169}$$

in which $\varepsilon = 1, 1/2, 0$ depending on whether P is inside Ω^*, on Γ^* or outside Ω^*. In order to establish the function $\hat{w}$, Eq. (2.168) is written using complex variables as [29]

$$16\frac{\partial^4 \hat{w}}{\partial z^2 \partial \bar{z}^2} = f(z, \bar{z}), \quad z = x + iy,\ \bar{z} = x - iy,\ i = \sqrt{-1} \tag{2.170}$$

Consecutive integration of Eq. (2.170) yields the function $\hat{w}$.

2.4.4 Internal Supports

Due to functional requirements, it is often necessary to support a plate not only on its boundary but also on columns (point supports) and/or on load-bearing walls (line supports) in the interior of the plate. At these interior supports the deflection is prescribed while the corresponding reaction is unknown. In order to treat this problem, Eqs. (2.127)-(2.129) must be supplemented by loading terms of the form given by Eqs. (2.156) and (2.160) for point and line supports, respectively. In this case the forces F and $f^*(s)$ denote the unknown point and line support reactions. To evaluate these unknown reactions additional equations are required. These equations can be provided by collocating Eq. (2.140) at the point and line support nodes, where the deflection is prescribed.

2.4.5 Evaluation of the corner forces

The solution of the set of Eqs. (2.131) and (2.138) theoretically gives the concentrated corner force R_k, if the displacement w_k is prescribed. However, due to the oscillatory behavior of the effective shear force in the neighborhood of the corner, the obtained corner force does not represent the actual one, except for the case where the adjacent to the corner edges are free. Moreover, it was observed that the solution is not influenced if Eq. (2.128) is not used when both adjacent edges are not free. In any event, the corner forces can be evaluated by numerical differentiation of the computed slope $w_{,n}$ along the boundary to obtain the twisting moment, whose discontinuity yields the corner force. Forward and backward derivative schemes are employed to evaluate the derivatives at the points adjacent to the corner. It was observed that the use of the value of the slope at the corner point resulting from the regularity corner conditions (see Table 2.1) improves the accuracy of the computed corner force. It should be noted that the corner force in the case of SS-F adjacent edges is zero [8]. The accuracy of the computed value of $w_{,n}$, hence of $w_{,nt}$, is considerably improved in the region of the corner if elements with C^1 continuity (Hermitian polynomials, see [1]) for the approximation of the displacement and linear elements for the other boundary quantities are employed.

Using the above described procedure, the uniformly loaded simply supported trapezoidal plate of Fig. 2.20 has been analyzed. Figure 2.21 shows the variation of $w_{,nt} = M_{nt}/D(1-\nu)$ along the boundary 1-2-3-4 for various values of the angle ϕ. The employed data are $E = 2.1 \times 10^7\,\mathrm{kN/m^2}$, $\nu = 0.3$, $h = 0.10\mathrm{m}$, $a = b + 8/\tan\phi$, $b = 4\mathrm{m(meters)}$. The values resulted using the program PLBECON (see Section 2.5) with N = 200 elements on each side. Finally, Fig. 2.22 shows the variation of the corner force $R = [\![M_{nt}]\!]$ at corner 1 versus angle ϕ as compared with those obtained with the FEM using the NASTRAN code. We observe that the corner force for $\phi < 60°$ is zero. For $\phi = 90°$ the BEM gives $R = 0.998$, the FEM $R = 0.97$ and the analytic $R = 1.04$ [14].

2.5 PLBECON PROGRAM FOR SOLVING THE PLATE EQUATION WITH CONSTANT BOUNDARY ELEMENTS

On the basis of the analysis and the numerical procedure presented in the previous sections a computer program has been written in FORTRAN language. The program employs constant elements for the discretization of the boundary integral equations. The shape of the plate is arbitrary and may have holes. All types of boundary conditions are encountered including internal point and line

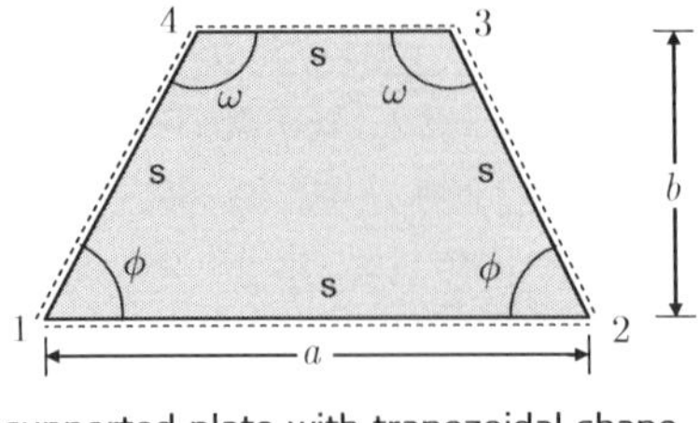

FIGURE 2.20 Simply supported plate with trapezoidal shape.

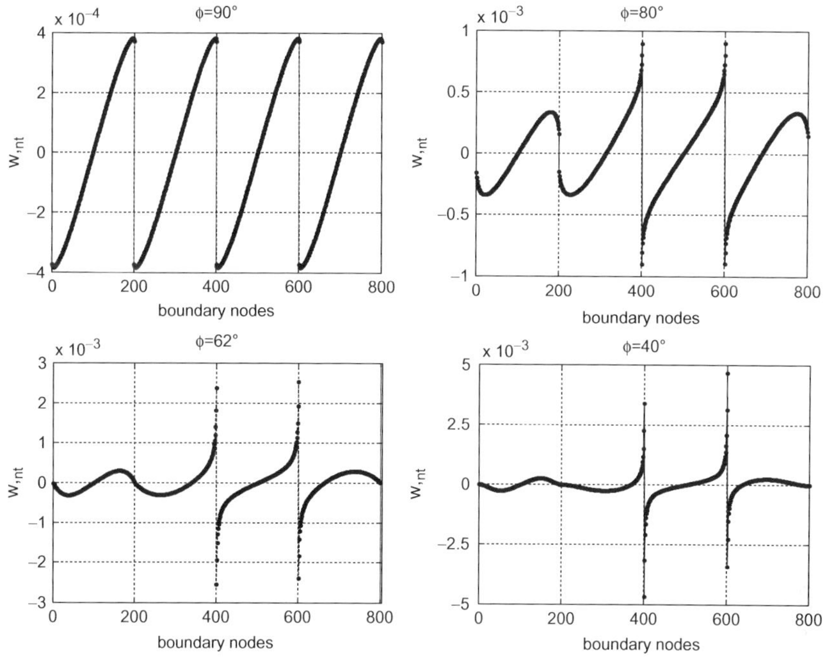

FIGURE 2.21 Distribution of $w_{,nt} = M_{nt}/D(1-\nu^2)$ on the boundary.

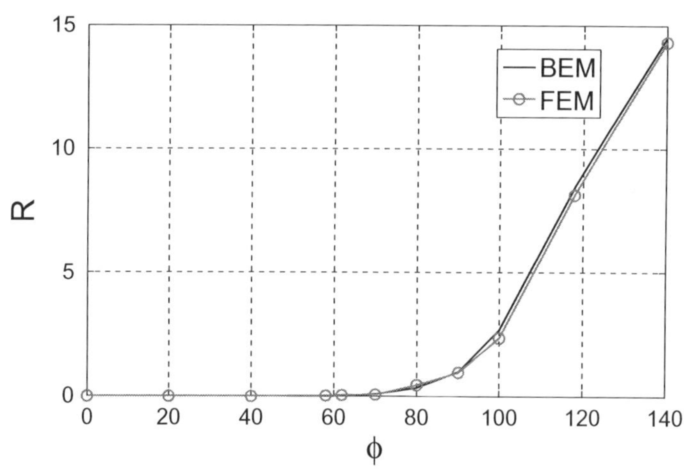

FIGURE 2.22 Variation of the corner force $R = [\![M_{nt}]\!]$ versus angle ϕ.

supports. Point supports can be placed at corner points and at isolated points at the free edges as well. In the latter case the point supports are treated as corner points. Regarding the loads, they may be distributed loads with linear variation, line loads, and concentrated loads. The structure of the PLBECON program is shown in the macro flowchart of Fig. 2.23. The preparation of the data is made using the program PLDATA. This program is written in MATLAB because its powerful graphics offers immediate check of the correctness of the input data. Both codes are available on this book's companion website.

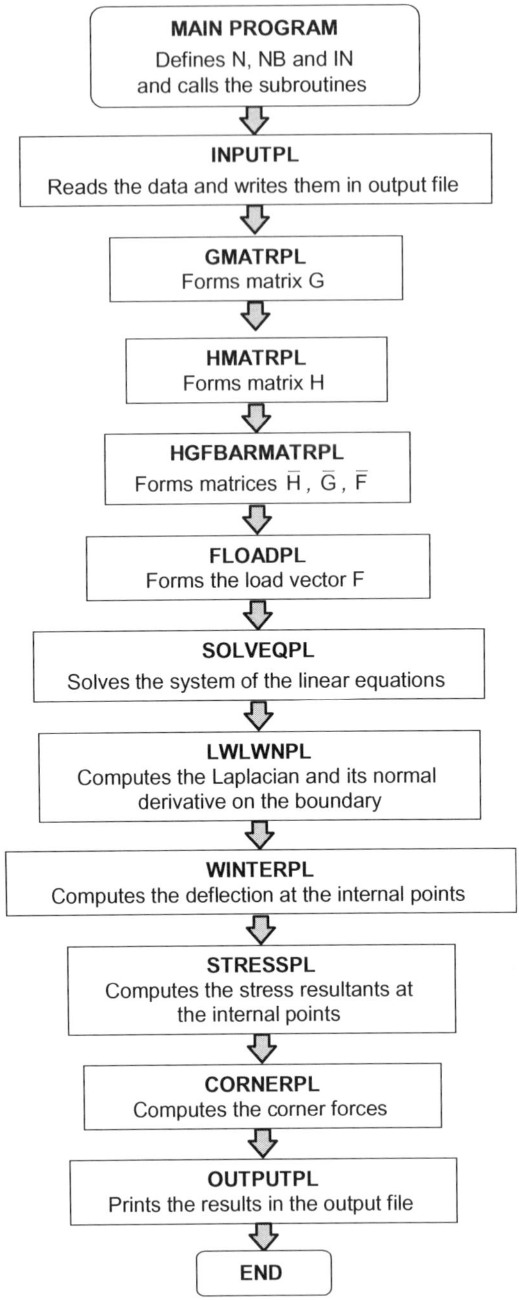

FIGURE 2.23 Macro flowchart of program PLBECON.

2.5.1 Main program

The main program defines the parameters N, NC, NFC, NB, and IN, which specify, respectively, the number of the boundary elements, the total number of the corner points, the number of the corner points with free adjacent sides, the number of boundaries, and the number of the internal points, where the solution and the stress resultants will be evaluated. Moreover, the main program defines the parameters: NPS (number of internal point supports), NLS (number of line supports), NELS (total number of elements on the line supports), NCL (number of concentrated loads), NLL (number of line loads), and NELL (total number of elements on line loads). Finally, it opens two files, the file INPUTFILE, which contains the data and the OUTPUTFILE file, in which the results are rendered. Subsequently, it calls the following 11 subroutines:

INPUTEPL	Reads the data from INPUTFILE.
GMATRPL	Forms the matrix $[G]$ defined by Eq. (2.133).
HMATRPL	Forms the matrix $[H]$ defined by Eq. (2.132).
HGFBARMATRPL	Forms the matrices $[\overline{H}]$, $[\overline{G}]$ and $[\overline{F}]$ defined by Eqs. (2.139).
FLOADPL	Forms the load vector $\{F\}=\{\boldsymbol{F}_1\ \boldsymbol{F}_2\ \boldsymbol{F}_3\}^T$ defined by Eq. (2.134).
SOLVEQPL	Solves the system of linear Eqs. (2.131) and (2.138) using Gauss elimination and forms the vectors $\{W\}$, $\{WN\}$, $\{MN\}$, $\{VN\}$, $\{WC\}$ and $\{RC\}$ of the boundary quantities.
LWLWNPL	Computes the Laplacian $\{LW\}$ and its normal derivative $\{LWN\}$ on the boundary by solving the coupled integral Eqs. (2.121) and (2.122).
WINTERPL	Computes the deflections w at the IN internal points using Eq. (2.119).
STRESSPL	Computes the stress resultants M_x, M_y, M_{xy}, Q_x, and Q_y at the IN internal points using Eqs. (2.19), (2.21), and (2.125).
CORNERPL	Computes the corner forces R_k.
OUTPUTEL	Writes the results in OUTPUTFILE.

The principal variables and matrices employed in the program are defined as follows:

IN	Number of internal points, where the displacements and the stress resultants are computed.
NB	Number of boundaries.
NC	Number of corners
NFC	Number of corners with F-F (F = free) adjacent edges.
NL	One-dimensional array with dimension NB containing the serial number of the last element of each boundary.
EL	Young's modulus of elasticity, E.
PN	Poisson's ratio, ν.

A1, A2, A3, B1, B2, B3 One dimensional arrays defining the type of the boundary condition at node I (I = 1,2,. . .,N).

CCODE	One-dimensional array specifying the support condition at corner K (K = 1,2,. . .,NC).
CURV	One-dimensional array containing the boundary curvature at node I (I = 1, 2,. . .,N).
XL, YL	One-dimensional arrays containing the x and y coordinates of the extreme points of the elements.
XC, YC	One-dimensional arrays containing the x and y coordinates of the corner points.
XM, YM	One-dimensional arrays containing the x and y coordinates of the boundary nodes.
XIN, YIN	One-dimensional arrays containing the x and y coordinates of the internal points, where the displacements and stress resultants are computed.
W, WN	One-dimensional arrays containing the boundary nodal values of w and w_n.
MN, VN	One-dimensional arrays containing the boundary nodal values of M_n and V_n.
WC, RC	One-dimensional arrays containing the corner displacements w_c and corner forces R_c.
WIN	One-dimensional array containing the computed values of the deflections w at the IN internal points.
MX, MY, MXY	One-dimensional arrays containing the computed values of the moments M_x, M_y and M_{xy} at the IN internal points.
QX, QY	One-dimensional arrays containing the computed values of the shear forces Q_x, Q_y at the IN internal points.

INPUTPL SUBROUTINE

The INPUTPL subroutine reads all the data required by the program using free FORMAT. The data have been written in INPUTFILE, to which the user has given a specific name as required by the main program. This file contains the following data:

1. *User's name*. One line containing the name of the user.
2. *Title*. One line containing the title of the program.
3. *Elastic constants*. Modulus of elasticity EL and Poisson's ratio PN.
4. The serial number of the last element of each boundary. NB integers forming the entries of the array NL.
5. *The coordinates of the extreme points of the boundary elements*. N couples of values consisting of the coordinates XL, YL of the extreme points of the boundary elements. They are read in the positive sense, i.e., counter-clockwise, on the external boundary and clockwise on the internal ones.
6. *The coordinates and the support conditions of the corner points*. NC couples of values consisting of the coordinates XC, YC of the corner points.

They are read together with the code number CCODE that specifies the type of the support condition:
CCODE = 0 corner at the end of a supported edge.
CCODE = 1 supported corner between two free edges. w_c^* is specified.
CCODE = 2 free corner between two free edges. R_c^* is specified.
CCODE = 3 elastically supported corner. k_T is specified.

7. *Boundary conditions.* N matrices with dimensions 7×1 containing the values of $\alpha 1$, $\alpha 2$, $\alpha 3$, $\beta 1$, $\beta 2$, $\beta 3$ and κ at the boundary nodal points. More specifically:
   $1, 0, w^*, 0, 1, M_n^*, \kappa$ for simply supported edge.
   $1, 0, w^*, 1, 0, w_n^*, \kappa$ for clamped edge.
   $0, 1, V_n^*, 0, 1, M_n^*, \kappa$ for free edge.
8. *Loading of the plate.* The program analyzes the plate subjected to (i) concentrated forces F (ii) line load $f = f(s)$ (iii) linearly varying distributed load $f = f_x x + f_y y + f_0$. The specified data are:
   (i) Concentrated forces. NCL triads of numbers representing the coordinates XCL, YCL of the point of application of the load and its value FCL.
   (ii) Line load. The vector NML containing NLL integer numbers representing the serial number of the last element of each line load. Moreover, NELL + NLL triads of numbers representing the coordinates XLL, YLL of the extreme points of the elements and the corresponding value of the line load FLL.
   (iii) Distributed load. The values F0, FX, FY representing the coefficients of the function $f = f_x x + f_y y + f_0$ are specified.
9. *Internal supports.* The plate may be supported at (i) internal line supports and (ii) internal point supports. The specified data are:
   (i) Line support. The vector NBL containing NLS integer numbers representing the serial number of the last element of each line support. Moreover, NELS + NLS couples consisting of the values of the coordinates XSL, YSL of the extreme points of the elements of the line supports.
   (ii) Point supports. NPS couples of values consisting of the coordinates XPS, YPS of the point supports.
10. *Coordinates of the internal points.* IN couples of values consisting of the coordinates XIN and YIN of the internal points, where the values of the deflections w and stress resultants M_x, M_y, M_{xy}, Q_x and Q_y are computed.

Finally, the subroutine INPUTPL writes the data in OUTPUTFILE, to which the user gives a specific name.

GMATRPL SUBROUTINE

The GMATRPL subroutine forms the matrix $[G]$ defined by Eq. (2.133). First, the submatrices $[v]^{ij}$, $[v]^{lj}$, $[v_1]^{ij}$, $[v]^{ik}$, $[v]^{lk}$, $[v_1]^{ik}$, $[v_n]^{ij}$, $[v_n]^{lj}$, $[v_{1n}]^{ij}$ $(i,j = 1,2,\ldots,N)$, $(l,k = 1,2,\ldots,N_c)$ are evaluated using the relevant Eqs. (2.130) and subsequently are placed in matrix $[G]$. The elements of the matrix $[G]^{ij}$ are line integrals of the fundamental solution and its derivatives. Except for the diagonal elements $[v_{1n}]^{ii}$, all other line integrals are regular and are evaluated using a four-point Gauss integration by calling the subroutines RLINTG and RLINTGC. The integrals $[v_{1n}]^{ii}$ exhibiting a logarithmic singularity are evaluated analytically using relation (2.144a).

HMATRPL SUBROUTINE

The HMATRPL subroutine forms the matrix $[H]$ defined by Eq. (2.132). First, the submatrices $[Vv]^{ij}, [Vv_1]^{lj}, [Vv_1]^{ij}, [\![Tv]\!]^{ik}, [\![Tv]\!]^{lk}, [\![Tv_1]\!]^{ij}, [Mv]^{ij}, [Mv]^{lj}, [Mv_1]^{ij}$ $(i,j=1,2,\ldots,N), (l,k=1,2,\ldots,N_c)$ are evaluated using the relevant Eqs. (2.130) and subsequently they are placed in matrix $[H]$. The elements of the matrix $[H]^{ij}$ are line integrals of the fundamental solution and its derivatives. The regular integrals are evaluated using a four-point Gauss integration by calling the subroutines RLINTH and RLINTHC, while the singular and hypersingular integrals are evaluated analytically using the appropriate relations derived in Section 2.4.2.

HGFBARMATRPL SUBROUTINE

The HGFBARMATRPL subroutine forms the matrices $[\overline{H}], [\overline{G}]$ and the vector $\{\overline{F}\}$ defined Eqs. (2.139).

FLOADPL SUBROUTINE

The FLOADPL subroutine forms the vectors $\{F_1\}^i, \{F_2\}^l, \{F_3\}^i$, $(i=1,2,\ldots,N)$, $(l=1,2,\ldots,N_C)$ defined by Eqs. (2.135) using Eqs. (2.156), (2.160), or (2.167) for concentrated load, line load, and linear varying distributed load, respectively.

SOLVEQPL SUBROUTINE

The SOLVEQPL subroutine solves the system of the linear equations (2.131) and (2.138) combined as single set of equations calling the subroutine LEQS and forms the vectors $\{w\}$, $\{w_c\}$, $\{w_n\}$, $\{V_n\}$, $\{R\}$, $\{M_n\}$ of the boundary quantities. The solution is obtained by Gauss elimination. The output parameter LSING takes the value LSING $=0$ if the coefficient matrix is regular, or LSING $=1$, if the coefficient matrix is singular. In the case of internal supports, the matrices $[G]$, $[H]$ and $[F]$ are augmented with those resulting from Eqs. (2.140) applied at the support points to obtain the unknown support reactions. Before solving the system the subroutine reorders the equations to include the reactions in the unknowns.

LWLWNPL SUBROUTINE

The LWLWNPL subroutine solves numerically the system of Eqs. (2.121) and (2.122) using the same discretization and computes the boundary nodal values of $\nabla^2 w$, $\nabla^2 w_{,n}$, which are employed in the subroutine WINTERPL.

WINTERPL SUBROUTINE

The WINTERPL subroutine computes the values of the displacement w and its derivatives at internal points adhering to the procedure described in Section 2.3.5.

STRESSPL SUBROUTINE

The STRESSPL subroutine computes the stress resultants M_x, M_y, M_{xy}, Q_x, Q_y at the internal points using Eqs. (2.19) and (2.21).

CORNERPL SUBROUTINE

The CORNERPL subroutine computes the concentrated corner forces at corner points with supported adjacent corners adhering to the procedure described in Section 2.4.5.

OUTPUTPL SUBROUTINE

The OUTPUTPL subroutine writes all the results in the output file.

2.6 EXAMPLES

Example 2.1

The program PLBECON has been used to analyze several plates with various shapes and boundary conditions subjected to a uniform load. Results for displacements and stress resultants at certain characteristic points of the plate are shown in Table 2.3 as compared with those obtained from existing analytical or other numerical solutions. The number of employed boundary elements is shown in parentheses. The FEM

TABLE 2.3 Deflections and stress resultants in plates of various shapes and boundary conditions under a uniform load. Upper value: Computed; Lower value: Existing analytical or other numerical

Shape and boundary conditions	$\frac{w^0}{(f_0 a^4/D)}$	$\frac{M_n^1}{(qa^2)}$	$\frac{V_n^1}{(qa)}$	$\frac{M_x^0}{(qa^2)}$
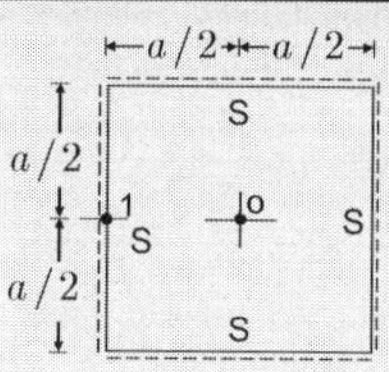	0.00407 (84) 0.00407 (404) 0.00405 (804)		-0.420 -0.420 -0.420	0.0479 0.0479 0.0479
	0.00406 [14]		-0.420 [14]	0.0479 [14]
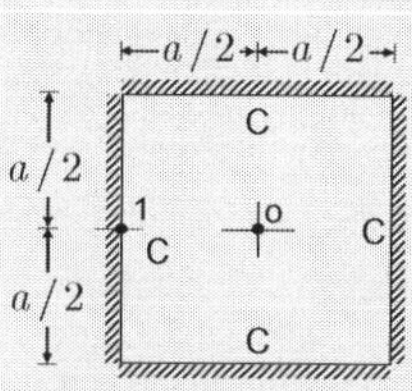	0.00127 (84) 0.00126 (404) 0.00126 (804)	-0.0514 -0.0513 -0.0513	-0.442 -0.441 -0.441	0.0229 0.0229 0.0229
	0.00126 [14]	-0.0513 [14]		0.0231 [14]
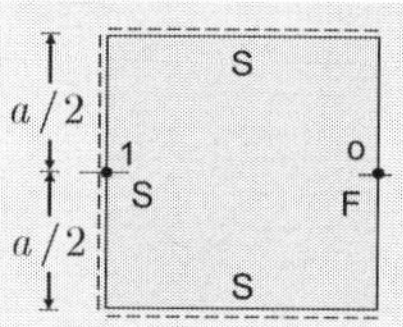	0.0128 (84) 0.0129 (404) 0.0130 (804)		-0.486 -0.487 -0.487	
	0.0129 FEM		-0.486 FEM	
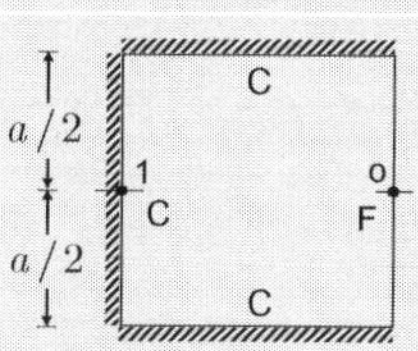	0.00294 (84) 0.00296 (404) 0.00297 (804)	-0.0563 -0.0563 -0.0563	-0.460 -0.460 -0.460	
	0.00296 FEM	-0.0561 FEM	-0.450 FEM	

Continued

TABLE 2.3 Deflections and stress resultants in plates of various shapes and boundary conditions under a uniform load. Upper value: Computed; Lower value: Existing analytical or other numerical—cont'd

Shape and boundary conditions	$\frac{w^0}{(f_0 a^4/D)}$	$\frac{M_n^1}{(qa^2)}$	$\frac{V_n^1}{(qa)}$	$\frac{M_x^0}{(qa^2)}$
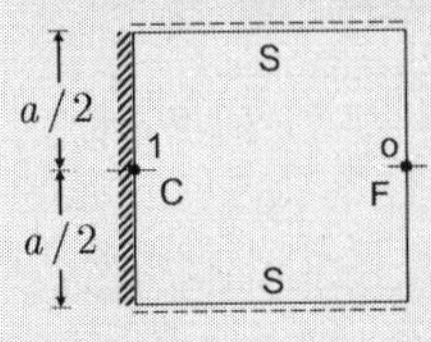	0.0112 (84) 0.0113 (404) 0.0114 (804) 0.0114(1204)	-0.118 -0.118 -0.118 -0.118	-0.709 -0.711 -0.712 -0.712	
	0.0113 [14]	-0.119 [14]	-0.696 FEM	
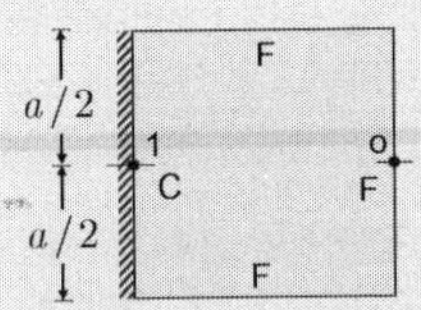	0.1309 (84) 0.1298 (484) 0.1297 (804)	-0.542 -0.538 -0.538	-1.218 -1.218 -1.230	
	0.129 FEM	-0.531 FEM	-1.161 FEM	
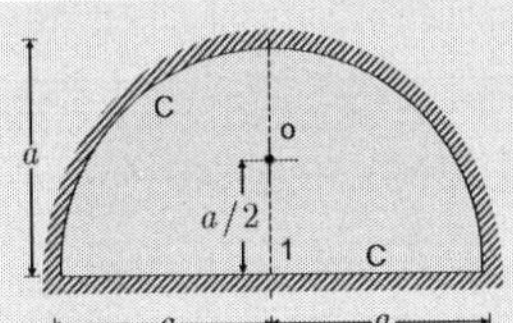	0.00202 (100) 0.00202 (250) 0.00202 (500)	-0.0731 -0.0731 -0.0731	-0.497 -0.497 -0.497	0.0193 0.0193 0.0193
	0.002021 [18]	-0.07 [18]	-0.491 [18]	0.021 FEM
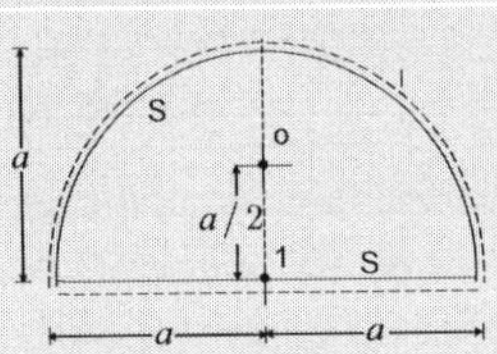	0.00812 (100) 0.00812 (250) 0.00812 (500)		-0.490 -0.491 -0.491	0.05149 0.05146 0.05146
	0.00812 FEM		-0.494 FEM	0.050 FEM
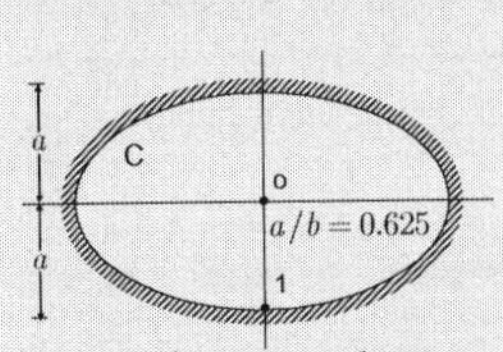	0.00606 (60) 0.00449 (100) 0.00449 (200)	-0.920 -0.0921 -0.0921	-0.4995 -0.4998 -0.4998	0.0318 0.0318 0.0318
	0.004499 [16]	-0.0921 [16]	-0.508 FEM	0.0318 [16]
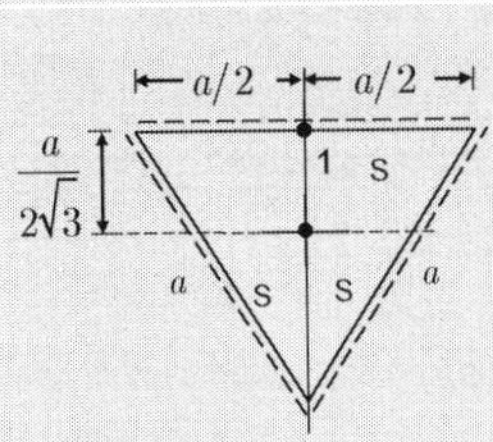	0.000578 (93) 0.000578(153) 0.000579(303)		-0.254 -0.254 -0.254	0.0180 0.0180 0.0180
	0.000582 FEM		-0.254 FEM	0.0183 FEM

solutions were obtained using the NASTRAN code with quadrilateral elements (1600 for square plates, 2962 for half-circle plates, 4884 for the elliptic plate, and 4972 for the triangular plate). The results demonstrate the efficiency and accuracy of the BEM with constant elements for plate bending analysis.

Example 2.2

In this example the triangular simply supported plate of Fig. 2.24 with $a = 2.5\mathrm{m}$ is analyzed. The plate, which has thickness $h = 0.05\mathrm{m}$ and elastic constants $E = 2.1 \times 10^6 \mathrm{kN/m^2}$ and $\nu = 0.3$, is subjected to hydrostatic load $f = 10(a - y)$. The solution is obtained with $\mathrm{N} = 10\overline{\mathrm{N}}$ elements, of which $4\overline{\mathrm{N}}$ are placed on edge 1-2 and $3\overline{\mathrm{N}}$ on each of the other two edges. The input data are produced with MATLAB program PLDATA_2.2 available on this book's companion website. The data and the results are given below as printed by the PLBECON in RESULTS_2.2, also available on the companion website. In Table 2.4, the values of the quantities w, M_x, M_y, Q_y at point $P(a, a/5)$ are presented for various values of N as compared with those resulting from FEM (NASTRAN) with 10693 quadrilateral elements. Fig. 2.25 shows the variations of w, M_x, M_y and Q_y along the line $x = a$ (N = 300) as compared with those obtained with the FEM. The variations of V_n and M_{nt} along the boundary are shown in Fig. 2.26. We observe that the oscillatory behavior of V_n at corner 3 does not permit the determination of the concentrated corner force, while the distribution of M_{nt} yields $R_3 = [\![M_{nt}]\!] = 4.06$. We see that there is no discontinuity of M_{nt} at the other corners with angle $= 45° < 60°$ as it was anticipated (see Fig 2.22). Finally, the contour curves of w, M_x, M_{xy}, M_y, Q_x and Q_y are shown in Fig. 2.27 through Fig. 2.32.

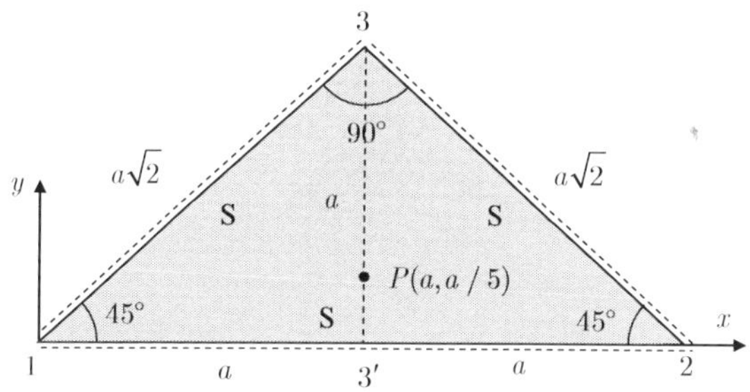

FIGURE 2.24 Triangular simply supported plate in Example 2.2.

EXAMPLE 2.2 (DATA and RESULTS)
****

```
J.T. KATSIKADELIS
Example 2.2

ELASTIC CONSTANTS: ELASTIC MODULUS = .2100E+07   POISSON RATIO =  .30
PLATE THICKNESS=  .05

NUMBER OF BOUNDARIES          1
BOUNDARY: 1     ELEMENTS: 300
TOTAL No OF BOUNDARY ELEMENTS: 300

COORDINATES OF THE MIDPOINTS OF THE BOUNDARY ELEMENTS
 POINT        XM                  YM
```

1	.20833E-01	.00000E+00
2	.62500E-01	.00000E+00
3	.10417E+00	.00000E+00
4	.14583E+00	.00000E+00
5	.18750E+00	.00000E+00
6	.22917E+00	.00000E+00
7	.27083E+00	.00000E+00
8	.31250E+00	.00000E+00
9	.35417E+00	.00000E+00
10	.39583E+00	.00000E+00
11	.43750E+00	.00000E+00
12	.47917E+00	.00000E+00
13	.52083E+00	.00000E+00
....		
297	.97222E-01	.97222E-01
298	.69444E-01	.69444E-01
299	.41667E-01	.41667E-01
300	.13889E-01	.13889E-01

CORNER POINTS

CORNER	XC	YC	KNOWN VALUE
1	.00000E+00	.90000E+01	CORNER NEXT TO SUPPORTED EDGES
2	.50000E+01	.00000E+00	CORNER NEXT TO SUPPORTED EDGES
3	.25000E+01	.25000E+01	CORNER NEXT TO SUPPORTED EDGES

BOUNDARY CONDITIONS

ELEMENT	A1	A2	A3	B1	B2	B3	CURV
1	.100E+01	.000E+00	.000E+00	.000E+00	.100E+01	.000E+00	.000E+00
2	.100E+01	.000E+00	.000E+00	.000E+00	.100E+01	.000E+00	.000E+00
3	.100E+01	.000E+00	.000E+00	.000E+00	.100E+01	.000E+00	.000E+00
4	.100E+01	.000E+00	.000E+00	.000E+00	.100E+01	.000E+00	.000E+00
5	.100E+01	.000E+00	.000E+00	.000E+00	.100E+01	.000E+00	.000E+00
6	.100E+01	.000E+00	.000E+00	.000E+00	.100E+01	.000E+00	.000E+00
7	.100E+01	.000E+00	.000E+00	.000E+00	.100E+01	.000E+00	.000E+00
8	.100E+01	.000E+00	.000E+00	.000E+00	.100E+01	.000E+00	.000E+00
9	.100E+01	.000E+00	.000E+00	.000E+00	.100E+01	.000E+00	.000E+00
10	.100E+01	.000E+00	.000E+00	.000E+00	.100E+01	.000E+00	.000E+00
11	.100E+01	.000E+00	.000E+00	.000E+00	.100E+01	.000E+00	.000E+00
12	.100E+01	.000E+00	.000E+00	.000E+00	.100E+01	.000E+00	.000E+00
13	.100E+01	.000E+00	.000E+00	.000E+00	.100E+01	.000E+00	.000E+00
....							

SURFACE LOAD: FX= .000E+00 FY= -.100E+02 F0= .250E+02

****

The system has been solved regularly

****

RESULTS

BOUNDARY NODES

NODE	X	Y	W	Wn	Vn	Mn
1	.20833E-01	.00000E+00	.00000E+00	-.11288E-05	-.21322E+00	.00000E+00
2	.62500E-01	.00000E+00	.00000E+00	-.42419E-04	.12386E+00	.00000E+00
3	.10417E+00	.00000E+00	.00000E+00	-.17001E-03	-.17916E+00	.00000E+00
4	.14583E+00	.00000E+00	.00000E+00	-.41353E-03	-.44831E+00	.00000E+00
5	.18750E+00	.00000E+00	.00000E+00	-.79322E-03	-.80283E+00	.00000E+00
6	.22917E+00	.00000E+00	.00000E+00	-.13232E-02	-.11835E+01	.00000E+00
7	.27083E+00	.00000E+00	.00000E+00	-.20133E-02	-.15918E+01	.00000E+00
8	.31250E+00	.00000E+00	.00000E+00	-.28697E-02	-.20178E+01	.00000E+00
9	.35417E+00	.00000E+00	.00000E+00	-.38958E-02	-.24573E+01	.00000E+00
10	.39583E+00	.00000E+00	.00000E+00	-.50929E-02	-.29065E+01	.00000E+00

```
11 .43750E+00 .00000E+00  .00000E+00 -.64602E-02 -.33625E+01 .00000E+00
12 .47917E+00 .00000E+00  .00000E+00 -.79956E-02 -.38229E+01 .00000E+00
13 .52083E+00 .00000E+00  .00000E+00 -.96954E-02 -.42857E+01 .00000E+00
..........................................................................
 CORNER POINTS

CORNER   XC          YC          WC          R
 1     .00000E+00 .90000E+01 .00000E+00  .20537E-01
 2     .50000E+01 .00000E+00 .00000E+00  .14774E-02
 3     .25000E+01 .25000E+01 .00000E+00 -.40600E+01

 DISPLACEMENTS AND SLOPES AT INTERNAL POINTS
 POINT     X            Y            W            Wx            Wy

  1   .25000E+01   .10000E+00   .11937E-01   -.69276E-13   .11770E+00
  2   .25000E+01   .20000E+00   .23388E-01   -.12215E-13   .11066E+00
  3   .25000E+01   .30000E+00   .33948E-01    .82326E-14   .10001E+00
  4   .25000E+01   .40000E+00   .43299E-01    .20349E-13   .86628E-01
  5   .25000E+01   .50000E+00   .51209E-01    .29525E-13   .71306E-01
  6   .25000E+01   .60000E+00   .57520E-01    .37403E-13   .54774E-01
..........................................................................

 BENDING AND TWISTING MOMENTS AT INTERNAL POINTS
  NODE     X            Y            Mx           My            Mxy

  1   .25000E+01   .10000E+00   .58879E+00   .12559E+01   .16156E-10
  2   .25000E+01   .20000E+00   .11128E+01   .23012E+01   .50157E-11
  3   .25000E+01   .30000E+00   .15684E+01   .31289E+01   .24712E-11
  4   .25000E+01   .40000E+00   .19591E+01   .37551E+01   .17344E-11
  5   .25000E+01   .50000E+00   .22883E+01   .41959E+01   .14160E-11
  6   .25000E+01   .60000E+00   .25595E+01   .44671E+01   .12213E-11
..........................................................................

 SHEAR FORCES AT INTERNAL POINTS
  NODE     X            Y            Qx            Qy

  1   .25000E+01   .10000E+00   -.22059E-09   .13221E+02
  2   .25000E+01   .20000E+00   -.38763E-10   .10947E+02
  3   .25000E+01   .30000E+00   -.10776E-10   .88224E+01
  4   .25000E+01   .40000E+00   -.39458E-11   .68476E+01
  5   .25000E+01   .50000E+00   -.97251E-12   .50224E+01
  6   .25000E+01   .60000E+00    .86196E-12   .33469E+01
.......................................................
*******************************************************************************
```

TABLE 2.4 Displacement w and stress resultants M_x, M_y, Q_y at point $(a, a/5)$ of the triangular plate in Example 2.2 for various values of N

	BEM					FEM
	N=150	N=200	N=400	N=600	N=800	
w	0.05296	0.05108	0.05123	0.05125	0.05125	0.05190
M_x	2.3756	2.2820	2.2896	2.2904	2.2906	2.3133
M_y	4.4269	4.1790	4.1990	4.2009	4.2013	4.2375
Q_y	4.6568	5.0476	5.0175	5.0144	5.0139	4.9606

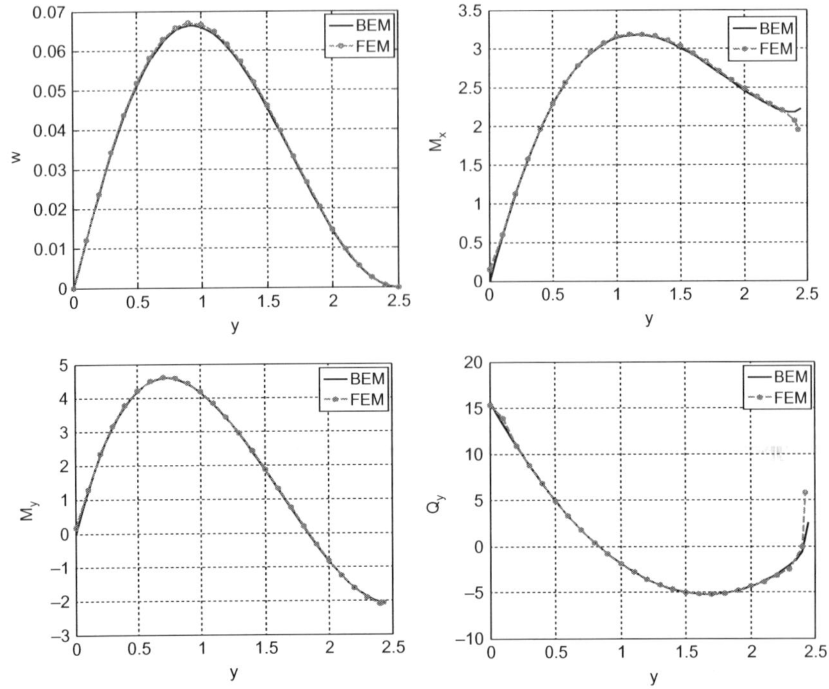

FIGURE 2.25 Distribution of w, M_x, M_y and Q_y along the line $x = 2.5$ in Example 2.2.

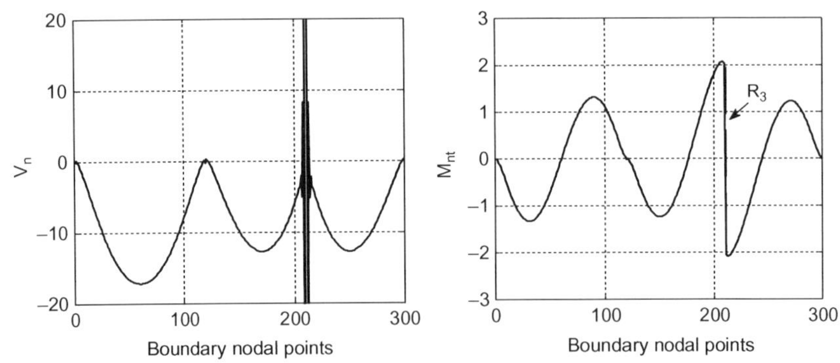

FIGURE 2.26 Distribution of V_n and M_{nt} on the boundary in Example 2.2.

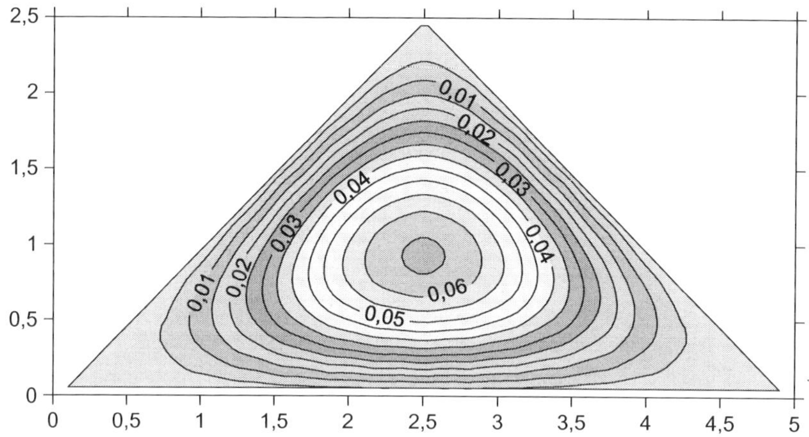

FIGURE 2.27 Contours of the deflection surface w in Example 2.2.

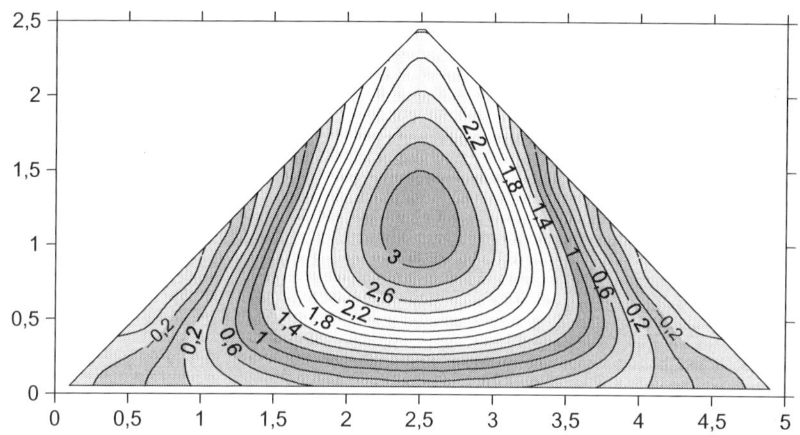

FIGURE 2.28 Contours of the bending moment M_x in Example 2.2.

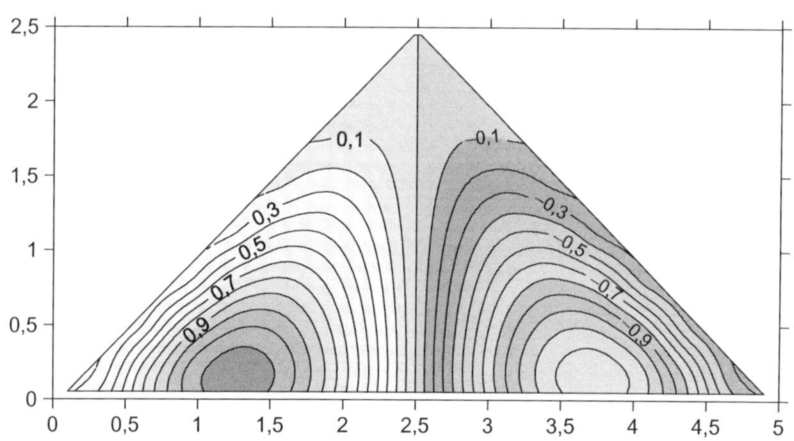

FIGURE 2.29 Contours of the twisting moment M_{xy} in Example 2.2.

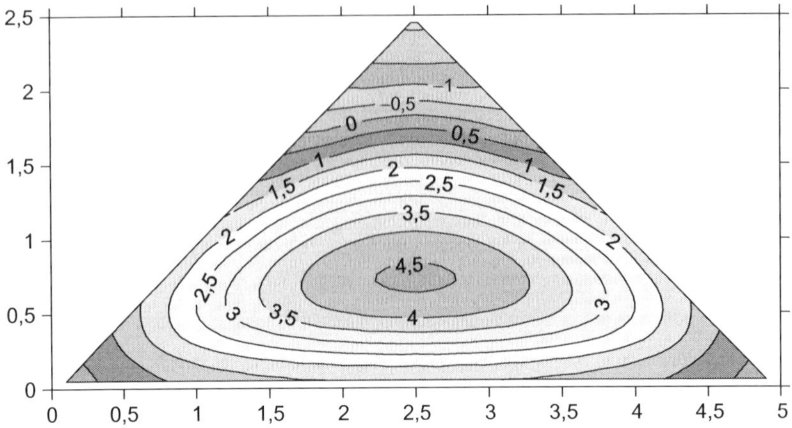

FIGURE 2.30 Contours of the bending moment M_y in Example 2.2.

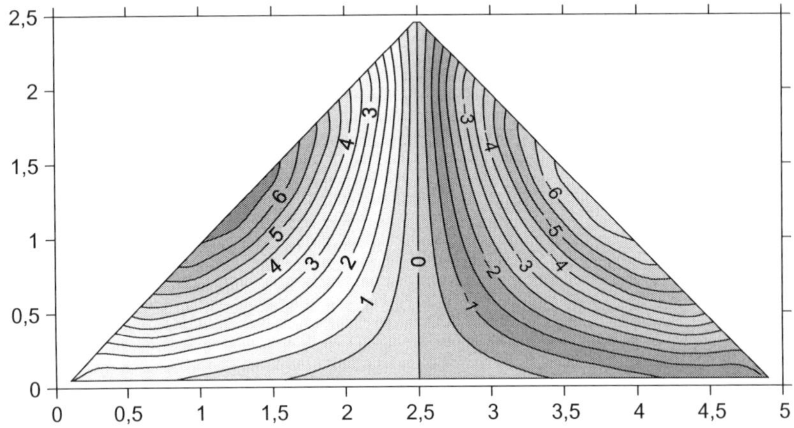

FIGURE 2.31 Contours of the shear force Q_x in Example 2.2.

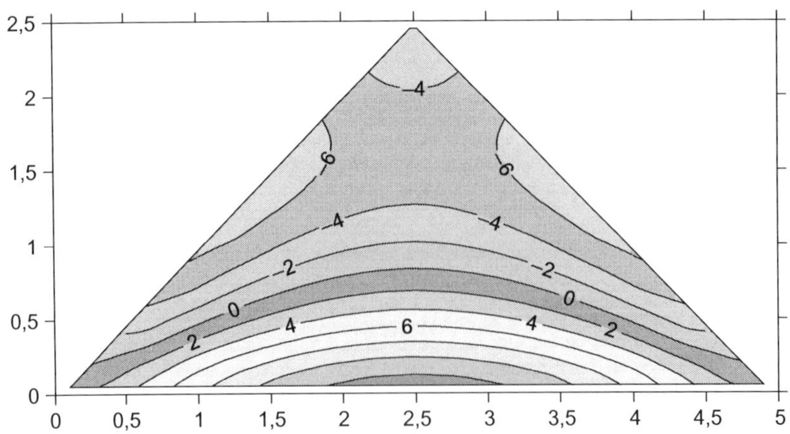

FIGURE 2.32 Contours of the shear force Q_y in Example 2.2.

Example 2.3

A polygonal plate with thickness $h = 0.05\mathrm{m}$ is analyzed in this example. The geometry of the plate and the boundary conditions are shown in Fig. 2.33. The plate is subjected to the uniform load $f_0 = 1\,\mathrm{kN/m^2}$. The material constants are $E = 2.1 \times 10^6 \mathrm{kN/m^2}$, $\nu = 0.3$. This example illustrates the capability of the program PLBECON to analyze plates with a complicated shape under mixed boundary conditions. The results were obtained with N=426 boundary elements distributed on the edges as follows: N1=70 on 1-2, N2=58 on 2-3, N3=50 on 3-4; N4=42 on 4-5, N5=19 on 5-6, N6=38 on 6-7, N7=59 on 7-8, and N8=90 on 8-1. The data file was created by the program PLDATA_2.3. The results are given in the file RESULTS_2.3, as they are printed by the program PLBECON. The files are available on the book's companion website. Fig. 2.34 presents the distributions of w,$w_{,x}$,M_x,M_{xy},Q_x and Q_y along the line $y = 2.5$ in comparison with those resulting from the FEM. Fig. 2.35 shows the variation of the boundary quantities w,$w_{,n}$,M_n,V_n (computed and prescribed) along the boundary, while Fig. 2.36 shows the variation of M_{nt} with the values of the corner forces. Finally, the contour curves of w,M_x,M_{xy},M_y,Q_x and Q_y are shown in Figs. 2.37 through 2.42.

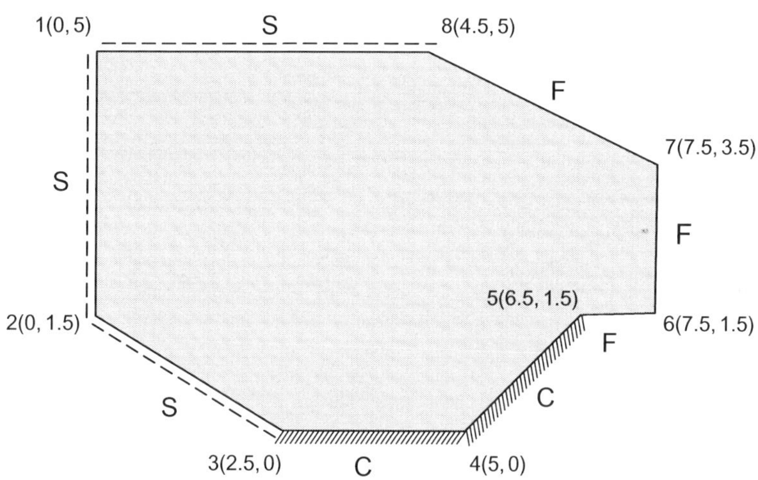

FIGURE 2.33 Polygonal plate with mixed boundary conditions in Example 2.3.

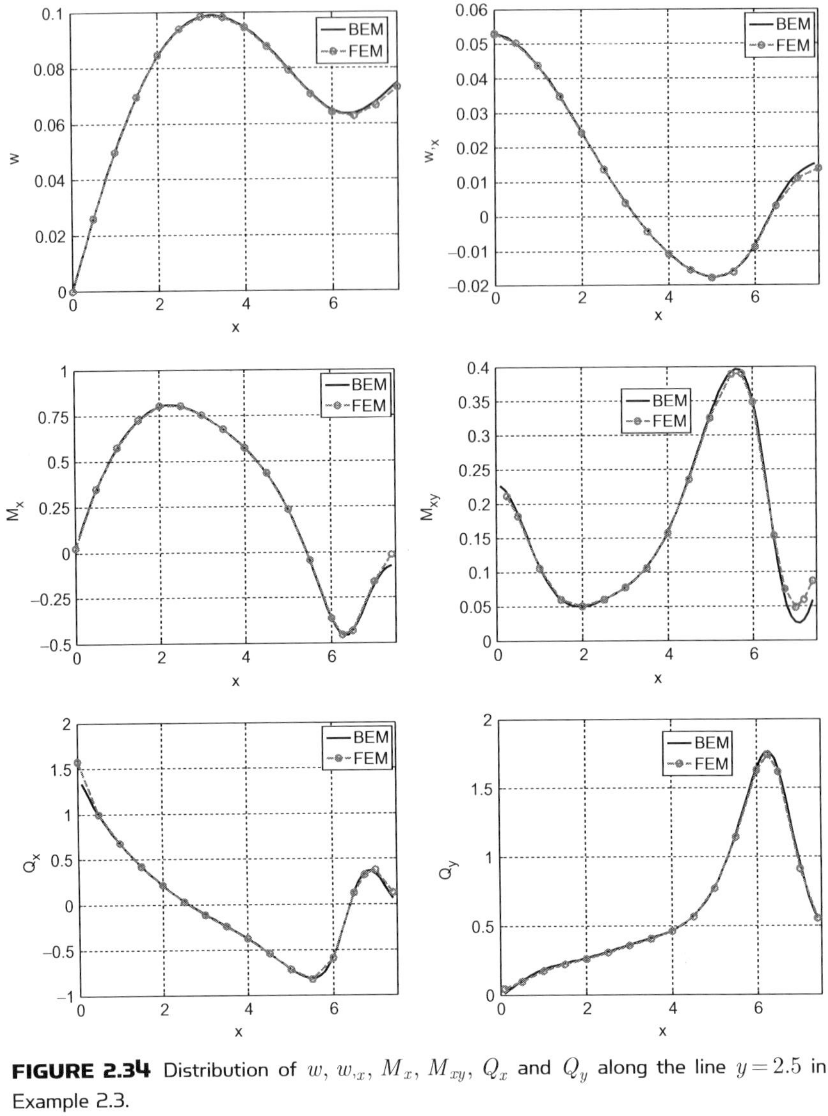

FIGURE 2.34 Distribution of w, $w_{,x}$, M_x, M_{xy}, Q_x and Q_y along the line $y = 2.5$ in Example 2.3.

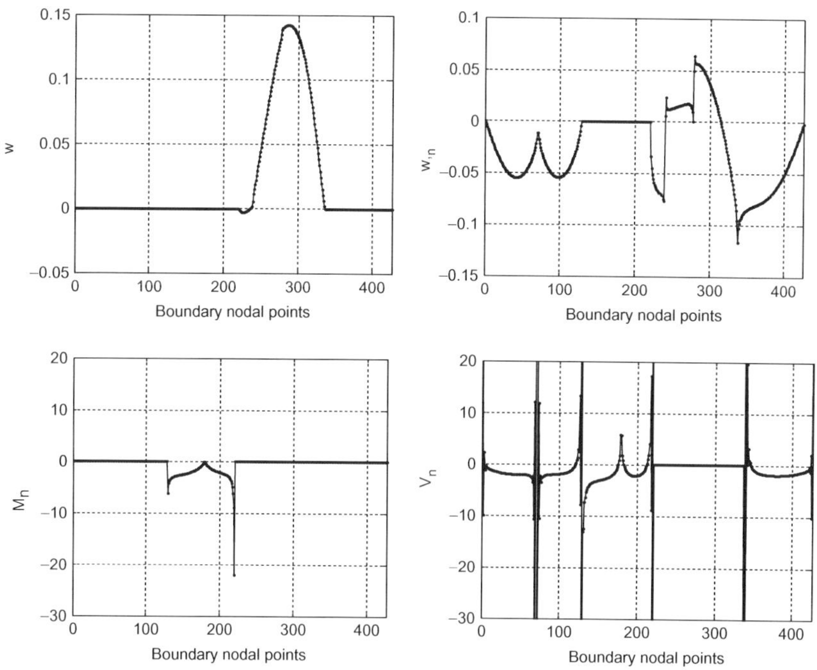

FIGURE 2.35 Distribution of w, $w_{,n}$, M_n and V_n on the boundary in Example 2.3.

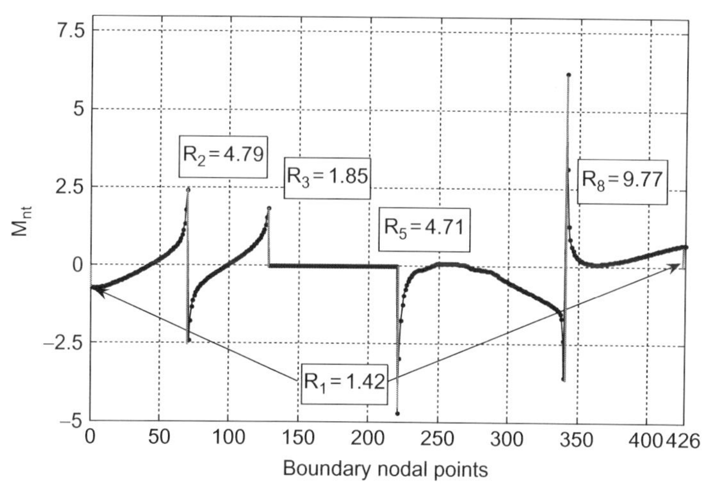

FIGURE 2.36 Distribution of M_{nt} on the boundary in Example 2.3.

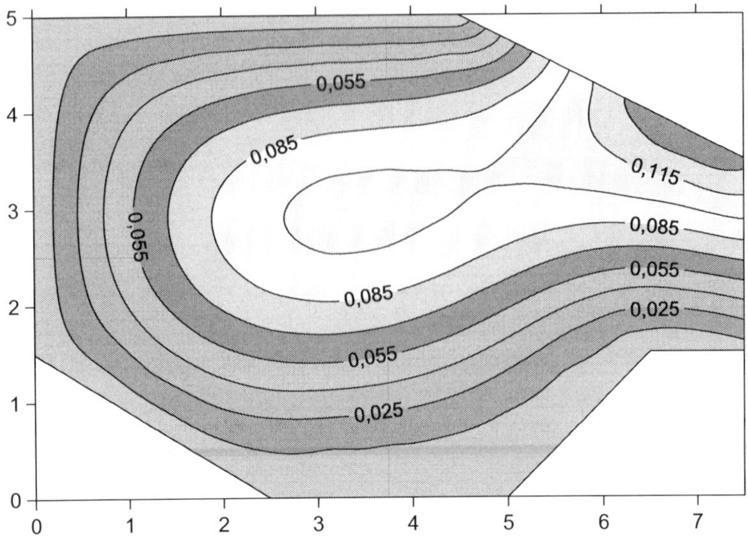

FIGURE 2.37 Contours of the deflection surface w in Example 2.3.

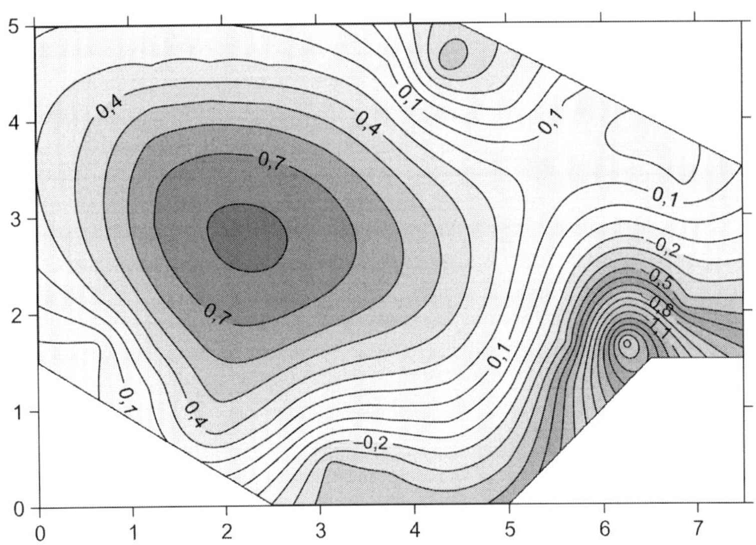

FIGURE 2.38 Contours of the bending moment M_x in Example 2.3.

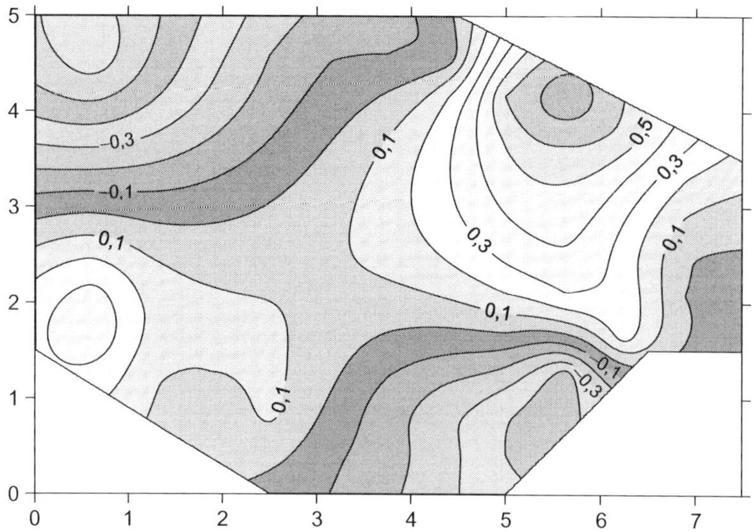

FIGURE 2.39 Contours of the twisting moment M_{xy} in Example 2.3.

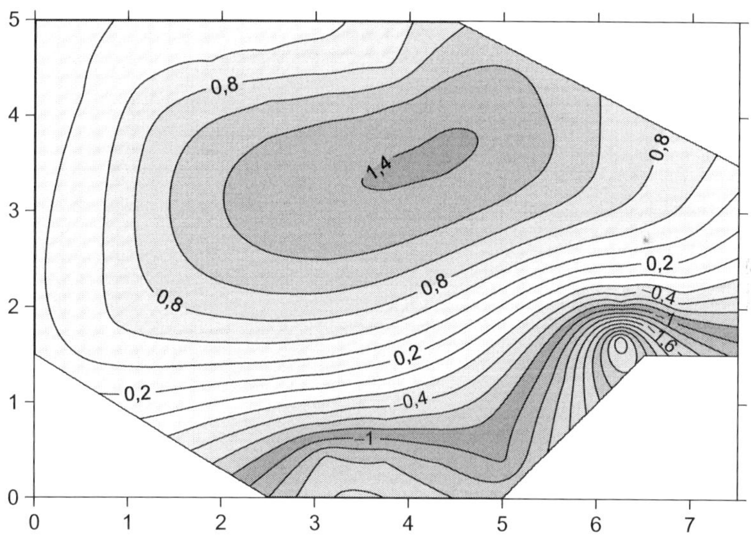

FIGURE 2.40 Contours of the bending moment M_y in Example 2.3.

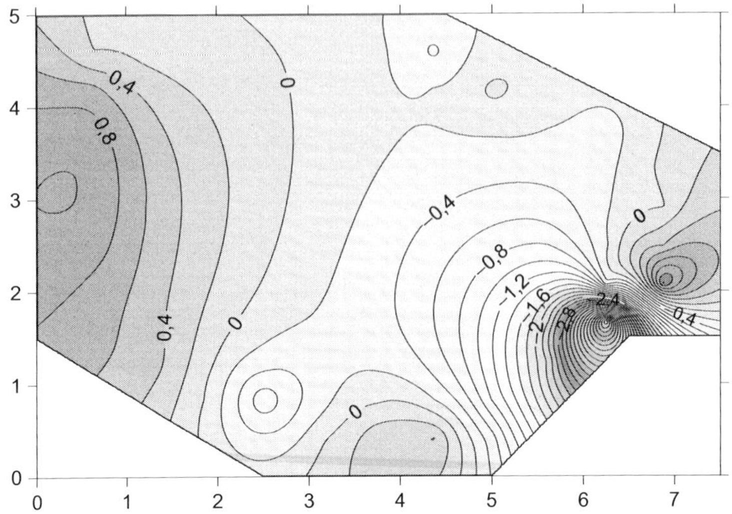

FIGURE 2.41 Contours of the shear force Q_x in Example 2.3.

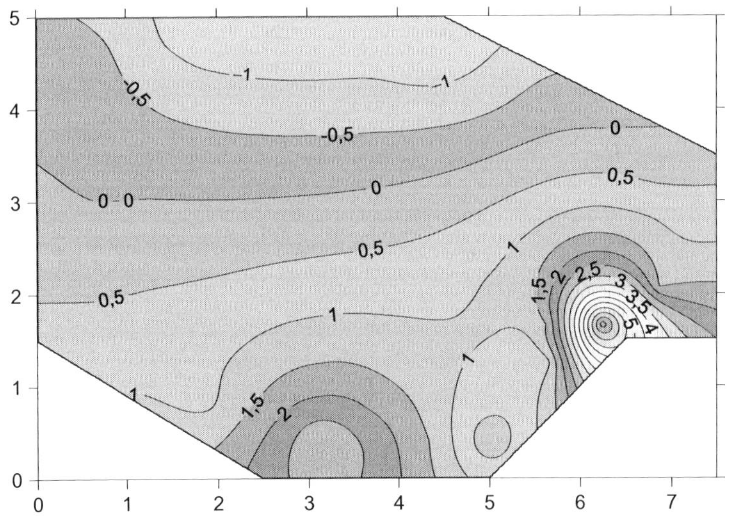

FIGURE 2.42 Contours of the shear force Q_y in Example 2.3.

Example 2.4

As a fourth example, a slab, which is usually encountered in buildings, is analyzed. The geometry of the plate and the support conditions are shown in Fig. 2.43. The plate having a thickness $h=0.1\,\mathrm{m}$ and material constants $E=2.1\times 10^6\mathrm{kN/m^2}$, $\nu=0.2$ is loaded by the uniform load $f_0=1\mathrm{kN/m^2}$ distributed on the whole surface of the plate and the line load $f^*=1\mathrm{kN/m}$ along the free edges of the hole. The results were obtained with N = 1328 boundary elements distributed on the plate edges as follows: N1 = 148 on 1-2 and 5-6, N2 = 60 on 2-3 and 4-5, N3 = 120 on 3-4 and 6-1, N4 = 144 on 7-8 and 9-10, and N5 = 192 on 8-9 and 10-7. The data file is produced by the PLDATA_2.4 program. The results are given in the file RESULTS_2.4 as they are printed by the PLBECON program. The files are available on book's companion website. Fig. 2.44 presents the distributions of w, M_x, M_y and Q_x along the line $y=5.0$, while Fig. 2.45 presents the distributions of M_x, M_{xy}, M_y and Q_y along the line $x=12.5$ in comparison with those resulting from FEM. In Fig. 2.46 the variation of the quantities w, $w,_n$, M_n and V_n (computed and prescribed) along the boundary are shown. Finally, the contour curves of w, M_x, M_{xy}, M_y, Q_x and Q_y are shown in Figs. 2.47 through 2.52. The computed reactions at the point supports are $P_3=P_4=26.16\mathrm{kN}$ and $P_{11}=P_{12}=8.41\mathrm{kN}$. The corresponding values obtained with the FEM are 26.55 and 8.40.

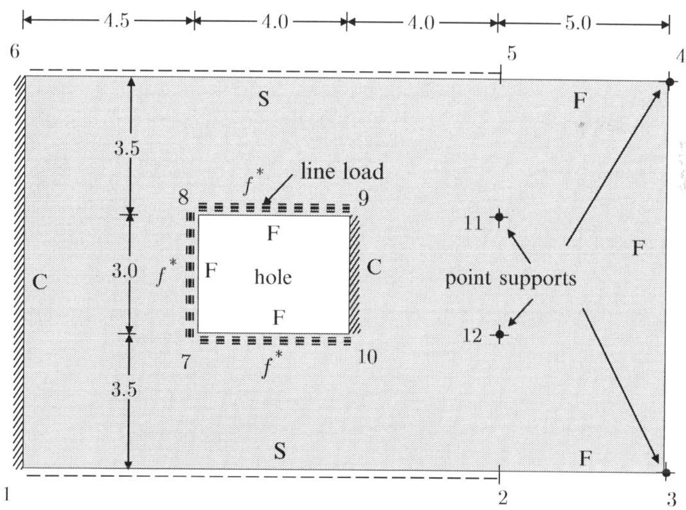

FIGURE 2.43 Plate with a hole and point supports in Example 2.4.

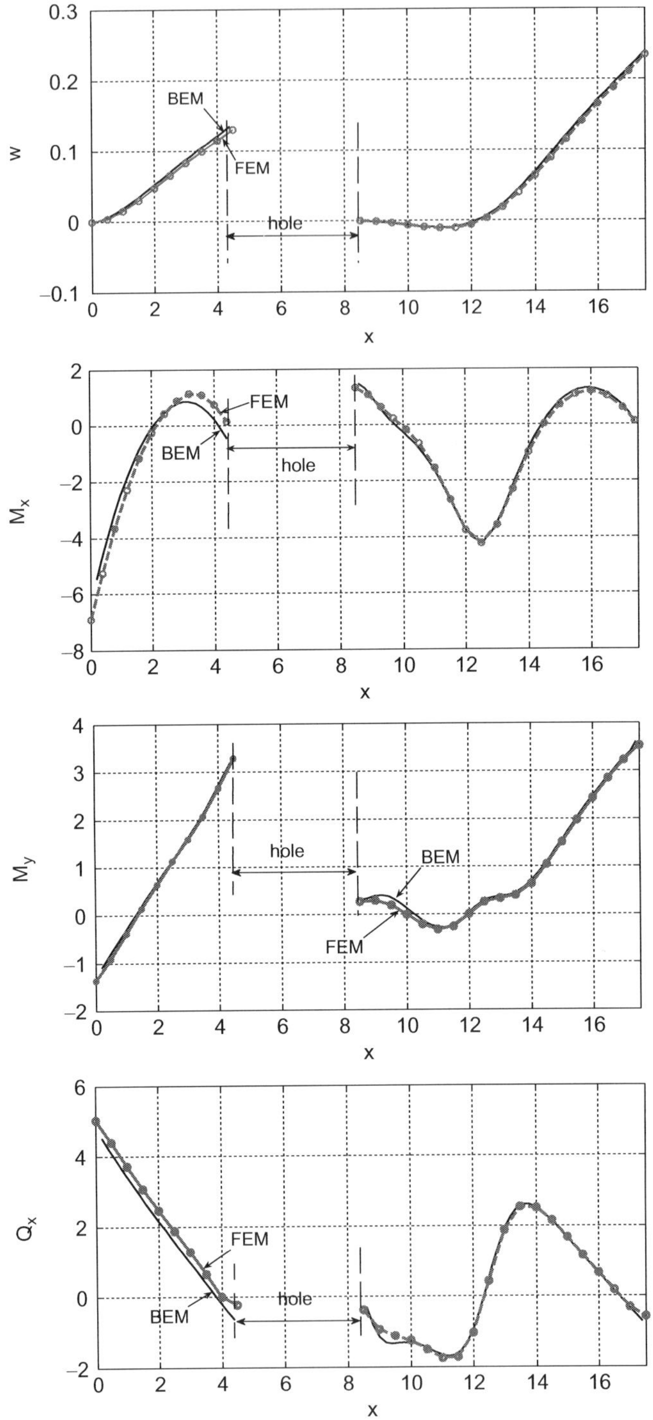

FIGURE 2.44 Distribution of w, M_x, M_y and Q_x along the line $y = 5.0$ in Example 2.4.

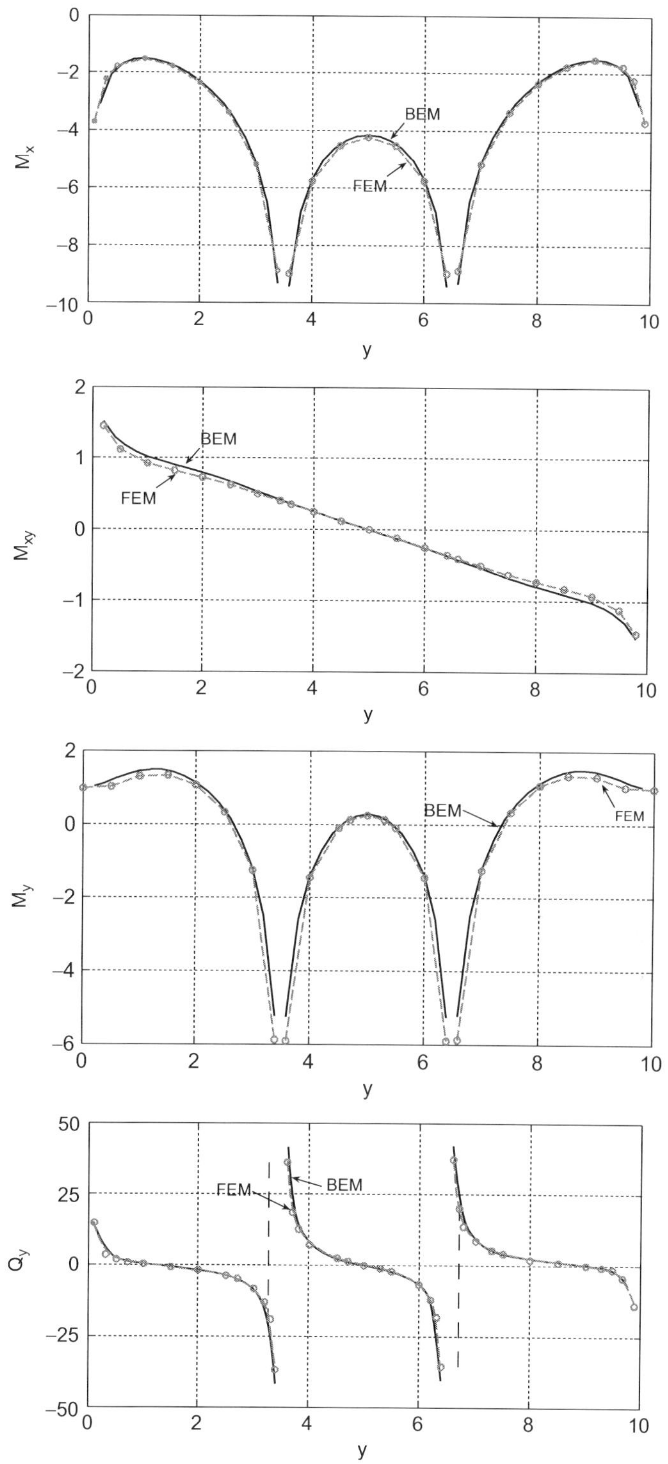

FIGURE 2.45 Distribution of M_x, M_{xy}, M_y and Q_y along the line $x = 12.5$ in Example 2.4.

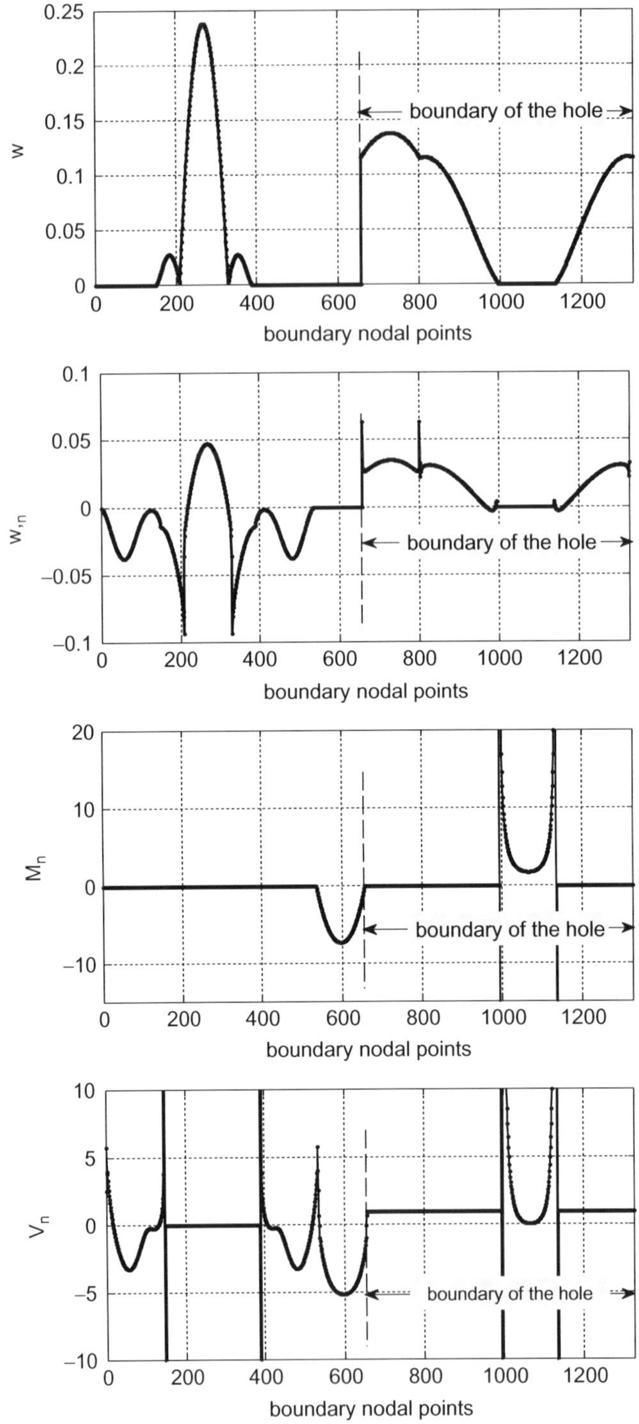

FIGURE 2.46 Distribution of w, $w_{,n}$, M_n and V_n on the boundary in Example 2.4.

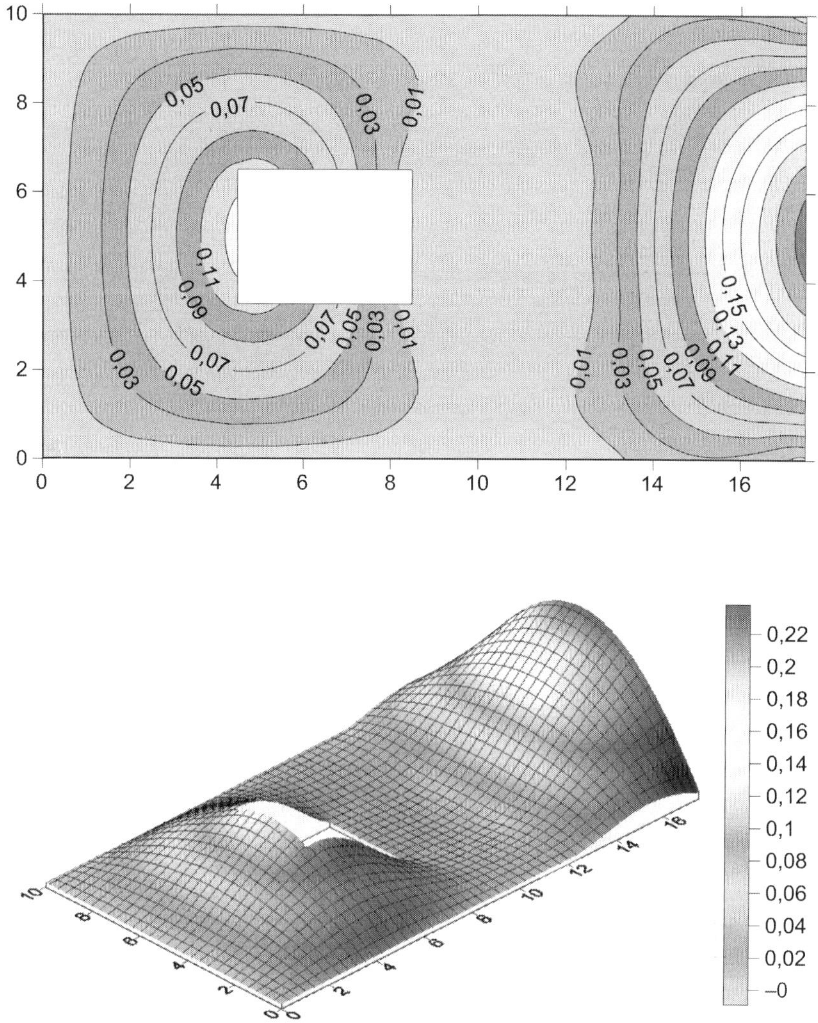

FIGURE 2.47 Deflection surface w in Example 2.4. Contours and 3-D view.

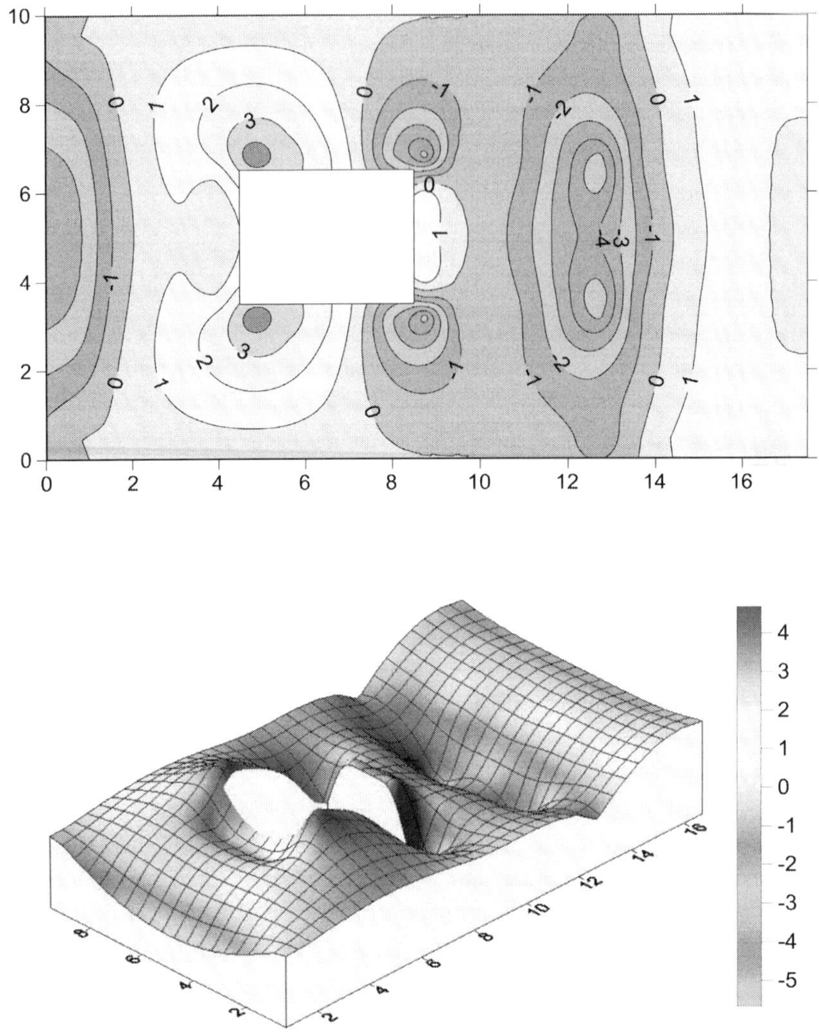

FIGURE 2.48 Bending moment M_x in Example 2.4. Contours and 3-D view.

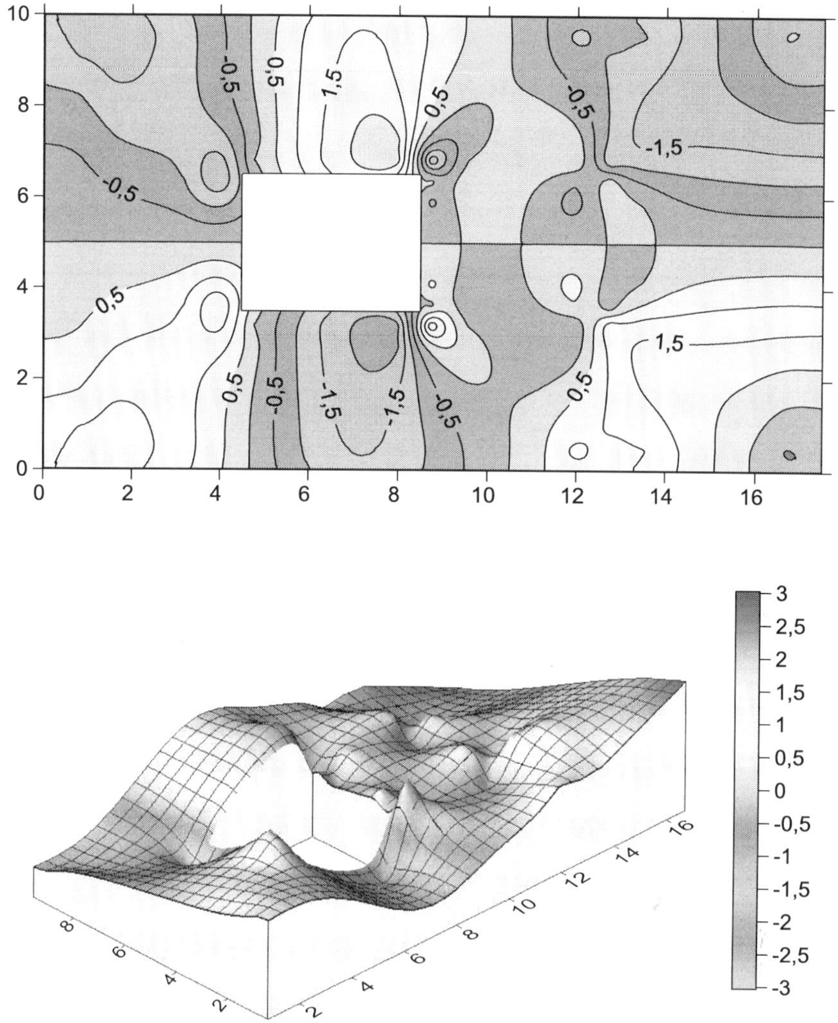

FIGURE 2.49 Twisting moment M_{xy} in Example 2.4. Contours and 3-D view.

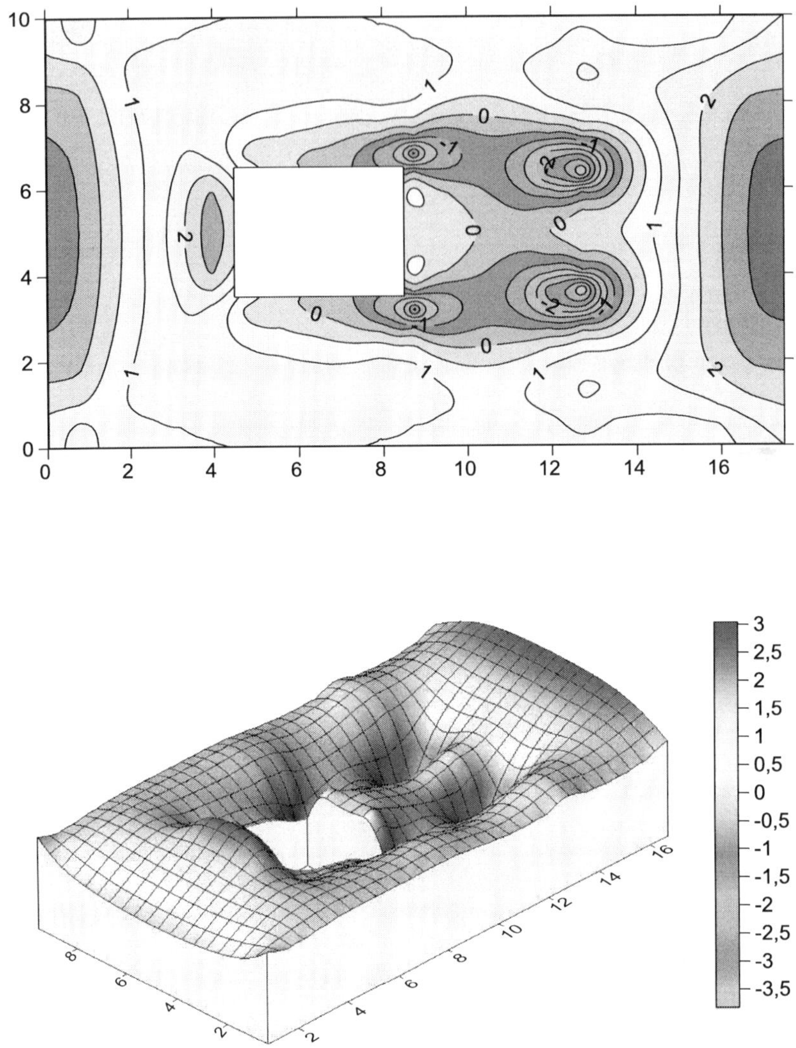

FIGURE 2.50 Bending moment M_y in Example 2.4. Contours and 3-D view.

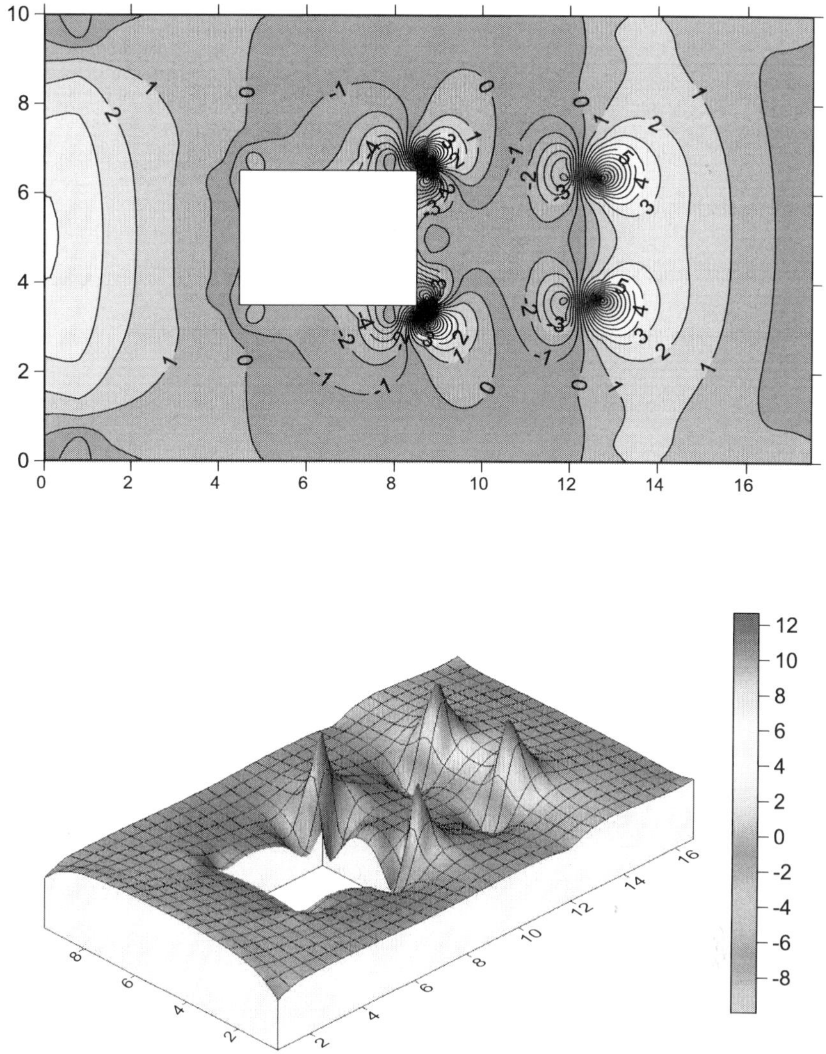

FIGURE 2.51 Shear force Q_x in Example 2.4. Contours and 3-D view.

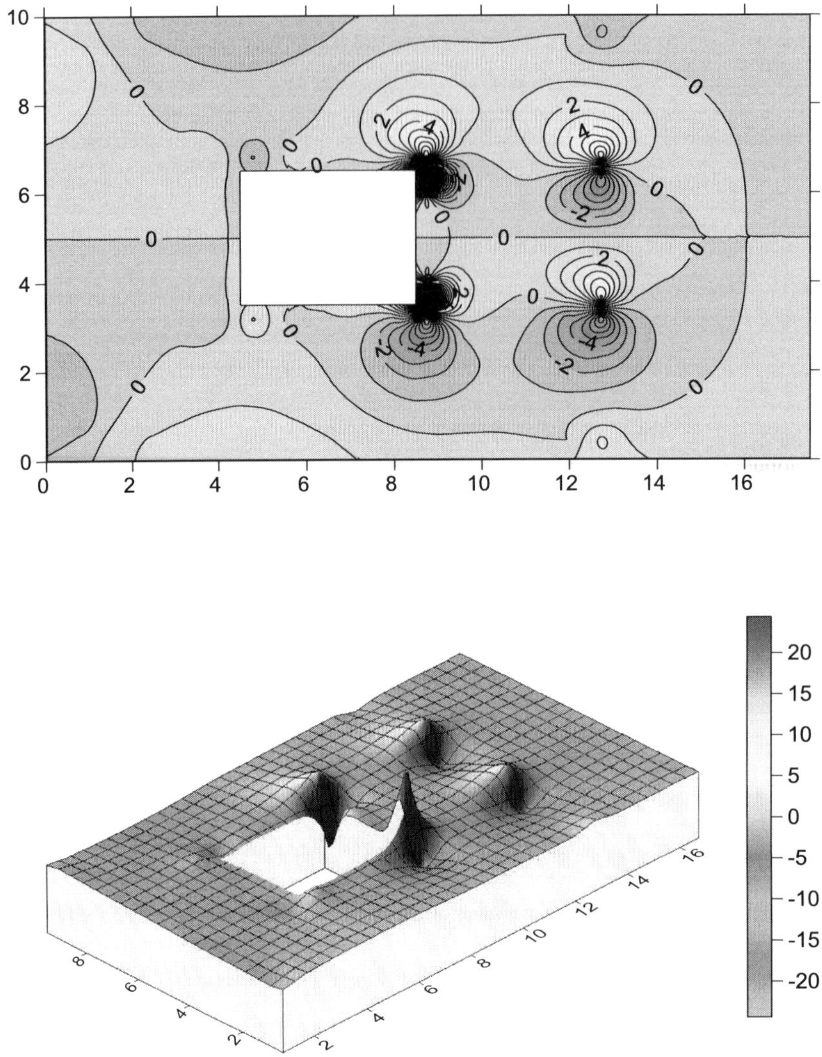

FIGURE 2.52 Shear force Q_y in Example 2.4. Contours and 3-D view.

Example 2.5

As a last example, a plate, whose boundary is partially curvilinear (circular), is analyzed. The geometry of the plate and the support conditions are shown in Fig. 2. 53. The plate with a thickness $h = 0.2\,\mathrm{m}$ and elastic constants $E = 2.1 \times 10^6 \mathrm{kN/m^2}$ and $\nu = 0.2$ is subjected to the uniform load $f = 1\,\mathrm{kN/m^2}$. The results were obtained with N = 900 boundary elements distributed as follows: N1 = 150 on the edges 1-2 and 3-4, N2 = 300 on the circular edge 2-3, N3 = 100 on the edge 4-1, and N4 = 200 on the boundary of the circular hole. The data file is produced with the program PLDATA_2.5. The results are given in the file RESULTS_2.5, as they are printed by the PLBECON program. The files are available on the book's companion website. Fig. 2.54 shows the variation of w,M_x,M_y and Q_x along the line $y = 5.0$ in comparison with those obtained with the FEM. In Fig. 2.55, the variations of the quantities w,$w_{,n}$ and V_n (computed and prescribed) along the boundary are shown. Finally, Figs. 2.56 through 2.61 present the contour curves of w,M_x,M_{xy},M_y,Q_x and Q_y.

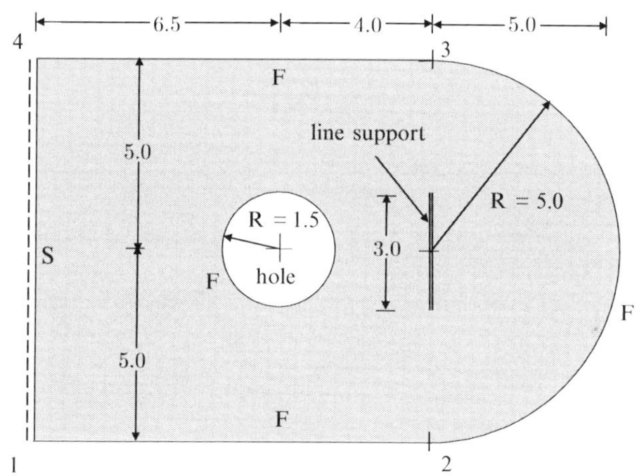

FIGURE 2.53 Plate with a hole and partially curvilinear boundary in Example 2.5.

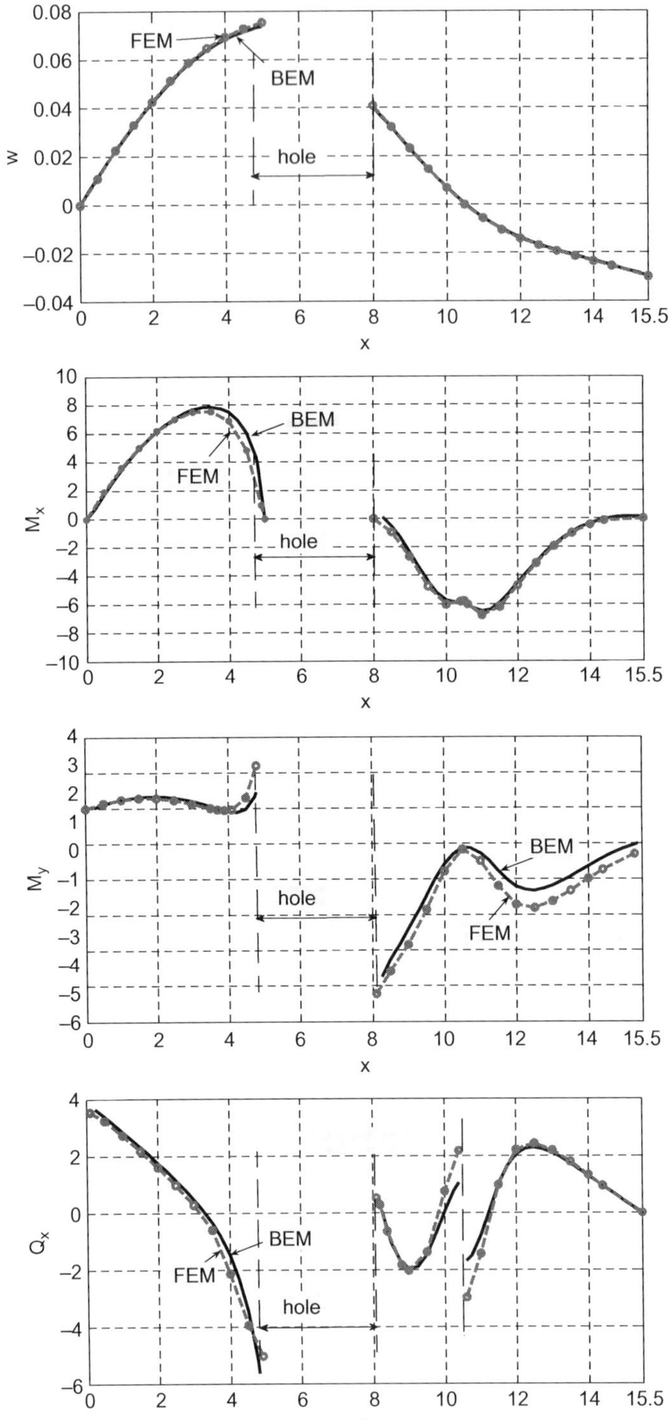

FIGURE 2.54 Distribution of w, M_x, M_y and Q_x along the line $y = 5.0$ in Example 2.5.

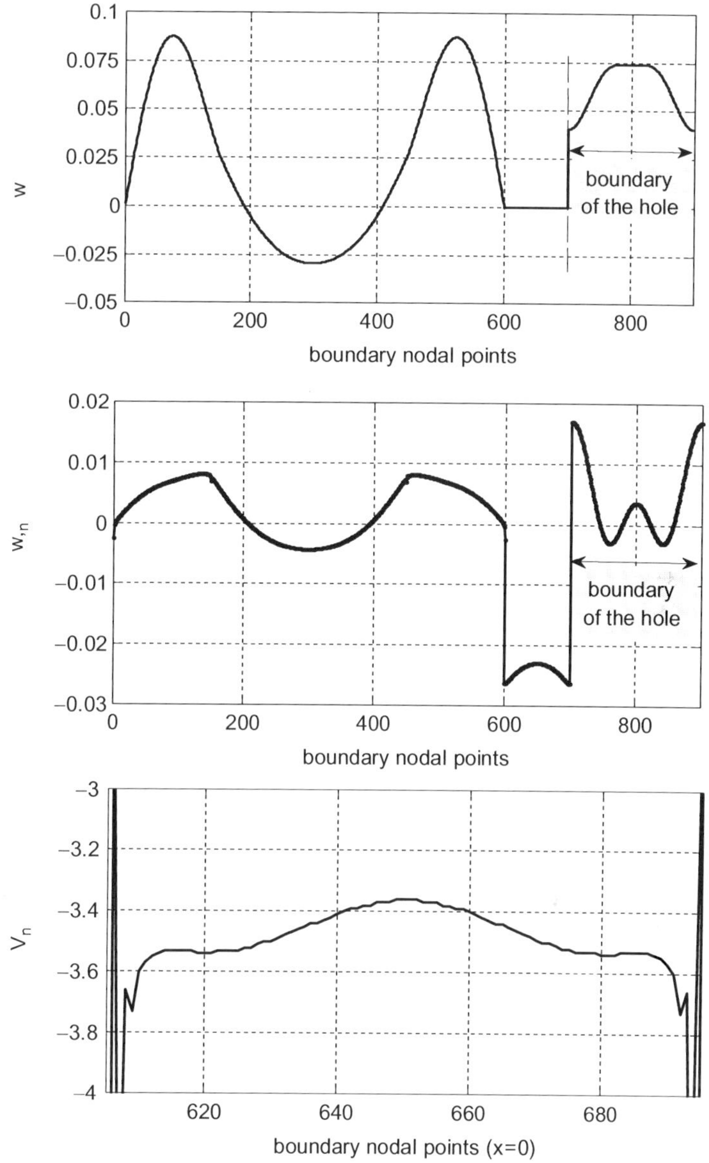

FIGURE 2.55 Distribution of w, $w_{,n}$ and V_n on the boundary in Example 2.5.

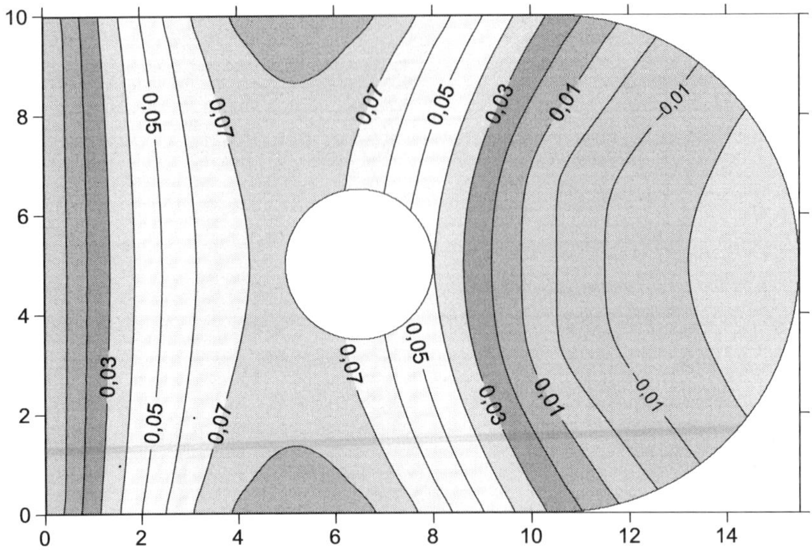

FIGURE 2.56 Contours of the deflection surface w in Example 2.5.

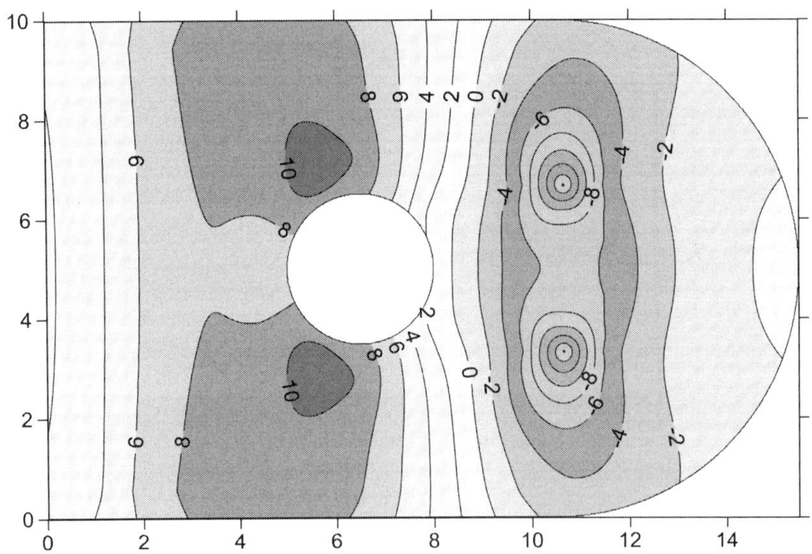

FIGURE 2.57 Contours of the bending moment M_x in Example 2.5.

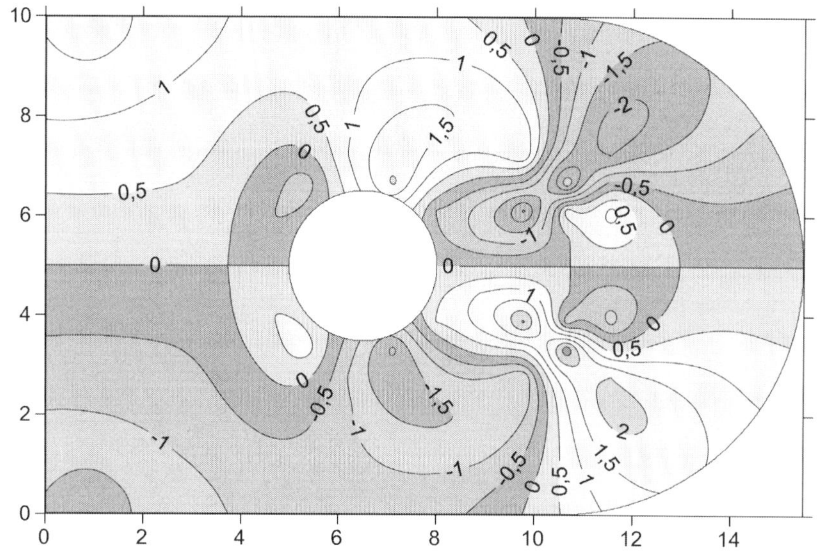

FIGURE 2.58 Contours of the twisting moment M_{xy} in Example 2.5.

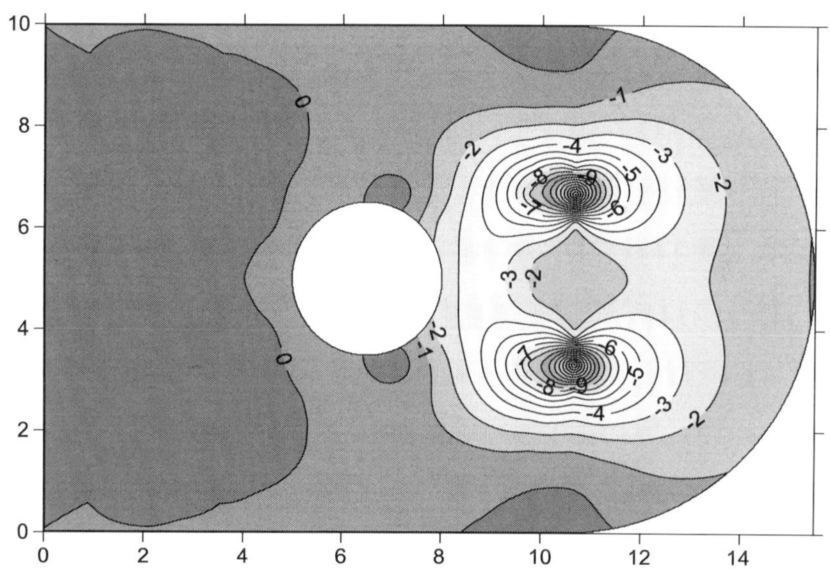

FIGURE 2.59 Contours of the bending moment M_y in Example 2.5.

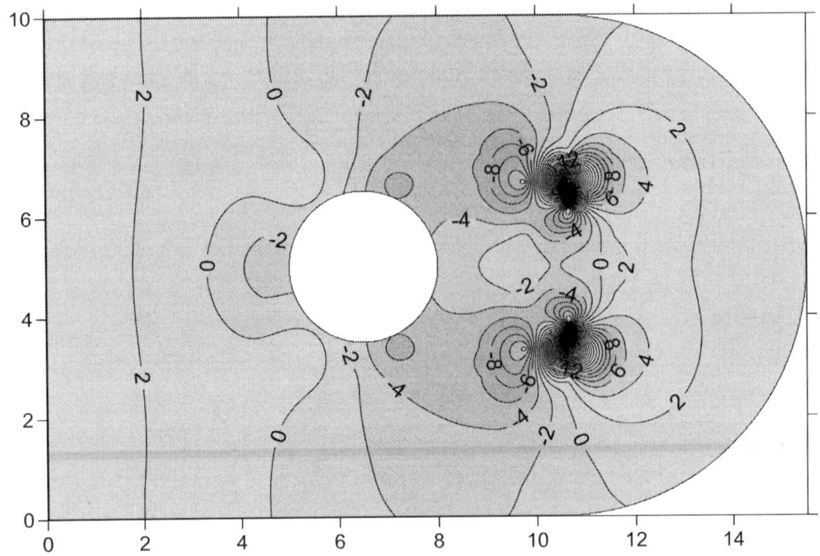

FIGURE 2.60 Contours of the shear force Q_x in Example 2.5.

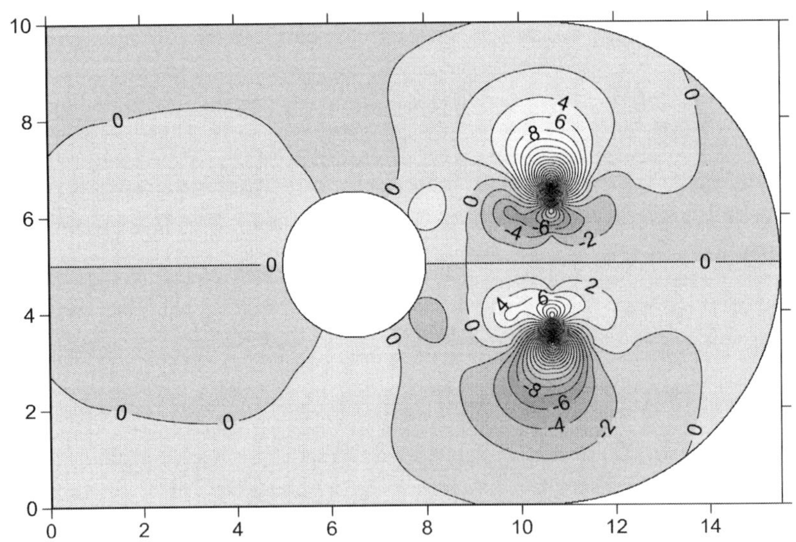

FIGURE 2.61 Contours of the shear force Q_y in Example 2.5.

2.7 REFERENCES

[1] J.T. Katsikadelis, Boundary Elements, Theory and Applications, Elsevier, UK, 2001.

[2] J.T. Katsikadelis, Special methods for plate analysis, in: D.E. Beskos (Ed.), Boundary Element Analysis of Plates and Shells, Springer-Verlag, Berlin, 1991, pp. 221–311.

[3] S. Bergman, M. Schiffer, Kernel Functions and Elliptic Differential Equations in Mathematical Physics, Academic Press, New York, 1953.

[4] D.J. Forbes, A.R. Robinson, Numerical Analysis of Elastic Plates and Shallow Shells by an Integral Equation Method, in: Structural Research SeriesUniversity of Illinois, Urbana, 1969 Report No. 345.

[5] G. Bezine, Boundary integral equations for plate flexure with arbitrary boundary conditions, Mech. Res. Comm. 5 (1978) 197–206.

[6] J.T. Katsikadelis, C.V. Massalas, G.J. Tzivanidis, An integral equation solution of the plane problem of the theory of elasticity, Mech. Res. Comm. 4 (1977) 199–208.

[7] M. Stern, A general boundary integral formulation for the numerical solution of plate bending problems, Int. J. Solid. Struct. 15 (1979) 769–782.

[8] F. Hartmann, R. Zotemantel, The direct boundary element method for plate bending, Int. J. Numer. Meth. Eng. 23 (1986) 2049–2069.

[9] F. Hartmann, Introduction to Boundary Elements, Springer-Verlag, Berlin, 1987.

[10] M.H. Aliabadi, in: The Boundary Element Method, vol. 2, John Wiley & Sons, UK, 2002.

[11] D.E. Beskos (Ed.), Boundary Element Analysis of Plates and Shells, Springer-Verlag, Berlin, 1991.

[12] H. Tottenham, The boundary element method for plates and shells, in: P.K. Banerjee, R. Butterfield (Eds.), Developments in Boundary Element Methods-1, Applied Science, London, 1979, pp. 173–205.

[13] M. Stern, Static analysis of beams, plates and shells, in: D.E. Beskos (Ed.), Boundary Element Methods in Structural Analysis, ASCE, New York, 1989, pp. 41–64.

[14] S. Timoshenko, S. Woinowsky-Krieger, Theory of Plates and Shells, second ed., McGraw-Hill Book Company, New York, 1959.

[15] K. Girkmann, Flächentragwerke, sixth ed., Springer-Verlag, Wien, 1963.

[16] A.E.H. Love, A Treatise on the Mathematical Theory of Elasticity, fourth ed., Dover Publications, New York, 1944.

[17] L.H. Donnell, Beams, Plates and Shells, McGraw-Hill, New York, 1976.

[18] R. Sillard, Theory and Analysis of Plates, Classical and Numerical Methods, Prentice Hall, New Jersey, 1974.

[19] V. Panc, Theories of Elastic Plates, Noordhoff International Publishing, Leyden, 1975.

[20] C.M. Wang, J.N. Reddy, K.H. Lee, Shear Deformable Beams and Plates, Elsevier Science, UK, 2000.

[21] R. Kienzler, On consistent plate theories, Arch. Appl. Mech. 72 (2002) 229–247.

[22] V.V. Novozhilov, Foundations of the Nonlinear Theory of Elasticity, Gray Lock Press, Rochester, NY, 1953.

[23] W. Pietraszkiewicz, Lagrangian description and incremental formulation in the nonlinear theory of thin shells, Int. J. Nonlin. Mech. 19 (2) (1984) 115–140.

[24] K. Washizu, Variational Methods in Elasticity and Plasticity, Pergamon Press, Oxford, 1975.

[25] F.B. Hildebrand, Methods of Applied Mathematics, Dover Publications, Inc., New York, 1965.

[26] J.T. Katsikadelis, The Analysis of Plates on Elastic Foundation by the Boundary Integral Equation Method, (1982), Dissertation in partial fulfillment for the Degree of Doctor of Philosophy (PhD) in Applied Mechanics at the Polytechnic University of New York, New York.

[27] M.L. Williams, Surface stress singularities resulting from various boundary conditions in angular corners of plates under bending, in: Proc. 1st U.S. National Congress of Applied Mechanics, 1951, pp. 325–392, Chicago.

[28] H. Melzer, R. Rannacher, Spannungskonzentrationen in Eckpunkten der Kirchhoffschen Platte, Bauingenieur 55 (1980) 181–184.

[29] J.T. Katsikadelis, A.E. Armenakas, A new boundary equation solution to the plate problem, ASME, J. Appl. Mech. 56 (1989) 364–374.

[30] J.B. Paiva, L.O. Neto, Cubic approximation for the transverse displacement in BEM for elastic plates analysis, Eng. Anal. Bound. Elem. 28 (2004) 869–880.
[31] E.S. Ventsel, Indirect boundary element method for plate bending analysis, Int. J. Numer. Meth. Eng. 40 (1997) 1597–1610.
[32] J. Hadamard, Lectures on Cauchy's Problem in Linear Partial Differential Equations, Yale University Press, New Haven, 1923.

PROBLEMS

2.1 Derive Eqs. (2.49).

2.2 Prove Eq. (2.50).

2.3 Using the procedure described in Section 2.2.5 derive the differential equation for a plate subjected to membrane forces N_x, N_y, N_{xy} in addition to the transverse load.

2.4 Using the procedure described in Section 2.2.5 derive the differential equation for moderately large deflection analysis of plates (von Kármán assumption).

2.5 Derive the reciprocal identity for a plate resting on (i) a Winkler and (ii) a biparametric elastic foundation (Pasternak).

2.6 Starting from Eq. (2.119) derive Eqs. (2.121) and (2.122).

2.7 Starting from Eq. (2.119) find the integral representation of $w_{,n}$ at (i) regular points and (ii) corner points.

2.8 Evaluate the singular line integrals arising when the BEM with linear elements is employed.

2.9 Given $f = 2(x^2 + y^2)^2$, convert the domain integrals $F_1(p) = \int_\Omega f(Q)v(p,Q)fd\Omega_Q$, $F_2(p) = \int_\Omega f(Q)v_1(p,Q)fd\Omega_Q$ to boundary line integrals on Γ.

2.10 Analyze the plates of Table 2.3 using the program PLBECON.

2.11 Analyze the plate of Fig. P2.1 using the program PLBECON. Data: uniform load $f_0 = 5\text{kN/m}$, $h = 0.10\,\text{m}$, $E = 2.1 \times 10^6 \text{kN/m}^2$ and $\nu = 0.2$.

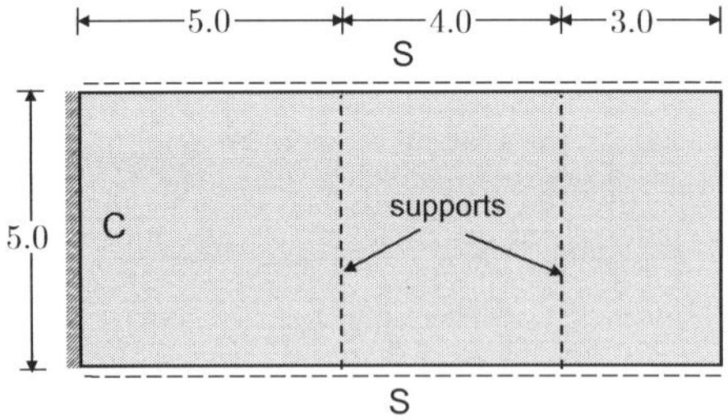

FIGURE P2.1 Continuous plate

2.12 Analyze the plate of Fig. P2.2 using the program PLBECON. Data: uniform load $f_0 = 2\text{kN/m}^2$, line load $f^* = 5\text{kN/m}$, concentrated loads $P = 10\text{kN}$, $h = 0.20\,\text{m}$, $E = 2.1 \times 10^6 \text{kN/m}^2$ and $\nu = 0.2$. The boundary of the circular hole is supported elastically with $k_T = 5000\text{kN/m}$ and $k_R = 0$.

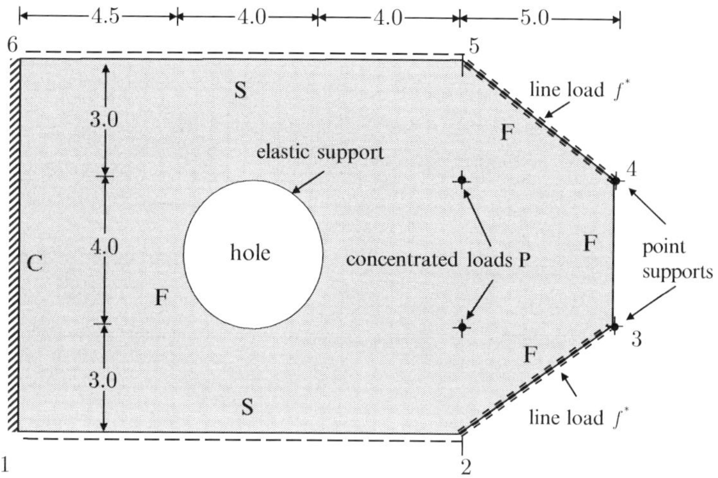

FIGURE P2.2 Plate with a hole and point supports.

Chapter | three

BEM for Other Plate Problems

CHAPTER OUTLINE

The Boundary Element Method for Plate Analysis

3.1 INTRODUCTION

In the previous chapter, the BEM was presented for the solution of the classical plate bending problem, namely, of the problem that is described with the nonhomogeneous partial differential equation for the biharmonic operator. In structural analysis, however, we come across more complicated plate problems, such as bending of plates under combined action of transverse and membrane (inplane) forces, plate buckling problems, plates on elastic foundation, large deflections of plates, postbuckling response of plates, plates with variable thickness, anisotropic plates, thick plates, and others. All these plate problems are governed by fourth-order partial differential equations, which are of elliptic type for static problems and include inertia terms beside the fourth-order spatial derivatives for dynamic problems. These equations, beside the fourth-order derivatives, also contain lower-order derivatives, while the coefficients may be variable. As we know, the BEM can be applied only when the fundamental solution of the differential equation is known or can be established.

Unfortunately, this is not feasible for the complete fourth-order partial differential equation, but only for some special forms of the governing equation, which describe particular plate problems, e.g., plates on Winkler or on biparametric elastic foundation, plates under uniform membrane force. The fundamental solution, whenever it can be established, is different for each plate problem and is expressed by special mathematical functions (e.g., Kelvin functions, Bessel functions of the second kind) whose manipulation to derive the boundary integral equations is a difficult and tedious task. The whole procedure demands special care for the evaluation of the derivatives, their limits on the boundary and at corner points as well as for the numerical evaluation of the boundary singular and hypersingular integrals. The procedure becomes even more complicated for dynamic problems. Thus, the BEM for plates becomes a difficult problem of advanced applied mathematics rather than a useful engineering method for plate analysis.

These observations discouraged researchers from developing and using the BEM as a computational method for plate analysis in engineering praxis. Thus, very early on investigators looked for BEM formulations that would solve the general plate problem using the known simple fundamental solution of the biharmonic equation at the cost of introducing domain integrals, which evidently annul the pure boundary character of the BEM. Several such methods are reported in the literature. Among them we distinguish the DRM (dual reciprocity method) and the AEM (analog equation method) as the most efficient. The DRM applies only when a dominant linear operator with a known fundamental solution can be extracted from the governing equation, while the AEM is alleviated from any restriction and can solve efficiently not only linear but also nonlinear problems, both static and dynamic. It is this method that we will utilize to solve various plate problems in the next sections of this chapter and in all subsequent chapters of this book. The AEM is based on the *principle of the analog equation* [1], which reduces all known plate problems to the classic plate bending problem presented in Chapter 2.

3.2 PRINCIPLE OF THE ANALOG EQUATION

The principle of the analog equation (PAE) was introduced by Katsikadelis in 1994 [1] and was used in conjunction with the BEM to develop the Analog Equation Method, or AEM. The PAE can be stated as:

> *Any differential equation, be it linear or nonlinear, can be replaced by another differential equation of the same order under an unknown fictitious source. The substitute equation is termed the analog equation.*

To make this concept more concrete we illustrate it as follows.

We consider the boundary value problem

$$N(u) = g(\mathbf{x}), \quad \mathbf{x} \in \Omega \tag{3.1a}$$

$$B(u) = \overline{g}(\mathbf{x}), \quad \mathbf{x} \subset \Gamma \tag{3.1b}$$

where N and B are linear or nonlinear differential operators.

Let $u(\mathbf{x})$ be the solution of the problem (3.1a,b). If N^* is another differential operator, linear or nonlinear, of the same order with N and apply it to $u(\mathbf{x})$ we will obtain

$$N^*(u) = b(\mathbf{x}), \quad \mathbf{x} \in \Omega \tag{3.2}$$

where now $b(\mathbf{x})$ is an unknown function (source, load). Eq. (3.2) together with the boundary condition (3.1b) can give the solution of the original problem, provided that the source $b(\mathbf{x})$ is first established. Eq. (3.2) is the analog equation of the original problem and the boundary value problem (3.2), (3.1b) the *equivalent* or the *substitute problem*. The replacement of the actual problem by the substitute one expresses the principle of the analog equation.

The PAE applies also to a system of coupled differential equations. In this case the analog equations may constitute a set of uncoupled equations. The PAE applies to elliptic, hyperbolic, and parabolic differential equations of integer or fractional derivative. Obviously, the practical significance of the PAE for the BEM is very important as a linear equation with simple known fundamental solution can be chosen as the analog equation. Then the BEM is readily employed to solve the substitute problem and the solution of the original problem is obtained from the integral representation of the substitute problem. The fictitious source is first established using a BEM-based procedure. The principle is depicted in Fig. 3.1. Fig. 3.1a shows the deflection curve $u(x)$ of a beam with

The Principle of the Analog Equation [1]

Real problem	Substitute problem
$N(u) = g(\mathbf{x}), \quad \mathbf{x} \in \Omega$ $B(u) = \overline{g}(\mathbf{x}), \quad \mathbf{x} \in \Gamma$ $N(\ )$: linear or nonlinear operator	$N^*(u) = b(\mathbf{x}), \quad \mathbf{x} \in \Omega$ $B(u) = \overline{g}(\mathbf{x}), \quad \mathbf{x} \in \Gamma$ $N^*(\ )$: linear or nonlinear operator of the order of $N(\ )$

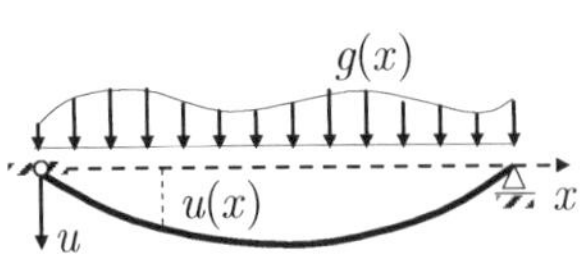

$[EI(x)u(x)_{,xx}]_{,xx} = g(x)$

(a)

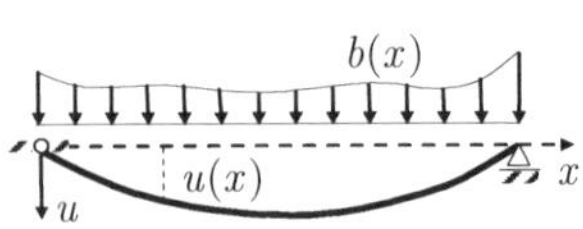

$u(x)_{,xxxx} = b(x)$

(b)

FIGURE 3.1 The principle of the analog equation.

variable stiffness $EI(x)$ under the load $g(x)$. According to the PAE, the $u(x)$ can result as the deflection curve of a beam with constant stiffness, say $EI(x)=1$, under an appropriate load $b(x)$. Obviously, the solution of the substitute beam problem is simple if the fictitious load $b(x)$ is first established. Some examples illustrating the PAE are presented in Tables 3.1, 3.2, and 3.3.

TABLE 3.1 Example 1: PAE applied to a linear problem

Real problem

$$(1+y^2)u_{,xx}+2xyu_{,xy}+(1+2x^2)u_{,yy}+xu_{,x}+yu_{,y}+u=7x^2-5xy+5y^2+4 \quad \text{in } \Omega$$

$$u=x^2-xy+y^2 \quad \text{on } \Gamma_u, \qquad u_{,n}=2x-y \quad \text{on } \Gamma_n$$

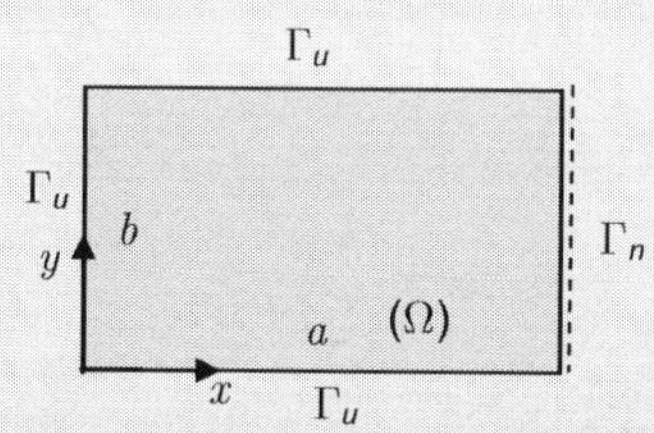

Substitute problem

$$\nabla^2 u=b(x,y) \quad \text{in } \Omega$$

$$u=x^2-xy+y^2 \quad \text{on } \Gamma_u, \qquad u_{,n}=2x-y \quad \text{on } \Gamma_n$$

If $b(x,y)=4$, both problems assume the same solution $u=x^2-xy+y^2$

TABLE 3.2 Example 2: PAE applied to a nonlinear problem

Real problem

$$u_{,xx}+u_{,yy}-\left(1+u_{,x}^2+u_{,y}^2\right)+5u=3(1+xy) \quad \text{in } \Omega$$

$$u=x^2-xy+y^2 \quad \text{on } \Gamma_u, \qquad u_{,n}^2=4x^2-4xy+y^2 \quad \text{on } \Gamma_n$$

The domain Ω is the rectangle of Example 1.

Substitute problem

$$\nabla^2 u=b(x,y) \quad \text{in } \Omega$$

$$u=x^2-xy+y^2 \quad \text{on } \Gamma_u, \qquad u_{,n}^2=4x^2-4xy+y^2 \quad \text{on } \Gamma_n$$

If $b(x,y)=4$, both problems assume the same solution $u=x^2-xy+y^2$

TABLE 3.3 Example 3: PAE applied to a system of two coupled equations

Real problem

$$\nabla^2 u + \frac{1+\nu}{1-\nu}(u_{,xx} + v_{,xy}) + \left(\frac{\lambda^*_{,x}}{\mu} + 2\frac{\mu_{,x}}{\mu}\right)u_{,x} + \frac{\mu_{,y}}{\mu}u_{,y} + \frac{\mu_{,y}}{\mu}v_{,x} + \frac{\lambda^*_{,x}}{\mu}v_{,y} = -\frac{f_x}{\mu} \quad \text{in } \Omega$$

$$\nabla^2 v + \frac{1+\nu}{1-\nu}(u_{,xy} + v_{,yy}) + \frac{\lambda^*_{,y}}{\mu}u_{,y} + \frac{\mu_{,x}}{\mu}u_{,y} + \frac{\mu_{,x}}{\mu}v_{,x} + \left(\frac{\lambda^*_{,y}}{\mu} + 2\frac{\mu_{,y}}{\mu}\right)v_{,y} = -\frac{f_y}{\mu} \quad \text{in } \Omega$$

$$u = xy + 0.2y^2,\ v = xy + 0.2x^2 \quad \text{on } \Gamma_u$$

$$t_x = \lambda^*(x+y) + 2\mu y,\quad v = xy + 0.2x^2 \quad \text{on } \Gamma_t$$

where

$$\lambda^* = \frac{\nu E(x,y)}{(1-\nu^2)},\quad \mu = \frac{E(x,y)}{2(1+\nu)}$$

$$f_x = E_0(1+x)[-0.1(1+x)(7+3\nu) - 2(\nu x + y)]/(1-\nu^2)$$

$$f_y = E_0(1+x)[-0.1(1+x)(7+3\nu) - 1.4(x+y)(1-\nu)]/(1-\nu^2)$$

$$E = E_0(1+x)^2,\quad E_0 = 2\times 10^5 \text{kN/m}^2,\quad \nu = 0.2$$

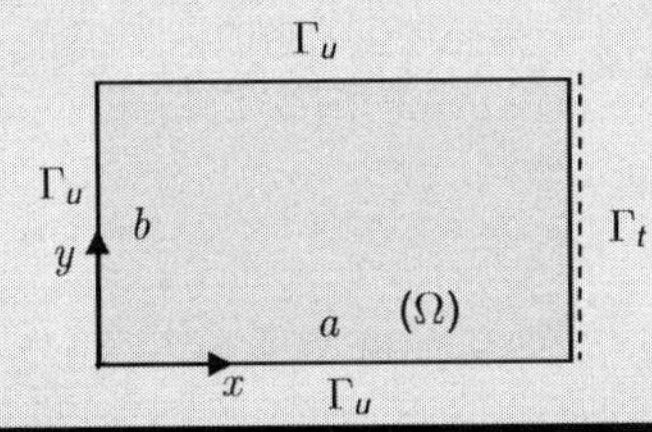

Substitute problem

$$\nabla^2 u = b_x(x,y),\quad \nabla^2 v = b_y(x,y) \quad \text{in } \Omega$$

$$u = xy + 0.2y^2,\quad v = xy + 0.2x^2 \quad \text{on } \Gamma_u$$

$$t_x = \lambda^*(x+y) + 2\mu y,\quad v = xy + 0.2x^2 \quad \text{on } \Gamma_t$$

If $b_x(x,y) = b_y(x,y) = 0.4$, both problems assume the same solution $u = xy + 0.2y^2,\ v = xy + 0.2x^2$

3.3 PLATE BENDING UNDER COMBINED TRANSVERSE AND MEMBRANE LOADS; BUCKLING

3.3.1 Problem statement

We consider a thin elastic plate of uniform thickness h occupying the two-dimensional multiply connected domain Ω of the xy plane whose boundary $\Gamma = \cup_{i=0}^{K}\Gamma_i$ may not be smooth (Fig. 3.2). The plate is subjected to the transverse load $f(x,y)$ and to the membrane loads $p_x(x,y)$, $p_y(x,y)$ acting in the middle plane.

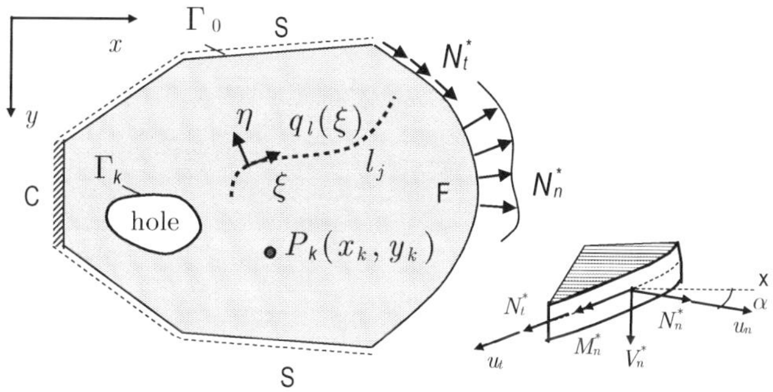

FIGURE 3.2 Plate geometry, notation, and various types of support (C = clamped edge, S = simply supported edge, F = free edge).

The transverse load may generally consist of the distributed load $q(x, y)$, the line loads $q_l(\xi)$ and the concentrated loads P_k. It can be represented as

$$f(\mathbf{x}) = q(\mathbf{x}) + q_l(\xi)\delta(\eta) + \sum_k P_k\delta(\mathbf{x} - \mathbf{x}_k), \ \mathbf{x}:(x, y) \tag{3.3}$$

where $\delta(\eta)$ is the one-dimensional delta function along the local η axis and $\delta(\mathbf{x} - \mathbf{x}_k)$ the two-dimensional delta function. The boundary of the plate may be elastically supported on springs having transverse stiffness $k_T(s)$ and rotational stiffness $k_R(s)$. The corner points k may be supported in the transverse direction on isolated springs with stiffness $k_c^{(k)}$. Finally, the loads N_n^*, N_t^*, M_n^*, V_n^* may act along the boundary and the concentrated load R_k^* $(k = 1, 2, \ldots, K)$ at corner k. Due to this loading, the plate undergoes bending combined with membrane deformation. The equation governing the response of the plate for the problem at hand is derived systematically using the energy method.

The total potential energy Π of the plate is the sum of the strain energy U_b due to bending, the strain energy U_m due to membrane deformation, the elastic energy U_s due to the yielding of the support, and the potential V of the external loads.

Hence, the total potential energy is

$$\Pi = U_b + U_s + U_m + V \tag{3.4}$$

The strain energy U_b is given by Eq. (2.28) and U_s by Eq. (2.29). Namely,

$$U_b = \frac{D}{2}\int_\Omega \left[w_{,xx}^2 + w_{,yy}^2 + 2\nu w_{,xx}w_{,yy} + 2(1-\nu)w_{,xy}^2\right] d\Omega \tag{3.5}$$

$$U_s = \frac{1}{2}\int_\Gamma \left(k_T w^2 + k_R w_{,n}^2\right) ds + \frac{1}{2}\sum_k k_c^{(k)} w_k^2 \tag{3.6}$$

where $D = Eh^2/12(1-\nu^2)$ is the bending stiffness of the plate.

The membrane strain energy U_m is given by the relation

$$U_m = \int_V \left(\overline{\sigma}_x \overline{\varepsilon}_x + \overline{\sigma}_y \overline{\varepsilon}_y + \overline{\tau}_{xy} \overline{\gamma}_{xy}\right) dV \tag{3.7}$$

The stress components $\overline{\sigma}_x, \overline{\sigma}_y, \overline{\tau}_{xy}$ are due to the membrane forces N_x, N_y, N_{xy}, which result from the plane stress state in the plate under the action of the loads $p_x(x, y)$, $p_y(x, y)$ and the boundary forces N_n^*, N_t^*. Since the problem is linear, the plane stress is not influenced by the bending, while the bending is affected by the plane stress. The stress components are evaluated from the relations

$$\overline{\sigma}_x = \frac{N_x}{h} = \frac{C}{h}\left(u_{,x} + \nu v_{,y}\right) \tag{3.8a}$$

$$\overline{\sigma}_y = \frac{N_y}{h} = \frac{C}{h}\left(\nu u_{,x} + v_{,y}\right) \tag{3.8b}$$

$$\overline{\tau}_{xy} = \frac{N_{xy}}{h} = \frac{C}{h}\frac{1-\nu}{2}\left(u_{,y} + v_{,x}\right) \tag{3.8c}$$

where $C = Eh/(1-\nu^2)$ is the membrane stiffness of the plate and u, v the membrane displacements at the middle surface resulting from the solution of the following boundary value problem (see [2], Chapter 7):

$$\nabla^2 u + \frac{1+\nu}{1-\nu}\left(\frac{\partial^2 u}{\partial x^2} + \frac{\partial^2 v}{\partial x \partial y}\right) + \frac{1}{Gh} p_x = 0 \quad \text{in } \Omega \tag{3.9a}$$

$$\nabla^2 v + \frac{1+v}{1-v}\left(\frac{\partial^2 u}{\partial x \partial y} + \frac{\partial^2 v}{\partial y^2}\right) + \frac{1}{Gh} p_y = 0 \quad \text{in } \Omega \tag{3.9b}$$

$$N_n = N_n^* \quad \text{or} \quad u_n = u_n^* \quad \text{on } \Gamma \tag{3.10a}$$

$$N_t = N_t^* \quad \text{or} \quad u_t = u_t^* \quad \text{on } \Gamma \tag{3.10b}$$

where u_n, u_t are the normal and the tangential displacements at the boundary.

The strain components $\overline{\varepsilon}_x, \overline{\varepsilon}_y, \overline{\gamma}_{xy}$ are obtained from Eqs. (2.6), which by virtue of Eqs. (2.8) become

$$\overline{\varepsilon}_x = -w_{,xx} z + u_{,x} + \frac{1}{2} w_{,x}^2 \tag{3.11a}$$

$$\overline{\varepsilon}_y = -w_{,yy} z + v_{,y} + \frac{1}{2} w_{,y}^2 \tag{3.11b}$$

$$\overline{\gamma}_{xy} = -2z w_{,xy} + u_{,y} + v_{,x} + w_{,x} w_{,y} \tag{3.11c}$$

Introducing Eqs. (3.8) and (3.11) in (3.7) and integrating through the thickness of the plate we obtain

$$U_m = \int_\Omega \left[N_x \left(u_{,x} + \frac{1}{2} w_{,x}^2 \right) + N_y \left(v_{,y} + \frac{1}{2} w_{,y}^2 \right) + N_{xy} \left(u_{,y} v_{,x} + w_{,x} w_{,y} \right) \right] d\Omega \tag{3.12}$$

Finally, the potential of the external forces is given by the relation

$$V = -\int_\Omega \left(fw + p_x u + p_y v \right) d\Omega - \int_\Gamma \left(N_n^* u_n + N_t^* u_t + V_n^* w - M_n^* w_{,n} \right) ds - \sum_k R_k^* w^{(k)} \tag{3.13}$$

The plate equilibrium requires that the variation of the total potential energy vanishes, $\delta\Pi = 0$. Thus, working as in Section 2.2.5 we obtain the differential equation of the problem

$$D\nabla^4 w - \left(N_x w_{,xx} + 2N_{xy} w_{,xy} + N_y w_{,yy} \right) + p_x w_{,x} + p_y w_{,y} = f \quad \text{in } \Omega \tag{3.14}$$

and the boundary conditions

$$Vw + N_n w_{,n} + N_t w_{,t} + k_T w = V_n^* \quad \text{or} \quad w = w^* \quad \text{on } \Gamma \tag{3.15a}$$

$$Mw - k_R w_{,n} = M_n^* \quad \text{or} \quad w_{,n} = w_n^* \quad \text{on } \Gamma \tag{3.15b}$$

$$k_c^{(k)} w_k - [\![T_w]\!]_k = R_k^* \quad \text{or} \quad w_k = w_k^* \quad \text{at corner } k \tag{3.15c}$$

It is convenient for programming purposes to write Eqs. (3.15) in the form

$$\alpha_1 w + \alpha_2 (Vw + N_n w_{,n} + N_t w_{,t}) = \alpha_3 \quad \text{on } \Gamma \tag{3.16a}$$

$$\beta_1 w_{,n} + \beta_2 Mw = \beta_3 \quad \text{on } \Gamma \tag{3.16b}$$

$$c_{1k} w_k + c_{2k} [\![Tw]\!]_k = c_{3k} \quad \text{at corner } k \tag{3.16c}$$

Evidently, the conventional boundary conditions are obtained from Eqs. (3.16) by specifying appropriately the coefficients α_i, β_i and c_{ik} (see Section 2.2.6).

For example, an edge is:

Clamped ($w = 0$, $\partial w / \partial n = 0$) if

$$\alpha_1 = 1,\ \alpha_2 = 0,\ \alpha_3 = 0,\ \beta_1 = 1,\ \beta_2 = 0,\ \beta_3 = 0,\ c_{1k} = 1,\ c_{2k} = 0,\ c_{3k} = 0$$

Free ($M_n = 0$, $V_n + N_n w_{,n} + N_t w_{,t} = 0$) if

$$\alpha_1 = 0,\ \alpha_2 = 1,\ \alpha_3 = 0,\ \beta_1 = 0,\ \beta_2 = 1,\ \beta_3 = 0,\ c_{1k} = 0,\ c_{2k} = 1,\ c_{3k} = 0$$

Simply supported with transverse elastic support ($M_n=0,\ V_n+N_n w_{,n}+N_t w_{,t}+k_T w=0$) if

$$\alpha_1=k_T,\ \alpha_2=1,\ \alpha_3=0,\ \beta_1=0,\ \beta_2=1,\ \beta_3=0,\ c_{1k}=k_c^{(k)},\ c_{2k}=1,\ c_{3k}=0$$

Similarly, we obtain the values $\alpha_i, \beta_i, c_{ik}$ for any other type of admissible boundary conditions.

The application of the BEM to Eq. (3.14) requires the knowledge of the fundamental solution. In general, the coefficients $N_x(x,y)$, $N_y(x,y)$, $N_{xy}(x,y)$, $p_x(x,y)$ and $p_y(x,y)$ of the equation are variable and the determination of the fundamental solution is not feasible. It can be established only in certain simple cases, where the membrane force is uniform ($N_x=N_y=\pm N=\text{constant},\ N_{xy}=0,\ p_x=p_x=0$) [3–5]. This happens in rectangular and circular plates under uniform membrane loading along the boundary. The state of uniform membrane force is rather ideal, because the geometry and the boundary conditions result in variable membrane forces in the interior of the plate and therefore it cannot simulate realistic engineering problems. As we will next see, the boundary value problem (3.14), (3.16) can be solved successfully in its general case using the AEM with the known fundamental solution of the biharmonic equation.

3.3.2 AEM solution

3.3.2.1 AEM FOR THE PLATE EQUATION (3.14)

Let w be the solution of Eq. (3.14). The function w is four times continuously differentiable at points $\mathbf{x}:(x,y)\in\Omega$, except at points $\mathbf{x}_k$ and at points ξ on the lines, where the concentrated loads P_k and the line loads $q_l(\xi)$ are applied. There, the derivatives of order higher than second are discontinuous but they can be represented using the two-dimensional delta function at points $\mathbf{x}_k$ and the one dimensional delta function at points on l_j. According to the PAE, the differential operator that produces the analog equation should be of the fourth order. The simplest operator of this order with known fundamental solution is the biharmonic ∇^4 operator. Thus, the analog equation of (3.14) reads

$$\nabla^4 w=\overline{b}(\mathbf{x}) \tag{3.17}$$

Eq. (3.17) represents the plate equation with $D=1$ and load $\overline{b}(\mathbf{x})$.

The fictitious load $\overline{b}(\mathbf{x})$ should include line and concentrated loads beside the distributed load. This is possible if it has the form

$$\overline{b}(\mathbf{x})=b(\mathbf{x})+\beta(\xi)\delta(\eta)+\sum_k \gamma_k\delta(\mathbf{x}-\mathbf{x}_k) \tag{3.18}$$

where $b(\mathbf{x})$ is distributed on the whole plate, $\beta(\xi)$ line load acting along the trace of the external line load ($\xi\in l,\ l=\cup l_j$) and γ_k concentrated load acting at point $\mathbf{x}_k$ of the external concentrated load P_k.

Equation (3.17) together with the boundary conditions (3.15a, b, c) is solved using the direct BEM for plates. The integral representation of the solution is given by Eq. (2.84), namely,

$$w(\mathbf{x}) = \int_\Omega v\bar{b}\,d\Omega + \int_\Gamma (vVw + w_{,n}Mv - v_{,n}Mw - wVv)\,ds - \sum_k (v[\![Tw]\!] - w[\![Tv]\!])_k \qquad \mathbf{x}\in\Omega \tag{3.19}$$

Moreover, the boundary integral equations are given by Eqs. (2.92) and (2.118), which for smooth boundary points are written as

$$\frac{1}{2}w(\mathbf{x}) = \int_\Omega v\bar{b}\,d\Omega + \int_\Gamma (vVw - wVv - v_{,n}Mw + w_{,n}Mv)\,ds - \sum_k (v[\![Tw]\!] - w[\![Tv]\!])_k \qquad \mathbf{x}\in\Gamma \tag{3.20a}$$

$$\frac{1}{2}w_{,\nu}(\mathbf{x}) = \int_\Omega v_1\bar{b}\,d\Omega + \int_\Gamma (v_1Vw - wVv_1 - v_{1,n}Mw + w_{,n}Mv_1)\,ds - \sum_k (v_1[\![Tw]\!] - w[\![Tv_1]\!])_k \qquad \mathbf{x}\in\Gamma \tag{3.20b}$$

where $v = v(\mathbf{x}, \mathbf{y})$ is the fundamental solution with $D = 1$, and $v_1 = v_{,\nu}(\mathbf{x}, \mathbf{y})$ its normal derivative at point $\mathbf{x}\in\Gamma$ (see Fig. 3.3). They are given by Eqs. (2.76a) and (2.97a), which now are written as

$$v = \frac{1}{8\pi} r^2 \ln r \tag{3.21a}$$

$$v_1 = \frac{1}{8\pi} r r_{,\nu}(2\ln r + 1) \tag{3.21b}$$

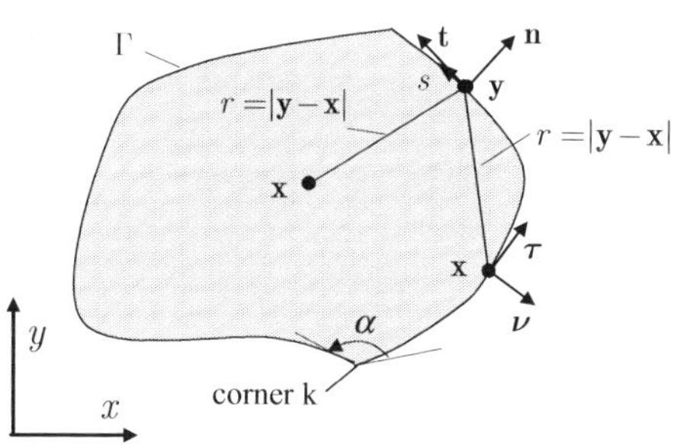

FIGURE 3.3 Notation.

In Eqs. (3.20a,b), the fictitious load $\overline{b}(\mathbf{x})$ is unknown, beside the unknown boundary quantities. It can be proved that(*)

$$\gamma_k = P_k/D \tag{3.22}$$

$$\beta(\xi) = q_l(\xi)/D \tag{3.23}$$

Hence, the loads $\beta(\xi) = q_l(\xi)/D$ and $\gamma_k = P_k/D$ are known and can be treated as described in Section 2.4.3. It remains to evaluate the load $b(\mathbf{x})$.

Equations (3.20a,b) constitute a system of two coupled boundary-domain integral equations, because they include the unknown quantity $b(\mathbf{x})$ in the domain integral. These integral equations can be solved numerically. The boundary integrals are approximated using boundary elements, while the domain integrals are approximated in the following two ways:

(a) *Domain discretization of* Ω

The domain Ω is divided into a finite number of elements over which the law of variation of $b(\mathbf{x})$ is assumed known, e.g., constant, linear, etc. Triangular elements with linear variation of $b(\mathbf{x})$ are very convenient, because, on the one hand, they give good accuracy and, on the other hand, the discretization in triangles is readily performed using the Delaunay triangulation. There are ready-to-use subroutines for this purpose (see the relevant MATLAB function). The program PLDATA available on this book's companion website, which prepares the data (see Section 2.5), can be also used for automatic domain triangulation, if it is provided with the MATLAB function `TRI=delaunay(X,Y)`. It should be noted that $b(\mathbf{x})$ is defined only in the interior of Ω and consequently the nodal points of the triangles adjacent to the boundary must be placed in the interior of the domain, (Fig. 3.4 and Fig. 3.5).

(*)Proof: Introducing Eqs. (3.3) and (3.18) in Eq. (3.14) yields

$$D\left[b(\mathbf{x}) + \beta(\xi)\delta(\eta) + \sum_k \gamma_k \delta(\mathbf{x} - \mathbf{x}_k)\right] - \left(N_x w_{,xx} + 2N_{xy} w_{,xy} + N_y w_{,yy}\right) \\ + p_x w_{,x} + p_y w_{,y} = q(\mathbf{x}) + q_l(\xi)\delta(\eta) + \sum_k P_k \delta(\mathbf{x} - \mathbf{x}_k) \tag{a}$$

Integrating Eq. (a) over a small circular domain Ω_ε with center $\mathbf{x}_k$, and radius ε, and letting $\varepsilon \to 0$ we obtain

$$\gamma_k = P_k/D \tag{b}$$

Further, integrating over the line l and taking into account that the terms $Db(\mathbf{x}) - \left(N_x w_{,xx} + 2N_{xy} w_{,xy} + N_y w_{,yy}\right) + p_x w_{,x} + p_y w_{,y}$ and $q(\mathbf{x})$ are continuous in Ω we obtain

$$\beta(\xi) = q_l(\xi)/D \tag{c}$$

We arrive at the same result if we take into account that points and lines are domains of zero measure when considered in two dimensions and the integrals of continuous functions vanish on them [6].

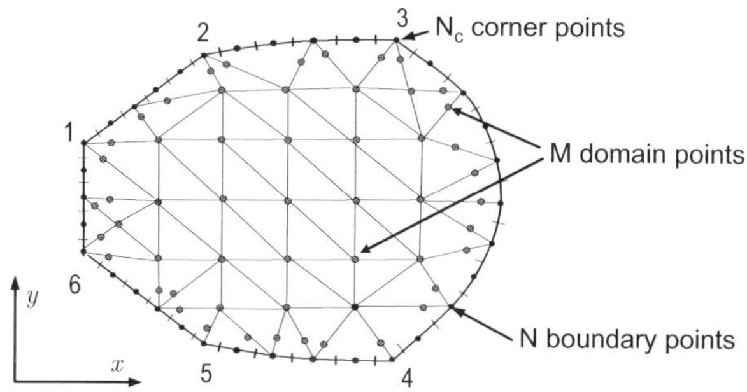

FIGURE 3.4 Domain discretization.

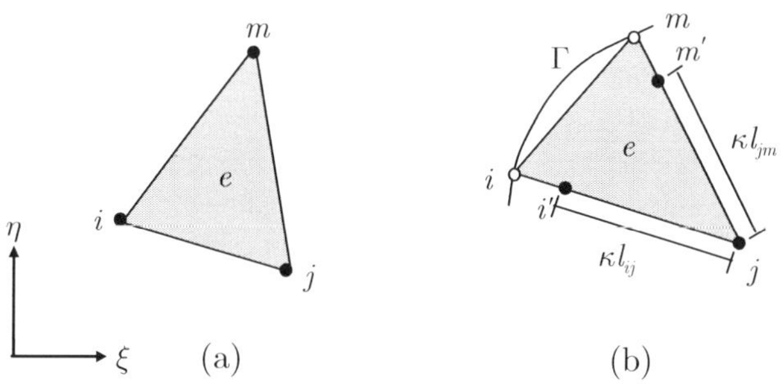

FIGURE 3.5 Triangular element (a) away from and (b) adjacent to the boundary.

(b) *Transformation of the domain integral to boundary integral on* Γ
In this method, the fictitious load $b(\mathbf{x})$ is approximated with radial basis functions (RBFs), namely,

$$b(\mathbf{x}) = \sum_{j=1}^{M} \alpha_j f_j(r), \quad r = |\mathbf{x} - \mathbf{x}_j| \tag{3.24}$$

where $f_j(r)$ are RBFs, $\mathbf{x}_j$ are M collocation points inside the domain, and a_j are M coefficients to be determined. There are several types of RBFs [7]. It has been shown that the RBFs of multiquadric type (multiquadrics) $f_j(r) = \sqrt{r^2 + c^2}$ give accurate results provided that an optimum value of the shape parameter c is employed. The determination of an optimum value is a difficult mathematical problem [8]. A computational method for the optimum value of c was developed by Katsikadelis [9]. This method is based on the minimization of the functional, which produces the governing differential equation as the Euler-Lagrange equation. In our problem, the functional is the total potential energy of the deformed plate.

Using Eq. (3.24), the domain integral in Eq. (3.20a) for $b(\mathbf{x})$ is written

$$\int_\Omega v b \, d\Omega = \sum_{j=1}^{M} a_j \int_\Omega v f_j(r) \, d\Omega \tag{3.25}$$

Further, application of Eq. (2.169) with $\varepsilon=½$, transforms the domain integrals in the right-hand side of Eq. (3.25) to the boundary line integrals

$$\int_\Omega v f_j(r)\,d\Omega = \frac{1}{2}\hat{w}_j(P) + \int_\Gamma \left(v\frac{\partial}{\partial n}\nabla^2\hat{w}_j - \hat{w}_j\frac{\partial}{\partial n}\nabla^2 v - v_{,n}\nabla^2\hat{w}_j + \hat{w}_{j,n}\nabla^2 v \right) d\Omega \tag{3.26}$$

where $\hat{w}_j$ is a particular solution of the equation

$$\nabla^4\hat{w}_j = f_j(r) \tag{3.27}$$

It should be noted that a particular solution of Eq. (3.27) can be always established (see Eqs. (2.169) and (2.170)).

A similar procedure transforms the domain integral in Eq. (3.20b), which involves the kernel $v_1(\mathbf{x},\mathbf{y})$.

The second method renders the BEM a boundary-only method in the sense that the discretization and the evaluation of the integrals are confined to the boundary. Thus, all the advantages of a boundary method are maintained. This technique has been successfully employed to solve plate problems [9–11]. However, because of the additional computational cost for the establishment of the optimal value of the shape parameter c, the method of the domain discretization is preferred. This is also dictated from the fact that nowadays the domain discretization can be performed, as mentioned, automatically with ready-to-use subroutines. Therefore, we limit the presentation to the AEM with domain discretization.

Equations (3.20a,b) are discretized by approximating the boundary integrals with constant elements and the domain integrals with linear triangular elements. This yields

$$\mathbf{H}\begin{Bmatrix}\mathbf{w}\\ \mathbf{w}_c\\ \mathbf{w}_n\end{Bmatrix} = \mathbf{G}\begin{Bmatrix}\mathbf{V}\\ \mathbf{R}\\ \mathbf{M}\end{Bmatrix} + \mathbf{A}\mathbf{b} + \frac{1}{D}\mathbf{F} \tag{3.28}$$

The vectors $\mathbf{w}$, $\mathbf{w}_c$, $\mathbf{w}_n$, $\mathbf{V}$, $\mathbf{R}$, $\mathbf{M}$ of the boundary quantities and the coefficient matrices $\mathbf{H}$, $\mathbf{G}$ are defined in Section 2.4.1. It should be noted that the coefficient matrices refer now to the analog equation, thus their evaluation should be made with $D=1$.

The matrix $\mathbf{A}$ and the vector $\mathbf{F}$ are given as

$$\mathbf{A} = \begin{bmatrix}\mathbf{A}_1\\ \mathbf{A}_2\\ \mathbf{A}_3\end{bmatrix}_{(2N+N_c)\times M}, \qquad \mathbf{F} = \begin{Bmatrix}\mathbf{F}_1\\ \mathbf{F}_2\\ \mathbf{F}_3\end{Bmatrix}_{(2N+N_c)\times 1} \tag{3.29}$$

The elements A_1^{ik}, A_2^{ik}, A_3^{ik} of the matrices $\mathbf{A}_1$, $\mathbf{A}_2$, $\mathbf{A}_3$ originate from the integration of the kernels of the domain integrals over the triangular elements. The integration procedure is presented in the next section. The vector $\mathbf{b}$ includes the M unknown values of the fictitious load at the interior domain points. Finally, the elements F_k^i of the vectors $\mathbf{F}_k$ represent the contribution of the point and line loads (see Eqs. (2.156) and (2.160)).

Further, applying the boundary conditions (3.16a,b) at the N boundary nodal points and (3.16c) at the N_c corner points, and approximating the tangential derivative $w_{,t}$ with finite differences, we obtain

$$\overline{\mathbf{H}}\left\{\begin{array}{c}\mathbf{w}\\ \mathbf{w}_c\\ \mathbf{w}_n\end{array}\right\}+\overline{\mathbf{G}}\left\{\begin{array}{c}\mathbf{V}\\ \mathbf{R}\\ \mathbf{M}\end{array}\right\}=\overline{\mathbf{F}} \tag{3.30}$$

where

$$\overline{\mathbf{H}}=\left[\begin{array}{ccc}\boldsymbol{\alpha}_1+\boldsymbol{\alpha}_2\mathbf{N}_t^*\mathbf{d} & \mathbf{0} & \boldsymbol{\alpha}_2\mathbf{N}_n^*\\ \mathbf{0} & \mathbf{c}_1 & \mathbf{0}\\ \mathbf{0} & \mathbf{0} & \boldsymbol{\beta}_1\end{array}\right],\quad \overline{\mathbf{G}}=D\left[\begin{array}{ccc}\boldsymbol{\alpha}_2 & \mathbf{0} & \mathbf{0}\\ \mathbf{0} & \mathbf{c}_2 & \mathbf{0}\\ \mathbf{0} & \mathbf{0} & \boldsymbol{\beta}_2\end{array}\right],\quad \overline{\mathbf{F}}=\left\{\begin{array}{c}\boldsymbol{\alpha}_3\\ \mathbf{c}_3\\ \boldsymbol{\beta}_3\end{array}\right\} \tag{3.31}$$

The multiplier D in matrix $\overline{\mathbf{G}}$ results from the fact that the boundary quantities $\mathbf{V}$, $\mathbf{R}$, $\mathbf{M}$ refer to the analog equation ($D=1$). Hence, $M_n=DMw, V_n=DVw$ and $R_k=D[\![Tw]\!]_k$.

The quantities appearing in Eq. (3.31) represent:

$\boldsymbol{\alpha}_1 :=N\times N$ diagonal matrix with values 0,1 or k_T
$\boldsymbol{\alpha}_2 :=N\times N$ diagonal matrix with values 0,1
$\boldsymbol{\alpha}_3 :=N\times 1$ vector with the nodal values w^* or V_n^*
$\boldsymbol{\beta}_1 :=N\times N$ diagonal matrix with values 0,1 or $-k_R$
$\boldsymbol{\beta}_2 :=N\times N$ diagonal matrix with values 0,1
$\boldsymbol{\beta}_3 :=N\times 1$ vector with the nodal values w_n^* or M_n^*
$\mathbf{c}_1 :=N_c\times N_c$ diagonal matrix with values 0,1 or $k_c^{(k)}$
$\mathbf{c}_2 :=N_c\times N_c$ diagonal matrix with values 0, -1
$\mathbf{c}_3 :=N_c\times 1$ vector with the nodal values of w_k^* or R_k^*
$\mathbf{N}_n^* :=N\times N$ diagonal matrix with the nodal values N_n^*
$\mathbf{N}_t^* :=N\times N$ diagonal matrix with the nodal values N_t^* and

$$\mathbf{d}=\frac{1}{2l}\begin{array}{|cccccccccccccc|l}
 &&&&&& j & j+1 &&&&&& & \\
 &&&&&& \downarrow & \downarrow &&&&&& & \\
-3 & 4 & -1 &&&&&&&&&&& & \\
-1 & 0 & 1 &&&&&&&&&&& & \\
 & -1 & 0 & 1 &&&&&&&&&& & \\
 && \ddots & \ddots & \ddots &&&&&&&&& & \\
 &&&& -1 & 0 & 1 &&&&&&& & \leftarrow j-1\\
 &&&& 1 & -4 & 3 &&&&&&& & \leftarrow j\\
 &&&&&&& -3 & 4 & -1 &&&& & \leftarrow j+1\\
 &&&&&&& -1 & 0 & 1 &&&& & \leftarrow j+2\\
 &&&&&&&&& \ddots & \ddots & \ddots && & \\
 &&&&&&&&&&& -1 & 0 & 1 & \\
 &&&&&&&&&&& 1 & -4 & 3 &
\end{array}$$

$l=$ the length of the element.

The matrix **d** results from the approximation of the tangential derivative $w_{,t} \equiv w_{,s}$ at the boundary nodal points with central differences. Forward and backward differences are employed at points adjacent to corner points. In the above matrix scheme, j and $j+1$ designate nodes adjacent to a corner (see [2], Section 6.2.2).

The values of α_i, β_i, c_i result from the specified boundary conditions (3.16).

Equations (3.28) and (3.30) constitute a system of $4N+2N_c$ equations, which is solved to give the boundary quantities $\mathbf{w}, \mathbf{w}_c, \mathbf{w}_n, \mathbf{V}, \mathbf{R}, \mathbf{M}$ in terms of the vector $\mathbf{b}$.

The final step of the AEM

Introducing Eq. (3.17) in Eq. (3.14) and taking into account Eqs. (3.3), (3.18), (3.22), and (3.23) we have

$$Db(\mathbf{x}) - \left(N_x w_{,xx} + 2N_{xy} w_{,xy} + N_y w_{,yy}\right) + p_x w_{,x} + p_y w_{,y} = q(\mathbf{x}) \qquad (3.32)$$

Now applying Eq. (3.32) at the M interior nodal points, we obtain

$$D\mathbf{b} - \left(\mathbf{N}_x \mathbf{w}_{,xx} + 2\mathbf{N}_{xy}\mathbf{w}_{,xy} + \mathbf{N}_y \mathbf{w}_{,yy}\right) + \mathbf{p}_x \mathbf{w}_{,x} + \mathbf{p}_y \mathbf{w}_{,y} = \mathbf{q} \qquad (3.33)$$

where $\mathbf{N}_x$, $\mathbf{N}_{xy}$, $\mathbf{N}_y$, $\mathbf{p}_x$, $\mathbf{p}_y$, and $\mathbf{q}$ are $M \times 1$ known vectors. They include the values of the membrane forces N_x, N_{xy}, N_y and the external loads p_x, p_y, q at the M interior nodal nodes, respectively. The vectors $\mathbf{w}_{,x}$, $\mathbf{w}_{,y}$, $\mathbf{w}_{,xx}$, $\mathbf{w}_{,xy}$, $\mathbf{w}_{,yy}$ include the values of the derivatives of the deflection w at the same points.(*) These vectors are unknown, but as we will see next they can be expressed in terms of $\mathbf{b}$.

The expressions of the derivatives can be derived by direct differentiation of the integral representation for the displacement (3.19). However, as it was explained in Section 2.3.5, the computational task is drastically reduced if the integral representation (2.85) is utilized, which for D = 1 is written as

$$w(\mathbf{x}) = \int_\Omega v\bar{b}\,d\Omega - \int_\Gamma \left(v\Psi - w\nabla^2 v_{,n} - v_{,n}\Phi + w_{,n}\nabla^2 v\right) ds \quad \mathbf{x} \in \Omega \qquad (3.34)$$

where it was set as $\Phi = \nabla^2 w$ and $\Psi = \nabla^2 w_{,n}$.

The integral equations that will be used to evaluate the boundary quantities Φ, Ψ are Eqs. (2.121) and (2.122), which are now written as

$$\frac{1}{2}w(\mathbf{x}) = \int_\Omega v\bar{b}\,d\Omega - \int_\Gamma \left(v\Psi - w\nabla^2 v_{,n} - v_{,n}\Phi + w_{,n}\nabla^2 v\right) ds \quad \mathbf{x} \in \Gamma \qquad (3.35)$$

$$\frac{1}{2}\Phi = \int_\Omega \nabla^2 v\bar{b}\,d\Omega - \int_\Gamma \left(\nabla^2 v\Psi - \nabla^2 v_{,n}\Phi\right) ds \quad \mathbf{x} \in \Gamma \qquad (3.36)$$

---

*For the simplicity of the presentation, the vectors in Eq. (3.33), and in subsequent appearances as well, should be understood as one-dimensional arrays and their product, e.g. $\mathbf{N}_x \mathbf{w}_{,xx}$, $\mathbf{p}_x \mathbf{w}_{,x}, \ldots$, as one-dimensional array, whose elements are the scalar product of the corresponding elements of the vectors (see corresponding operation in Matlab). In matrix algebra this operation is meaningful if the preceding array is understood as a diagonal matrix.

Using the same boundary and domain discretization and approximating the boundary integrals with constant elements and the domain integrals with linear triangular elements, Eqs. (3.35) and (3.36) give

$$\frac{1}{2}w^i = \sum_{k=1}^{M} B_1^{ik} b^k + \frac{1}{D}C_1^i - \sum_{j=1}^{N}[v]^{ij}\Psi^j + \sum_{j=1}^{N}\left[\nabla^2 v_n\right]^{ij} w^j + \sum_{j=1}^{N}[v_n]^{ij}\Phi^j - \sum_{j=1}^{N}\left[\nabla^2 v\right]^{ij} w_n^j \tag{3.37}$$

$$\frac{1}{2}\Phi^i = \sum_{k=1}^{M} B_2^{ik} b^k + \frac{1}{D}C_2^i - \left[\nabla^2 v\right]^{ij}\Psi^j + \left[\nabla^2 v_n\right]^{ij}\Phi^j \tag{3.38}$$

where

$$[v]^{ij} = \int_{\Gamma_j} v(\mathbf{x}_i, \mathbf{y})\, ds_y, \quad [v_n]^{ij} = \int_{\Gamma_j} \frac{\partial v(\mathbf{x}_i, \mathbf{y})}{\partial n_y}\, ds_y \tag{3.39a,b}$$

$$\left[\nabla^2 v\right]^{ij} = \int_{\Gamma_j} \nabla^2 v(\mathbf{x}_i, \mathbf{y})\, ds_y, \quad \left[\nabla^2 v_n\right]^{ij} = \int_{\Gamma_j} \frac{\partial}{\partial n_y}\nabla^2 v(\mathbf{x}_i, \mathbf{y})\, ds_y \tag{3.39c,d}$$

The coefficients B_1^{ij} and B_2^{ij} originate from the integration of the kernels over the triangular elements. The integration technique is presented in full detail in a later section. ${C_1}^i$, ${C_2}^i$ represent the contribution of the actual line and concentrated loads.

Equations (3.37) and (3.38) are written in matrix form

$$\mathbf{K}\begin{Bmatrix}\mathbf{\Psi}\\ \mathbf{\Phi}\end{Bmatrix} = \mathbf{M}\begin{Bmatrix}\mathbf{w}\\ \mathbf{w}_n\end{Bmatrix} + \mathbf{Bb} + \frac{1}{D}\hat{\mathbf{F}} \tag{3.40}$$

where

$$\mathbf{K} = \begin{bmatrix} [v]^{ij} & -[v_n]^{ij} \\ \left[\nabla^2 v\right]^{ij} & \frac{1}{2}\delta^{ij} - \left[\nabla^2 v_n\right]^{ij} \end{bmatrix}_{2N\times 2N} \tag{3.41}$$

$$\mathbf{M} = \begin{bmatrix} -\frac{1}{2}\delta^{ij} + \left[\nabla^2 v_n\right]^{ij} & -\left[\nabla^2 v\right]^{ij} \\ \mathbf{0} & \mathbf{0} \end{bmatrix}_{2N\times 2N} \tag{3.42}$$

$$\mathbf{B} = \begin{bmatrix} B_1^{ik} \\ B_2^{ik} \end{bmatrix}_{2N\times M}, \quad \hat{\mathbf{F}} = \begin{Bmatrix} C_1^i \\ C_2^i \end{Bmatrix}_{2N\times 1}, \quad \mathbf{\Psi} = \left[\Psi^i\right]_{N\times 1}, \quad \mathbf{\Phi} = \left[\Phi^i\right]_{N\times 1} \tag{3.43}$$

It should be noted that $B_1^{ik} = A_1^{ik}$ and $C_1^i = F_1^i$. The coefficients C_2^i are evaluated from the relation (see Section 2.4.3)

$$C_2^i = P_k \nabla^2 v(\mathbf{x}_i, \mathbf{x}_k) + \sum_{k=1}^{L} f_k \int_{L_k} \nabla^2 v(\mathbf{x}_i, s)\, ds \tag{3.44}$$

Introducing the expressions for $\mathbf{w}$ and $\mathbf{w}_n$ resulting from the solution of Eqs. (3.28) and (3.30) in Eq. (3.40), we obtain the boundary quantities $\boldsymbol{\Psi}, \boldsymbol{\Phi}$ in terms of the vector $\mathbf{b}$.

Further, differentiation of Eq. (3.34) gives

$$\begin{aligned} w_{,pqr}(\mathbf{x}) &= \int_\Omega v_{,pqr} \bar{b}\, d\Omega \\ &\quad - \int_\Gamma \left[v_{,pqr} \Psi - w\left(\nabla^2 v_{,n}\right)_{,pqr} - \left(v_{,n}\right)_{,pqr} \Phi + w_{,n}\left(\nabla^2 v\right)_{,pqr} \right] ds \end{aligned} \tag{3.45}$$

where p, q, r represent 0, x, y. Evidently, $w_{,pqr}$ designates any derivative up to the third order, e.g., $w_{,x00} = w_{,x}$, $w_{,xy0} = w_{,xy}$ etc. Note that $w_{,000} = w$. The derivatives of the fundamental solution are given by Eqs. (2.126) with $\mathrm{D}=1$.

After discretization Eq. (3.45) gives

$$\begin{aligned} \mathbf{w}_{,pqr} = \mathbf{A}_{1,pqr}\, \mathbf{b} + \mathbf{F}^i_{pqr} - [v]_{,pqr} \boldsymbol{\Psi} + \left[\nabla^2 v_n\right]_{,pqr} \mathbf{w} \\ + [v_n]_{,pqr} \boldsymbol{\Phi} - \left[\nabla^2 v\right]_{,pqr} \mathbf{w}_n \end{aligned}$$

or using the expressions of $\mathbf{w}$, $\mathbf{w}_n$, $\boldsymbol{\Psi}$, $\boldsymbol{\Phi}$ in terms of $\mathbf{b}$ we obtain

$$\mathbf{w}_{,pqr} = \mathbf{S}_{,pqr}\, \mathbf{b} + \mathbf{c} \tag{3.46}$$

Finally, introducing Eq. (3.46) into Eq. (3.33) yields the equation

$$\mathbf{T}\mathbf{b} = \mathbf{e} \tag{3.47}$$

where the matrix $\mathbf{T}$ with dimensions $M \times M$ and the vector $\mathbf{e}$ with dimensions $M \times 1$ are known. The solution of Eq. (3.47) gives the fictitious load $\mathbf{b}$. Once $\mathbf{b}$ is known, the boundary quantities $\mathbf{w}$, $\mathbf{w}_n$, $\boldsymbol{\Psi}$, $\boldsymbol{\Phi}$ can be computed. Then Eq. (3.45) is used to compute the solution of the original problem and its derivatives at the interior nodal points. The solution at any interior point $\mathbf{x}$ is computed from Eq. (3.34) using the same discretization. Finally, the stress resultants are evaluated from Eqs. (2.19) and (2.21). The derivation of the expression of the matrices $\mathbf{S}_{,pqr}$, $\mathbf{T}$, and the vectors $\mathbf{c}, \mathbf{e}$ is left to the reader as an exercise.

3.3.2.2 EVALUATION OF THE LINE AND DOMAIN INTEGRALS

The line integrals (3.39a) and (3.39b) are regular, while (3.39c) and (3.39d) exhibit logarithmic and Cauchy type singularity, respectively. Their evaluation is performed using the techniques presented in Section 2.4.2. Finally, all line integrals in Eq. (3.45) are regular, because it is always $r \neq 0$.

The domain integrals are evaluated over the N_e triangular elements. The linear variation of the fictitious load on the triangular element produces integrals of the form

$$I_e(\mathbf{x}) = \int_{\Omega e} K_n(r) b^e(\mathbf{y}) d\Omega_{\mathbf{y}}, \; r = |\mathbf{y} - \mathbf{x}|, \, \mathbf{x}(x, y), \; \mathbf{y}(\xi, \eta) \tag{3.48}$$

where

$$b^e(\mathbf{y}) = a_1^e + a_2^e \xi + a_3^e \eta, \qquad e = 1, 2, \ldots, N_e \tag{3.49}$$

The coefficients a_1^e, a_2^e, a_3^e are expressed in terms of the nodal values b_i, b_j, b_m, which define the plane of the linear distribution of $b(\mathbf{x})$ on the triangular element. Thus, referring to Fig. 3.5a we have

$$\begin{Bmatrix} a_1^e \\ a_2^e \\ a_3^e \end{Bmatrix} = \begin{bmatrix} 1 & \xi_i & \eta_i \\ 1 & \xi_j & \eta_j \\ 1 & \xi_m & \eta_m \end{bmatrix}^{-1} \begin{Bmatrix} b_i \\ b_j \\ b_m \end{Bmatrix} \tag{3.50}$$

and Eq. (3.49) becomes

$$b^e(\mathbf{y}) = \frac{1}{\Delta} \left[p_i(\xi, \eta) b_i + p_j(\xi, \eta) b_j + p_m(\xi, \eta) b_m \right] \tag{3.51}$$

in which

$$\Delta = \det \begin{bmatrix} 1 & \xi_i & \eta_i \\ 1 & \xi_j & \eta_j \\ 1 & \xi_m & \eta_m \end{bmatrix} \tag{3.52}$$

and

$$p_i(\xi, \eta) = \alpha_i + \beta_i \xi + \gamma_i \eta \tag{3.53a}$$

$$p_j(\xi, \eta) = \alpha_j + \beta_j \xi + \gamma_j \eta \tag{3.53b}$$

$$p_m(\xi, \eta) = \alpha_m + \beta_m \xi + \gamma_m \eta \tag{3.53c}$$

The coefficients of the polynomial $p_i(\xi, \eta)$ are given by the relations

$$\begin{aligned} \alpha_i &= \xi_j \eta_m - \xi_m \eta_j \\ \beta_i &= \eta_j - \eta_m \\ \gamma_i &= \xi_m - \xi_j \end{aligned} \tag{3.54}$$

The coefficients of the polynomials $p_j(\xi, \eta)$ and $p_m(\xi, \eta)$ are obtained from (3.54) by cyclic permutation of the subscripts i, j, m.

In elements adjacent to the boundary, the nodal points are placed at distances $(1-\kappa)l_{ij}$ and $(1-\kappa)l_{jm}$ from the boundary (see Fig. 3.5b). Therefore, it must be set $b_m = b'_m/\kappa$ and $b_i = b'_i/\kappa$ in Eq. (3.50).

By virtue of Eq. (3.51), the integral (3.48) becomes

$$I_e(\mathbf{x}) = \frac{b_i}{\Delta}\int_{\Omega e} K_n(r)p_i(\xi,\eta)\,d\xi\,d\eta + \frac{b_j}{\Delta}\int_{\Omega e} K_n(r)p_j(\xi,\eta)\,d\xi\,d\eta + \frac{b_m}{\Delta}\int_{\Omega e} K_n(r)p_m(\xi,\eta)\,d\xi\,d\eta \tag{3.55}$$

The kernels $K_n(r)$ are given as

$$K_1(r) = v = \frac{1}{8\pi}r^2\ln r \tag{3.56a}$$

$$K_2(r) = v_1 = \frac{1}{8\pi}r(2\ln r + 1)r_\nu \tag{3.56b}$$

$$K_3(r) = v_{,x} = \frac{1}{8\pi}r(2\ln r + 1)r_x \tag{3.56c}$$

$$K_4(r) = v_{,y} = \frac{1}{8\pi}r(2\ln r + 1)r_y \tag{3.56d}$$

$$K_5(r) = v_{,xx} = \frac{1}{8\pi}(2\ln r + 2r_x^2 + 1) \tag{3.56e}$$

$$K_6(r) = v_{,yy} = \frac{1}{8\pi}\left(2\ln r + 2r_y^2 + 1\right) \tag{3.56f}$$

$$K_7(r) = v_{,xy} = \frac{1}{4\pi}r_x r_y \tag{3.56g}$$

$$K_8(r) = \nabla^2 v = \frac{1}{2\pi}(\ln r + 1) \tag{3.56h}$$

$$K_9(r) = \nabla^2 v_{,x} = \frac{1}{2\pi}\frac{r_x}{r} \tag{3.56i}$$

$$K_{10}(r) = \nabla^2 v_{,y} = \frac{1}{2\pi}\frac{r_y}{r} \tag{3.56j}$$

When point $\mathbf{x}$ lies outside the element e, it is $r = |\mathbf{y}-\mathbf{x}| \neq 0$. Consequently, all kernels are regular and the domain integrals in Eq. (3.55) are evaluated using two-dimensional Gauss integration (see Appendix B). When point $\mathbf{x}$ lies on the element e, then it is $\lim_{\mathbf{y}\to\mathbf{x}} r = 0$ and the evaluation of the integrals requires special care. We can readily show that $\lim_{\mathbf{y}\to\mathbf{x}} K_n(r) = 0$ when $n = 1, 2, 3, 4$. Hence, the respective integrals in Eq. (3.55) are regular and can be evaluated using Gauss integration. The remaining kernels $(n = 5, 6, \ldots, 10)$ are singular and the evaluation of the domain integrals can be done by converting them to line integrals using Green's identity (1.16). Thus, setting

$$\nabla^2 \mathcal{K}_n(r) = K_n(r) \tag{3.57}$$

and applying Eq. (1.16) for the functions $\mathcal{K}_n(r)$ and $p_i(\xi,\eta)$, we obtain

$$\int_{\Omega e} K_n(r) p_i \, d\Omega = \int_{\Gamma e} \left(p_i \frac{\partial \mathcal{K}_n}{\partial n} - \mathcal{K}_n \frac{\partial p_i}{\partial n} \right) ds \tag{3.58}$$

Similar expressions are derived for $p_j(\xi,\eta)$, $p_m(\xi,\eta)$. The functions $\mathcal{K}_n(r)$, obtained as particular solutions of Eq. (3.57), are given in Appendix A.

Let k be the number of triangles having the node i as common vertex. Referring to Fig. (3.6a) we observe that the contribution of the domain integrals around i is converted into a line integral on the boundary $1-2-3-4-5-k-1$. Thus, we have

$$\sum_{e=1}^{k} \int_{\Omega e} K_n(r) p_i^{(e)} \, d\Omega = \sum_{e=1}^{k} \int_{le} \left(p_i^{(e)} \frac{\partial \mathcal{K}_n}{\partial n} - \mathcal{K}_n \frac{\partial p_i^{(e)}}{\partial n} \right) ds \tag{3.59}$$

When node i is near the boundary Γ, the contribution of the domain integrals is converted to a line integral on the boundary $1-2-3-4-1$ (see Fig. 3.6b). Thus, we have

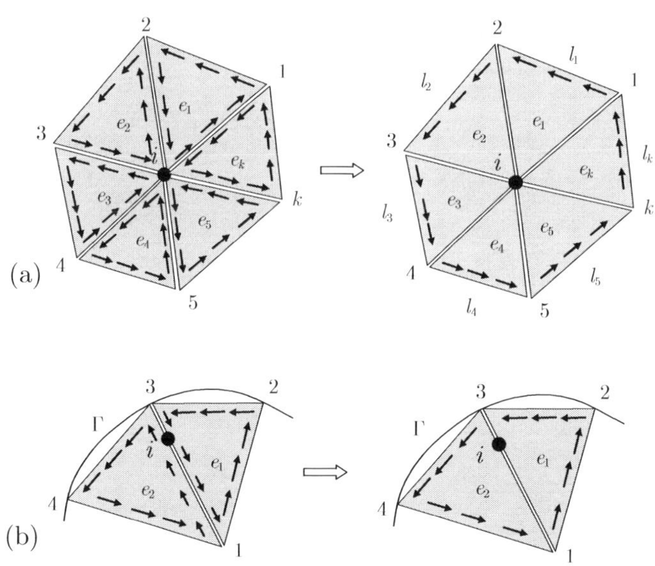

FIGURE 3.6 Node (a) away from and (b) adjacent to the boundary.

$$\sum_{e=1}^{2}\int_{\Omega e} K_n(r)p_i^{(e)}\,d\Omega = \int_{1-2-3}\left(p_i^{(1)}\frac{\partial \mathcal{K}_n}{\partial n} - \mathcal{K}_n\frac{\partial p_i^{(1)}}{\partial n}\right)ds + \int_{3-4-1}\left(p_i^{(2)}\frac{\partial \mathcal{K}_n}{\partial n} - \mathcal{K}_n\frac{\partial p_i^{(2)}}{\partial n}\right)ds \tag{3.60}$$

Obviously, the resulting line integrals are regular ($r=|\mathbf{y}-\mathbf{x}|\neq 0$) and can be evaluated using Gauss integration. This technique is also convenient to evaluate the domain integrals with regular kernels ($n=1, 2, 3, 4$).

3.3.2.3 AEM FOR THE PLANE STRESS PROBLEM

The boundary value problem (3.9), (3.10) can be solved using the BEM for the plane stress problem (see [2], Chapter 7) to evaluate the membrane forces N_x, N_y, N_{xy} involved in Eq. (3.33). Here, however, the AEM will be employed, because, on the one hand, the plane elasticity problem is highly simplified as it is reduced to two plane potential problems (Poison's equations), and on the other hand, because it will be used to solve the nonlinear and dynamic plate problems. Moreover, it allows writing a unified computer code for the numerical solution of linear, nonlinear, static as well as dynamic plate problems.

Equations (3.9) are of the second order. Hence, the Laplace operator can give two uncoupled analog equations that can be readily treated by the BEM. Thus, we have

$$\nabla^2 u = b^{(1)}(\mathbf{x}) \tag{3.61a}$$

$$\nabla^2 v = b^{(2)}(\mathbf{x}) \tag{3.61b}$$

where $b^{(1)}(\mathbf{x})$ and $b^{(2)}(\mathbf{x})$ are the two fictitious sources, unknown in the first instance.

The integral representation of the solution of Eq. (3.61a) is (see [2], Section 3.4)

$$\varepsilon u(\mathbf{x}) = \int_{\Omega} v^* b^{(1)}\,d\Omega - \int_{\Gamma}\left(v^* u_{,n} - v_{,n}^* u\right)ds \qquad \mathbf{x}\in\Omega\cup\Gamma \tag{3.62}$$

where $v^* = \ln r/2\pi$ is the fundamental solution of Eq. (3.61a), $r=|\mathbf{y}-\mathbf{x}|$, $\mathbf{x}\in\Omega\cup\Gamma$ and $\mathbf{y}\in\Gamma$; ε is the free term coefficient, which takes the values $\varepsilon=1$ if $\mathbf{x}\in\Omega$, $\varepsilon=1/2$ if $\mathbf{x}\in\Gamma$ (for smooth boundary) and $\varepsilon=0$ if $\mathbf{x}\notin\Omega\cup\Gamma$.

Using the same discretization with N constant boundary elements and linear triangular domain elements, Eq. (3.62) for points $\mathbf{x}\in\Gamma$ is written as

$$\frac{1}{2}u^i = \sum_{k=1}^{M} D^{ik} b_k^{(1)} - \sum_{j=1}^{N}[v^*]^{ij} u_n^j + \sum_{j=1}^{N}\left[v_n^*\right]^{ij} u^j, \quad i=1,2,\ldots,N \tag{3.63a}$$

Similarly, we obtain for the displacement v

$$\frac{1}{2}v^i = \sum_{k=1}^{M} D^{ik} b_k^{(2)} - \sum_{j=1}^{N} [v^*]^{ij} v_n^j + \sum_{j=1}^{N} [v_n^*]^{ij} v^j, \quad i = 1, 2, \ldots, N \tag{3.63b}$$

where it was set as

$$[v^*]^{ij} = \int_{\Gamma_j} v^*(\mathbf{x}_i, \mathbf{y})\, ds_y, \quad [v_n^*]^{ij} = \int_{\Gamma_j} \frac{\partial v^*(\mathbf{x}_i, \mathbf{y})}{\partial n_y}\, ds_y \tag{3.64a,b}$$

The numerical evaluation of the line integrals (3.64a,b) is presented in Section 2.4.2. The elements of the matrix $[D^{ik}]$ originate from the integration of $v^*(\mathbf{x}, \mathbf{y})$ over the triangular elements using the technique presented in Section 3.3.2.2. We recall that this kernel exhibits logarithmic singularity when $r \to 0$.

Equations (3.63a,b) are written in matrix form

$$\hat{\mathbf{H}}\begin{Bmatrix} \mathbf{u} \\ \mathbf{v} \end{Bmatrix} = \hat{\mathbf{G}}\begin{Bmatrix} \mathbf{u}_n \\ \mathbf{v}_n \end{Bmatrix} + \hat{\mathbf{D}}\begin{Bmatrix} \mathbf{b}^{(1)} \\ \mathbf{b}^{(2)} \end{Bmatrix} \tag{3.65}$$

where

$$\hat{\mathbf{H}} = \begin{bmatrix} -[v_n^*]^{ij} + \frac{1}{2}\delta_{ij} & \mathbf{0} \\ \mathbf{0} & -[v_n^*]^{ij} + \frac{1}{2}\delta_{ij} \end{bmatrix} \tag{3.66a}$$

$$\hat{\mathbf{G}} = \begin{bmatrix} -[v^*]^{ij} & \mathbf{0} \\ \mathbf{0} & -[v^*]^{ij} \end{bmatrix} \tag{3.66b}$$

$$\hat{\mathbf{D}} = \begin{bmatrix} [D^{ik}] \\ [D^{ik}] \end{bmatrix} \tag{3.66c}$$

$$(i, j = 1, 2, \ldots, N, \quad k = 1, 2, \ldots, M)$$

$\mathbf{u}$, $\mathbf{v}$, $\mathbf{u}_n$, $\mathbf{v}_n$ are $N \times 1$ vectors containing the values of the displacements u, v and their normal derivatives at the N boundary nodal points, while $\mathbf{b}^{(1)}$, $\mathbf{b}^{(2)}$ are $M \times 1$ vectors containing the values of the fictitious sources at the M interior nodal points.

It is convenient to write the boundary conditions (3.10) in the form

$$\gamma_1 u + \gamma_2 t_x = \gamma_3 \tag{3.67a}$$

$$\delta_1 v + \delta_2 t_y = \delta_3 \tag{3.67b}$$

in which the coefficients γ_1, γ_2 and δ_1, δ_2 take the values 0 or 1, while γ_3 takes the values u^* or t_x^* and δ_3 the values v^* or t_y^*. The quantities u^*, v^* and t_x^*, t_y^* result from the specified boundary quantities via the transformation relations

$$\begin{Bmatrix} u^* \\ v^* \end{Bmatrix} = \mathbf{R}^T \begin{Bmatrix} u_n^* \\ u_t^* \end{Bmatrix} \tag{3.68a}$$

$$\begin{Bmatrix} t_x^* \\ t_y^* \end{Bmatrix} = \frac{1}{h}\mathbf{R}^T \begin{Bmatrix} N_n^* \\ N_t^* \end{Bmatrix} \tag{3.68b}$$

where

$$\mathbf{R} = \begin{bmatrix} n_x & n_y \\ -n_y & n_x \end{bmatrix} = \begin{bmatrix} \cos\alpha & \sin\alpha \\ -\sin\alpha & \cos\alpha \end{bmatrix}, \quad \alpha = \measuredangle x, n \tag{3.69}$$

is the transformation matrix due to the rotation of axes by an angle $\alpha = \measuredangle x, n$.

The tractions t_x, t_y are given by the equations [see [2], Eqs. (7.37)]

$$\begin{aligned} \begin{Bmatrix} t_x \\ t_y \end{Bmatrix} &= C \begin{bmatrix} n_x & \dfrac{1-\nu}{2} n_y \\ \nu n_y & \dfrac{1-\nu}{2} n_x \end{bmatrix} \begin{Bmatrix} u_{,x} \\ u_{,y} \end{Bmatrix} + C \begin{bmatrix} \dfrac{1-\nu}{2} n_y & \nu n_x \\ \dfrac{1-\nu}{2} n_x & n_y \end{bmatrix} \begin{Bmatrix} v_{,x} \\ v_{,y} \end{Bmatrix} \\ &= C \begin{bmatrix} n_x & \dfrac{1-\nu}{2} n_y \\ \nu n_y & \dfrac{1-\nu}{2} n_x \end{bmatrix} \mathbf{R}^T \begin{Bmatrix} u_{,n} \\ u_{,t} \end{Bmatrix} + C \begin{bmatrix} \dfrac{1-\nu}{2} n_y & \nu n_x \\ \dfrac{1-\nu}{2} n_x & n_y \end{bmatrix} \mathbf{R}^T \begin{Bmatrix} v_{,n} \\ v_{,t} \end{Bmatrix} \end{aligned} \tag{3.70}$$

where $C = Eh/(1-\nu^2)$ is the membrane stiffness.

Applying Eqs. (3.67) at the N boundary nodal points yields

$$\begin{bmatrix} \boldsymbol{\gamma}_1 & \mathbf{0} \\ \mathbf{0} & \boldsymbol{\delta}_1 \end{bmatrix} \begin{Bmatrix} \mathbf{u} \\ \mathbf{v} \end{Bmatrix} + \begin{bmatrix} \boldsymbol{\gamma}_2 & \mathbf{0} \\ \mathbf{0} & \boldsymbol{\delta}_2 \end{bmatrix} \begin{Bmatrix} \mathbf{t}_x \\ \mathbf{t}_y \end{Bmatrix} = \begin{Bmatrix} \boldsymbol{\gamma}_3 \\ \boldsymbol{\delta}_3 \end{Bmatrix} \tag{3.71}$$

Introducing now Eqs. (3.70) into (3.71), replacing the tangential derivatives $u_{,t}$, $v_{,t}$ with finite differences, we obtain the boundary conditions in the form

$$\tilde{\mathbf{H}} \begin{Bmatrix} \mathbf{u} \\ \mathbf{v} \end{Bmatrix} + \tilde{\mathbf{G}} \begin{Bmatrix} \mathbf{u}_n \\ \mathbf{v}_n \end{Bmatrix} = \tilde{\mathbf{F}} \tag{3.72}$$

Equations (3.65) and (3.72) constitute a system of $4N$ equations whose solution gives the boundary quantities $\mathbf{u}$, $\mathbf{u}_n$, $\mathbf{v}$, $\mathbf{v}_n$ in terms of the vectors $\mathbf{b}^{(1)}$, $\mathbf{b}^{(2)}$ whose evaluation is achieved as follows.

Applying Eqs. (3.9a,b) at the M interior nodal points and taking into account Eqs. (3.61a,b), we have

$$\mathbf{b}^{(1)}+\frac{1+\nu}{1-\nu}\left(\mathbf{u}_{,xx}+\mathbf{v}_{,xy}\right)+\frac{1}{Gh}\mathbf{p}_x=0 \tag{3.73a}$$

$$\mathbf{b}^{(2)}+\frac{1+\nu}{1-\nu}\left(\mathbf{u}_{,xy}+\mathbf{v}_{,yy}\right)+\frac{1}{Gh}\mathbf{p}_y=0 \tag{3.73b}$$

where $\mathbf{p}_x$, $\mathbf{p}_y$ are known vectors including the values of the membrane loads at the M nodal points. The vectors $\mathbf{u}_{,xx}$, $\mathbf{u}_{,xy}$, $\mathbf{v}_{,xx}$, $\mathbf{v}_{,xy}$ contain the values of the derivatives of the displacements at the same nodal points. These vectors are unknown, but they can be expressed in terms of $\mathbf{b}^{(1)}$ and $\mathbf{b}^{(2)}$ as follows.

The integral representations of the derivatives of the displacements can be obtained by direct differentiations of Eq. (3.62). Thus, we have

$$u_{,pq}(\mathbf{x})=\int_\Omega v^*_{,pq}b^{(1)}\,d\Omega-\int_\Gamma\left(v^*_{,pq}u_{,n}-v^*_{,npq}u\right)ds\quad p,q=0,\,x,\,y \tag{3.74a}$$

Similarly, we obtain

$$v_{,pq}(\mathbf{x})=\int_\Omega v^*_{,pq}b^{(2)}\,d\Omega-\int_\Gamma\left(v^*_{,pq}v_{,n}-v^*_{,npq}v\right)ds\quad p,q=0,\,x,\,y \tag{3.74b}$$

Approximating the boundary integrals with constant elements and the domain integrals with linear triangular elements in Eqs. (3.74a,b) yields

$$\mathbf{u}_{,pq}=\mathbf{D}_{,pq}\mathbf{b}^{(1)}-[v^*]_{,pq}\mathbf{u}_n+\left[v_n^*\right]_{,pq}\mathbf{u} \tag{3.75a}$$

$$\mathbf{v}_{,pq}=\mathbf{D}_{,pq}\mathbf{b}^{(2)}-[v^*]_{,pq}\mathbf{v}_n+\left[v_n^*\right]_{,pq}\mathbf{v} \tag{3.75b}$$

Then replacing $\mathbf{u}$, $\mathbf{u}_n$, $\mathbf{v}$, $\mathbf{v}_n$ obtained from the solution of the system of Eqs. (3.65) and (3.72) in Eqs. (3.75a,b) gives

$$\mathbf{u}_{,pq}=\mathbf{U}^{(1)}_{,pq}\mathbf{b}^{(1)}+\mathbf{U}^{(2)}_{,pq}\mathbf{b}^{(2)}+\mathbf{e}_{,pq} \tag{3.76a}$$

$$\mathbf{v}_{,pq}=\overline{\mathbf{U}}^{(1)}_{,pq}\mathbf{b}^{(1)}+\overline{\mathbf{U}}^{(2)}_{,pq}\mathbf{b}^{(2)}+\overline{\mathbf{e}}_{,pq} \tag{3.76b}$$

in which $\mathbf{U}^{(1)}_{,pq}$, $\mathbf{U}^{(2)}_{,pq}$, $\overline{\mathbf{U}}^{(1)}_{,pq}$, $\overline{\mathbf{U}}^{(2)}_{,pq}$ are $M\times M$ known matrices and $\mathbf{e}_{,pq}$, $\overline{\mathbf{e}}_{,pq}$ known vectors. Their expressions are not given here as it is more convenient to derive them in the computer program for the numerical solution.

Finally, introducing the derivatives (3.76a,b) into Eqs. (3.73a,b) yields

$$\begin{aligned}\left[\mathbf{I}+\frac{1+\nu}{1-\nu}\left(\mathbf{U}^{(1)}_{,xx}+\overline{\mathbf{U}}^{(1)}_{,xy}\right)\right]\mathbf{b}^{(1)}+\frac{1+\nu}{1-\nu}\left(\mathbf{U}^{(2)}_{,xx}+\overline{\mathbf{U}}^{(2)}_{,xy}\right)\mathbf{b}^{(2)}=\\-\frac{1+\nu}{1-\nu}\left(\mathbf{e}_{,xx}+\overline{\mathbf{e}}_{,xy}\right)-\frac{1}{Gh}\mathbf{p}_x\end{aligned} \tag{3.77a}$$

$$\frac{1+\nu}{1-\nu}\left(\mathbf{U}^{(1)}_{,xy}+\overline{\mathbf{U}}^{(1)}_{,yy}\right)\mathbf{b}^{(1)})+\left[\mathbf{I}+\frac{1+\nu}{1-\nu}\left(\mathbf{U}^{(2)}_{,xy}+\overline{\mathbf{U}}^{(2)}_{,yy}\right)\right]\mathbf{b}^{(2)}=$$

$$-\frac{1+\nu}{1-\nu}\left(\mathbf{e}_{,xy}+\bar{\mathbf{e}}_{,yy}\right)-\frac{1}{Gh}\mathbf{p}_y \quad (3.77\text{b})$$

Equations (3.77a,b) are solved to give the fictitious forces $\mathbf{b}^{(1)}$, $\mathbf{b}^{(2)}$, which are used in Eqs. (3.76a,b) to yield the values of the derivatives at the M interior nodal points. The membrane forces $\mathbf{N}_x$, $\mathbf{N}_y$, $\mathbf{N}_{xy}$ involved in Eq. (3.33) are computed using Eqs. (3.8a,b,c). Thus, we have

$$\mathbf{N}_x=C\left(\mathbf{u}_{,x}+\nu\mathbf{v}_{,y}\right) \quad (3.78\text{a})$$

$$\mathbf{N}_y=C\left(\nu\mathbf{u}_{,x}+\mathbf{v}_{,y}\right) \quad (3.78\text{b})$$

$$\mathbf{N}_{xy}=C\frac{1-\nu}{2}\left(\mathbf{u}_{,y}+\mathbf{v}_{,x}\right) \quad (3.78\text{c})$$

3.3.3 Linear buckling of plates

The linear buckling of plates is the most usual case of a plate under inplane forces. The transverse load f is zero and the boundary conditions (3.15) are homogeneous, i.e., $w^*=w_n^*=w_k^*=0$ or $V_n^*=M_n^*=R_k^*=0$. The buckling in the plate is caused by the membrane loads p_x, p_y and/or nonhomogeneous membrane boundary conditions, i.e., $\left(N_n^*,\, N_t^*\right)\neq 0$ or $\left(u_n^*,\, u_t^*\right)\neq 0$. First, the membrane loading is applied, which produces the membrane forces N_x, N_{xy}, N_y. Then the loading increases gradually with a multiplying parameter λ. Thus, Eq. (3.14) becomes

$$D\nabla^4 w-\lambda\left[\left(N_x w_{,xx}+2N_{xy}w_{,xy}+N_y w_{,yy}\right)-p_x w_{,x}-p_y w_{,y}\right]=0 \quad (3.79)$$

Consequently, we have an eigenvalue problem for the plate equation, namely, we seek the values of the parameter λ for which Eq. (3.79) has a nontrivial solution. Application of the AEM reduces Eq. (3.79) to

$$D\mathbf{b}-\lambda\left[\left(\mathbf{N}_x\mathbf{w}_{,xx}+2\mathbf{N}_{xy}\mathbf{w}_{,xy}+\mathbf{N}_y\mathbf{w}_{,yy}\right)-\mathbf{p}_x\mathbf{w}_{,x}-\mathbf{p}_y\mathbf{w}_{,y}\right]=0 \quad (3.80)$$

Further, using Eqs. (3.46) to replace the derivatives, an equation of the following form is obtained

$$(\mathbf{A}-\lambda\mathbf{B})\mathbf{b}=0 \quad (3.81)$$

for the evaluation of the fictitious load $\mathbf{b}$. Evidently, Eq. (3.81) represents a generalized eigenvalue problem of the linear algebra. The eigenvalues λ_i give the critical values of the *buckling parameter*, while the eigenvectors $\mathbf{b}_i$ are used in

Eq. (3.46) ($p=q=r=0$) to compute the *buckling modes*. It should be noted that in the case of a free edge the parameter λ is involved in the boundary condition (3.15a), hence it appears in matrix $\overline{\mathbf{H}}$ (see Eq. (3.31)), namely, the elements $\alpha_2\mathbf{N}_n^*$ and $\alpha_2\mathbf{N}_t^*\mathbf{d}$ become $\alpha_2\lambda\mathbf{N}_n^*$ and $\alpha_2\lambda\mathbf{N}_t^*\mathbf{d}$ ($\alpha_2=1$). The consequence of this is that the matrices $\mathbf{A},\mathbf{B}$ in Eq. (3.81) depend on λ and thus the algebraic eigenvalue problem becomes nonlinear. We can cope with this problem by establishing the critical values as the zeroes of the equation $f(\lambda)=|\mathbf{A}(\lambda)-\lambda\mathbf{B}(\lambda)|=0$. Another technique is to use the dynamic criterion (see Example 4.3).

3.3.4 Examples

EXAMPLE 3.1

In this example, the rectangular simply supported plate of Fig. 3.7 is analyzed. The plate, which has a thickness $h=0.15\text{m}$, is subjected to the transverse uniform load $f=20\text{kN/m}^2$ and the membrane boundary load $N_n^*=2000\text{kN/m}$ as shown in the figure. The elastic constants of the material are $\nu=0.2$ and $E=30\times10^6\text{kN/m}^2$. The solution was obtained with $N=326$ boundary elements and $M=95$ domain nodal points resulting from 138 linear triangular elements. The variation of displacement w and bending moment M_x along the line $y=2.5$ is shown in Fig. 3.8. The results are compared with those obtained from an analytical solution [12].

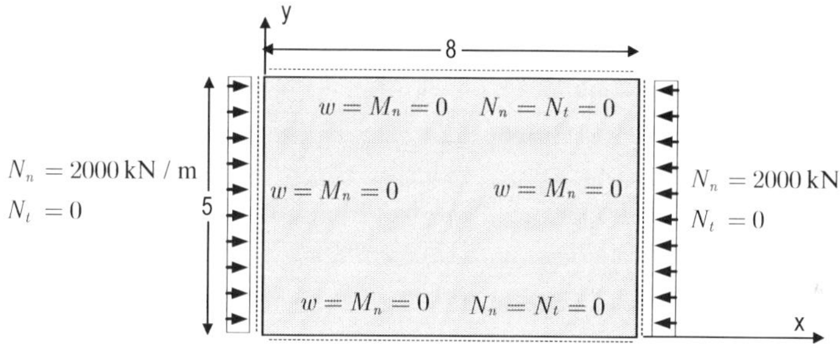

FIGURE 3.7 Simply supported rectangular plate in Example 3.1.

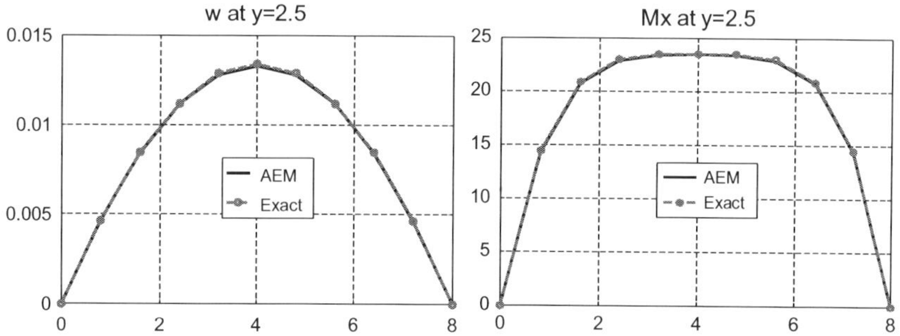

FIGURE 3.8 Displacement w and bending moment M_x along the line $y=2.5$ of the plate in Example 3.1.

EXAMPLE 3.2

In this example the buckling of the square plate shown in Fig. 3.9 is studied. The edge 1-4 is simply supported, while all other edges are free ($M_n^* = V_n^* = 0$). The plate has also a point support at its center. The plate, which has a thickness $h = 0.08\,\mathrm{m}$ and elastic constants $E = 3 \times 10^7 \mathrm{kN/m^2}$ and $\nu = 0.2$, is loaded by the uniform membrane load N_n^*. The solution is obtained with N = 200 boundary elements and M = 112 domain nodal points resulting from 174 linear triangular elements. Fig. 3.10 shows the three first buckling modes together with the respective critical values of the parameter λ. The values obtained from a FEM solution with 1600 quadrilateral elements are given in parentheses.

EXAMPLE 3.3

As a second example of plate buckling analysis, a plate with a complicate geometry and boundary conditions is chosen. The plate and the support conditions are shown in Fig. 3.11. The supported edges are free to move in the plane of the plate. The plate, which has a thickness $h = 0.2\,\mathrm{m}$ and elastic constants $E = 2.1 \times 10^7 \mathrm{kN/m^2}$ and $\nu = 0.2$, is loaded by the uniform membrane force P along the edges 1-2 and 3-4. The results were obtained with N = 900 boundary elements and M = 236 domain nodal points resulting from 368 linear triangular elements. The boundary and domain discretization are shown in Fig. 3.12. The boundary elements are distributed as follows: N1 = 150 on each of the edges 1-2 and 3-4, N2 = 300 on the circular edge 2-3, N3 = 100 on the edge 4-1, and N4 = 200 on the circumference of the circular hole. The contours and 3-D views of the computed membrane forces N_x, N_y, N_{xy} with $P = 1$ are shown in Figs. 3.13 through 3.15. Fig. 3.16 presents the first three buckling modes with the respective critical values of the parameter λ. The corresponding values resulting from a FEM solution with 3695 quadrilateral elements are given in parentheses.

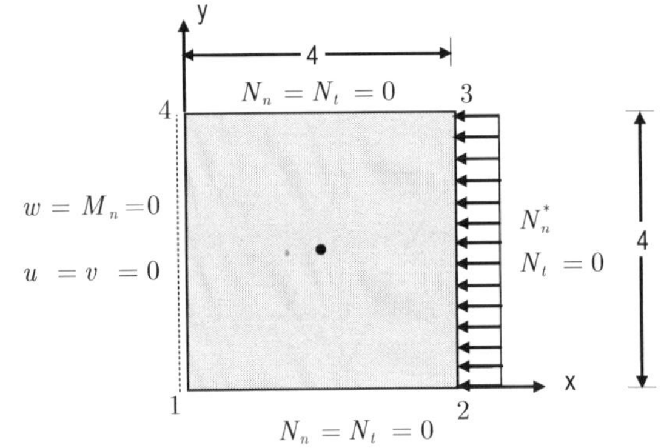

FIGURE 3.9 Square plate with boundary conditions in Example 3.2.

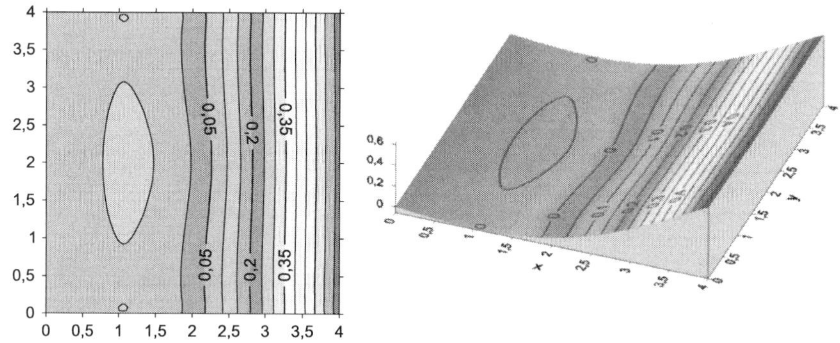

1st buckling mode. λ_1 = 442.57 (439.87).

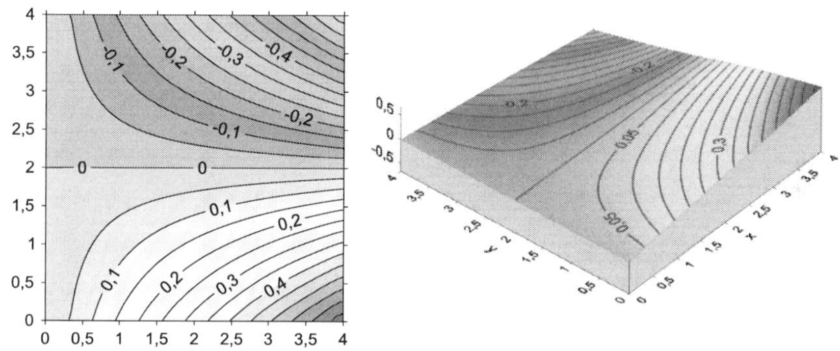

2nd buckling mode. λ_2 = 1447.35 (1507.46).

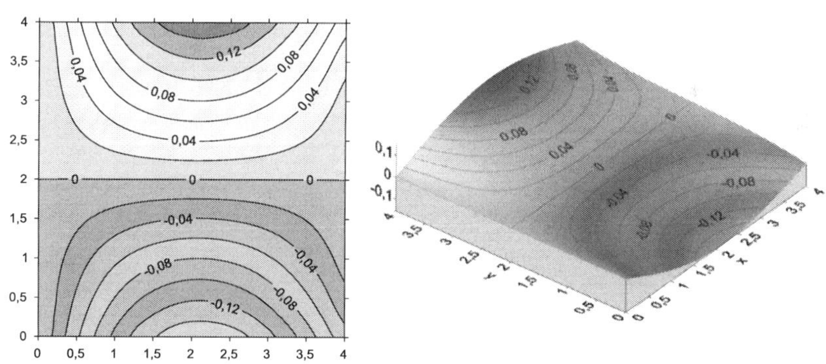

3rd buckling mode. λ_3 = 2299.34 (2246.27).

FIGURE 3.10 Buckling modes of the square plate in Example 3.2.

3.4 PLATES ON ELASTIC FOUNDATION

3.4.1 Foundation models

The bending of plates on elastic foundation is related to the solution of many engineering problems of considerable practical importance. Floating bridges

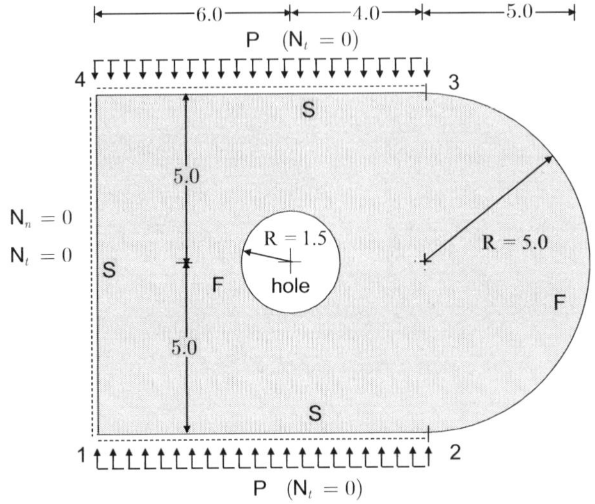

FIGURE 3.11 Plate with complicate boundary in Example 3.3.

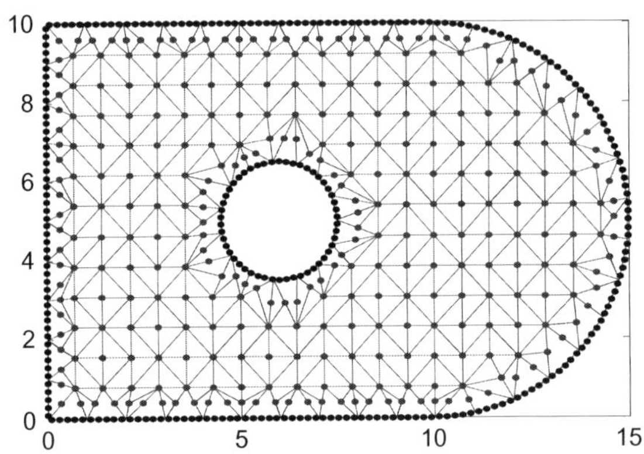

FIGURE 3.12 Boundary and domain discretization ($N=900$, $M=236$) in Example 3.3.

and ice sheets, foundation slabs of buildings, reinforced concrete pavements and airfield runways, to name few, are well-known direct applications. The governing equation of the problem results by including the foundation reaction in the differential equation of the plate as an additional external loading. Hence, Eq. (2.42) becomes

$$D\nabla^4 w + p_s = f \quad \text{in } \Omega \tag{3.82}$$

where $p_s = p_s(x, y)$ is the reaction of the elastic subgrade.

The expression of $p_s(x, y)$ is unknown. The inherent complexity in the behavior of real foundations has led to the development of many idealized models of foundation behavior. The idealized models prove to be particularly useful in the analysis of soil-structure interaction problems. The results of such

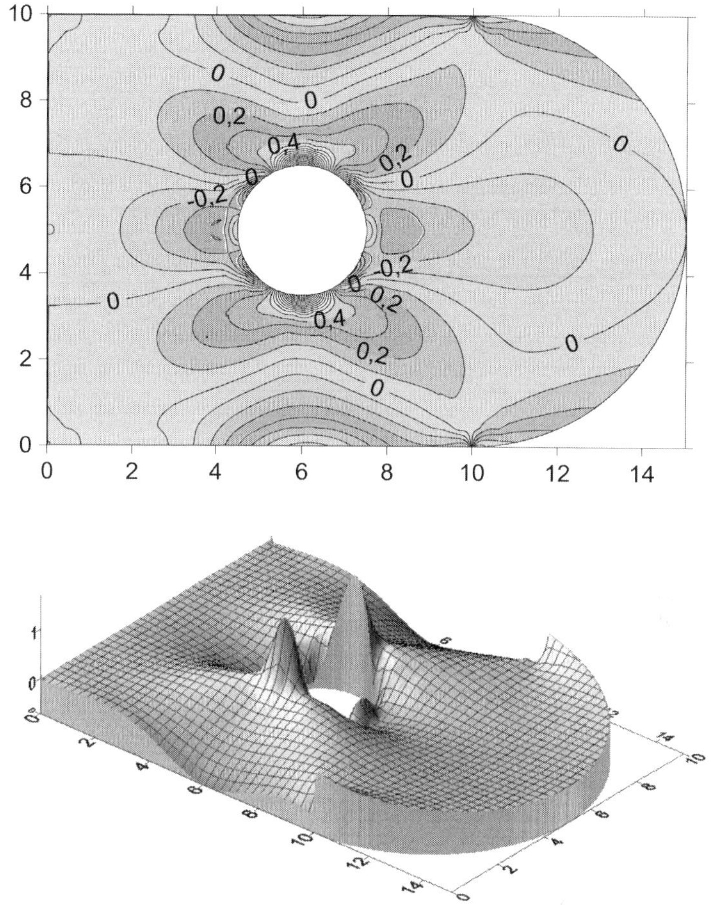

FIGURE 3.13 Membrane force N_x $(\mathrm{P}=1)$ in Example 3.3. Contours and 3-D view.

analysis provide information that can be used for the design of plates and evaluation of the stress and deformation state in the foundation medium. In these models it is assumed that the subgrade reaction depends on the plate deflection and its derivatives [13], namely,

$$p_s = p_s\big(x,\ y,\ w,\ w_{,x},\ w_{,y},\ w_{,xy},\ \ldots\big) \tag{3.83}$$

If the subgrade material is homogeneous and linear elastic, Eq. (3.83) takes the form

$$p_s = k_0 w + k_1 \nabla^2 w + k_2 \nabla^4 w + \ldots \tag{3.84}$$

with $k_0 > 0,\ k_1 < 0,\ k_2 > 0,\ \ldots$.

Models that have been proposed and are commonly utilized for the analysis of plates on elastic foundation are shown in Table 3.4.

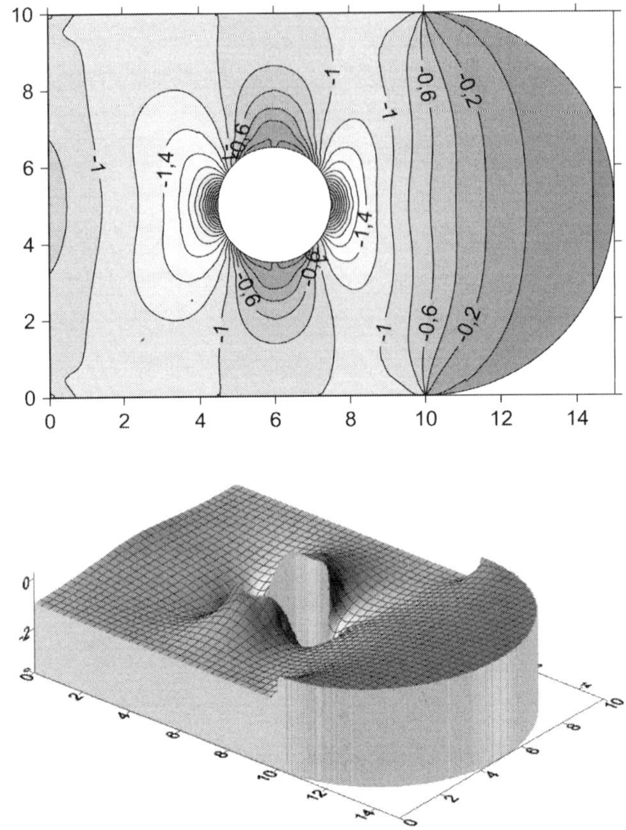

FIGURE 3.14 Membrane force $N_y\,(\mathrm{P}=1)$ in Example 3.3. Contours and 3-D view.

The simplest foundation model was suggested by Winkler [14] who assumed linear relation between p_s and w

$$p_s = kw \tag{3.85}$$

where k represents the proportionality constant referred to as modulus of the subgrade reaction. The mechanical model describing this response is a layer of closely spaced linear independent springs of stiffness k without any shear coupling. This simple model, though useful for the analysis of many engineering problems, exhibits shortcomings, especially at the free boundary of the plate. This situation created the need for the development of more accurate pressure displacement relations. A number of these models have been discussed by Kerr [15,16] and Selvadurai [17]. In the development of these models the investigators proceeded in three ways: (1) they started with the Winkler model and assumed some kind of interaction between the spring elements in order to make the model more realistic (Filonenko-Borodich [18]; Hetenyi [19]; Pasternak [20]; Kerr [15]), (2) they started with the three-dimensional continuum and then

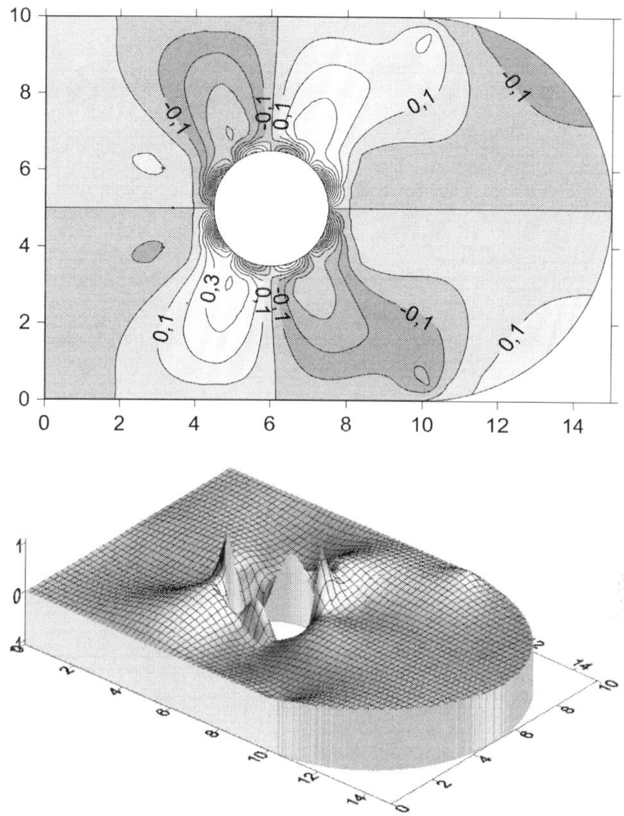

FIGURE 3.15 Membrane force N_{xy} $(\mathrm{P}=1)$ in Example 3.3. Contours and 3-D view.

introduced simplifications with respect to the displacements and/or stresses (Reissner [21]; Vlasov and Leontiev [22]), and (3) they assumed a form of dependence of the contact pressure on the displacements and then made formal series expansion of the adopted function (Ratzersdorfer [23]; Kerr [16]; Bharatha and Levinson [13]).

In general, the methods generating foundation models by using formal expansions reduce to two forms [16]:

$$p_s = Lw \tag{3.86}$$

and

$$L_1 p_s = L_2 w \tag{3.87}$$

where L, L_1 and L_2 are linear differential operators containing derivatives of even order. The first-order approximation of both expressions (3.86) and (3.87) represents the compressibility in the Winkler sense, Eq. (3.85), while the next higher-order approximation includes the effect of shear interaction

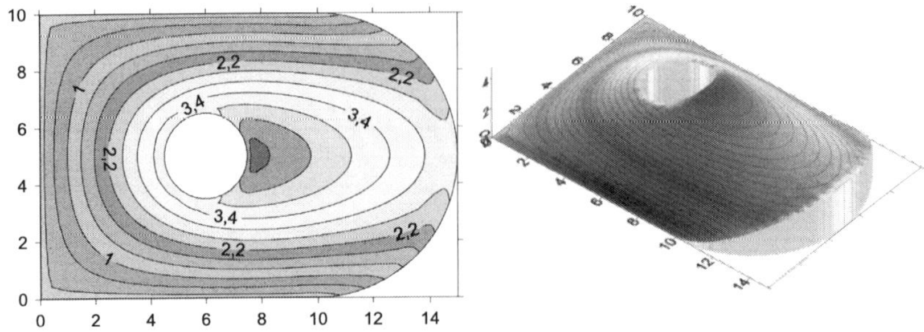

1st buckling mode. $\lambda_1 = 2610.12\,(2605.20)$.

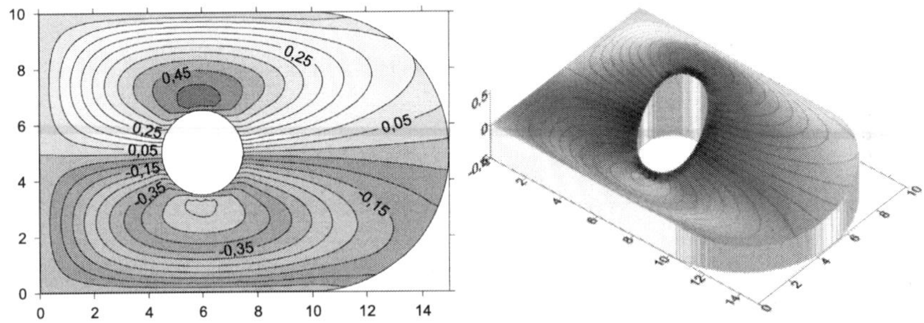

2nd buckling mode. $\lambda_2 = 7194.59\,(7394.03)$.

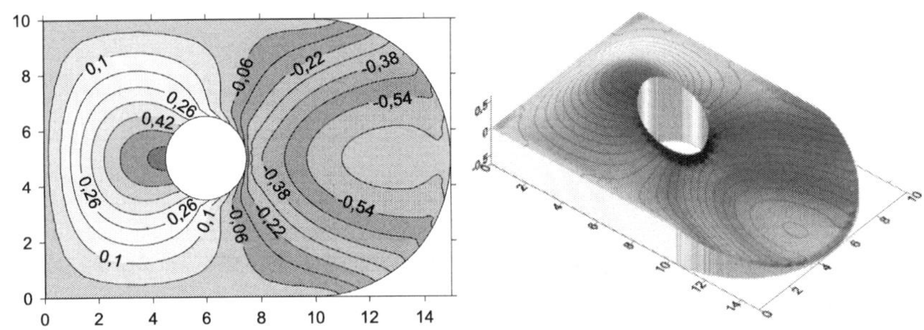

3rd buckling mode. $\lambda_3 = 10451.07\,(10565.90)$.

FIGURE 3.16 Buckling modes of the plate in Example 3.3.

between the spring elements as in the Pasternak model. This is realized by inserting a shear layer with modulus G between plate and springs (see Table 3.4) and is mathematically expressed by

$$p_s = kw - G\nabla^2 w \tag{3.88}$$

The first publications on BEM solutions for plates on elastic foundation appeared in the mid-1980s. The most significant works for plates on Winkler foundation are those by Katsikadelis and Armenakas [24,25], Costa and Brebbia [26] and for plates on biparametric foundation those by Katsikadelis and Kallivokas [27,28], Balas et al. [29]. In these investigations, the solution has been

TABLE 3.4 Foundation models

Mathematical model	Mechanical model
Winkler: $p_s = kw$	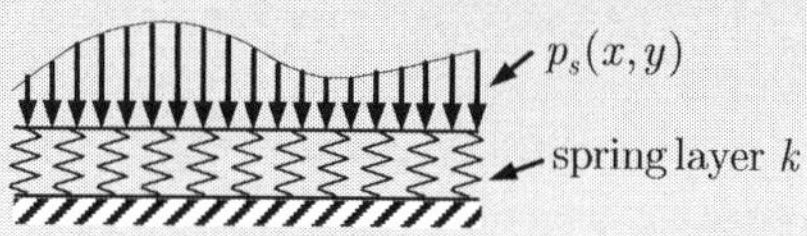
Pasternak: $p_s = kw - G\nabla^2 w$	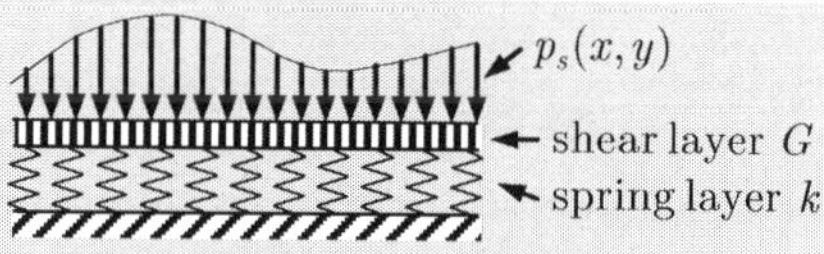
Levinson-Bharatha: $p = k_0 w - k_1 \nabla^2 w + k_2 \nabla^4 w$ with $k_0, k_1, k_2 > 0$	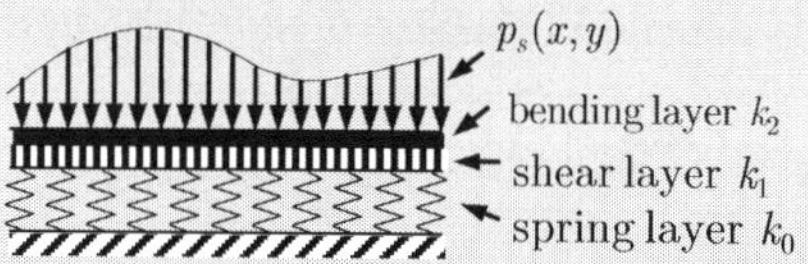
Kerr: $\left(1 + \frac{k}{c}\right) p_s - \frac{G}{c} \nabla^2 p_s = kw - G\nabla^2 w$	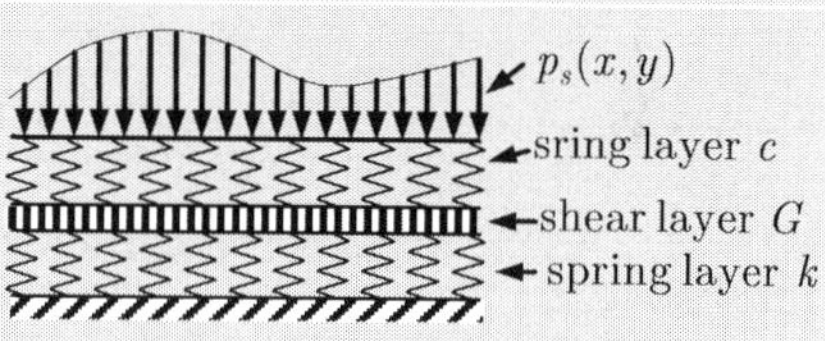

obtained by the direct BEM using the fundamental solution of the governing equation for the derivation of the boundary integral equations. The fundamental solution is either the zero-order Kelvin function (Winkler foundation [24,25]) or the real part of the zero-order complex Hankel function (Pasternak foundation [27,28]). Puttonen and Varpasuo [30] utilized a Fourier-Bessel integral representation for the fundamental solution.

The analytical manipulations of these fundamental solutions are tedious and lead to complicated expressions, whose numerical treatment requires considerable computational task and special care. Besides, the plate and foundation rigidities are included in the argument of the kernels. Consequently, a change in the rigidities changes the problem entirely as all matrices need to be recomputed. Moreover, it is not possible to treat problems in which the foundation rigidity is not constant (non-homogeneous foundation) or depends nonlinearly on the deflection. Therefore, although the BEM approaches presented by the previous researchers are very elegant, their merits and practical significance is diminished due to the mentioned shortcomings. To overcome these disadvantages D/BEM solutions have been developed by some investigators [31–35]. In this approach the fundamental solution of the biharmonic equation is employed and the subgrade reaction is treated as unknown external loading.

Moreover, BEM solutions have been developed to analyze plates on nonlinear nonhomogeneous elastic foundation with unilateral contact [36] as well as BEM-based messless methods for plates on biparametric elastic foundation [37,38]. The Pasternak type biparametric model with experimental determination of the parameters approximates adequately the response of the elastic subgrade. Thus, difficult approximations based on the half space modeling can be avoided.

3.4.2 AEM solution

The AEM provides an efficient solution method for Eq. (3.82) for any function $p_s = p_s(x, y, w, w_{,x}, w_{,y}, w_{,xy}, \ldots)$ representing the subgrade reaction, which may depend linearly or nonlinearly on the deflection w and its derivatives.

The application of the AEM as described in Section 3.3.2.1 converts Eq. (3.82) into the equation

$$D\mathbf{b} + \mathbf{p}_s(\mathbf{w}, \mathbf{w}_{,x}, \mathbf{w}_{,y}, \mathbf{w}_{,xy}, \ldots) = \mathbf{f} \tag{3.89}$$

which by virtue of Eq. (3.46) results in an equation of the form

$$D\mathbf{b} + \mathbf{p}_s(\mathbf{b}) = \mathbf{f} \tag{3.90}$$

Equation (3.90) is a linear or nonlinear algebraic equation depending on the function (3.83), whose solution gives the vector **b** of the nodal values of the fictitious load. Evidently, once **b** is known, the deflections and the derivatives are computed from Eq. (3.46). Special care should be given in deriving the boundary conditions for free edges, when the mechanical model includes shear or bending layer [28].

3.4.3 Examples

The computer program PLBECON is modified to include the subgrade reaction and is used to analyze the plates on elastic foundation in the examples that follow.

EXAMPLE 3.4

In this example, the square plate of Fig. 3.17 is analyzed. The plate has dimensions $2a \times 2a$ and rests on a Pasternak-type biparametric elastic foundation $p_s = kw - G\nabla^2 w$ and is loaded by a uniform load f. Beside the boundary support, the plate is also supported on four unyielding piles represented by point supports. The results were obtained with $N = 200$ boundary elements and $M =$ 89 domain nodal points resulting from 136 linear triangular elements. Table 3.5 shows the reactions $\overline{R} = R/fa^2$ at the point supports. Fig. 3.18 shows the deflection surface in 3-D view, while Fig. 3.19 gives the contours of the deflection surface w, the subgrade reaction p_s and the bending moments M_x, M_y. The

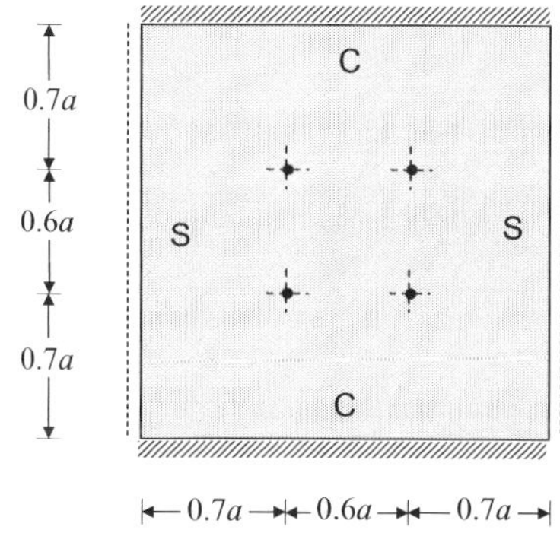

FIGURE 3.17 Square plate on Pasternak-type biparametric elastic foundation with four unyielding internal point supports in Example 3.4.

TABLE 3.5 Reactions $\overline{R} = R/fa^2(\nu = 0.3)$ at the point supports in Example 3.4

	Winkler				Pasternak	
	$\lambda=0$ $s=0$	$\lambda=3$ $s=0$	$\lambda=4$ $s=0$	$\lambda=5$ $s=0$	$\lambda=5$ $s=5$	$\lambda=5$ $s=7$
AEM	0.441	0.416	0.371	0.305	0.298	0.290
Ref. [37]	0.441	0.416	0.371	0.306	0.295	0.283
FEM	0.441	0.415	0.371	0.306		

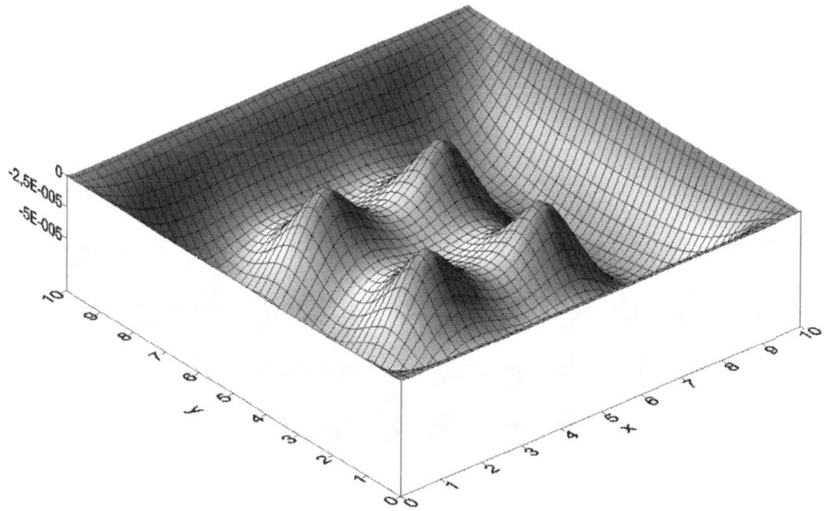

FIGURE 3.18 Deflection surface w in Example 3.4.

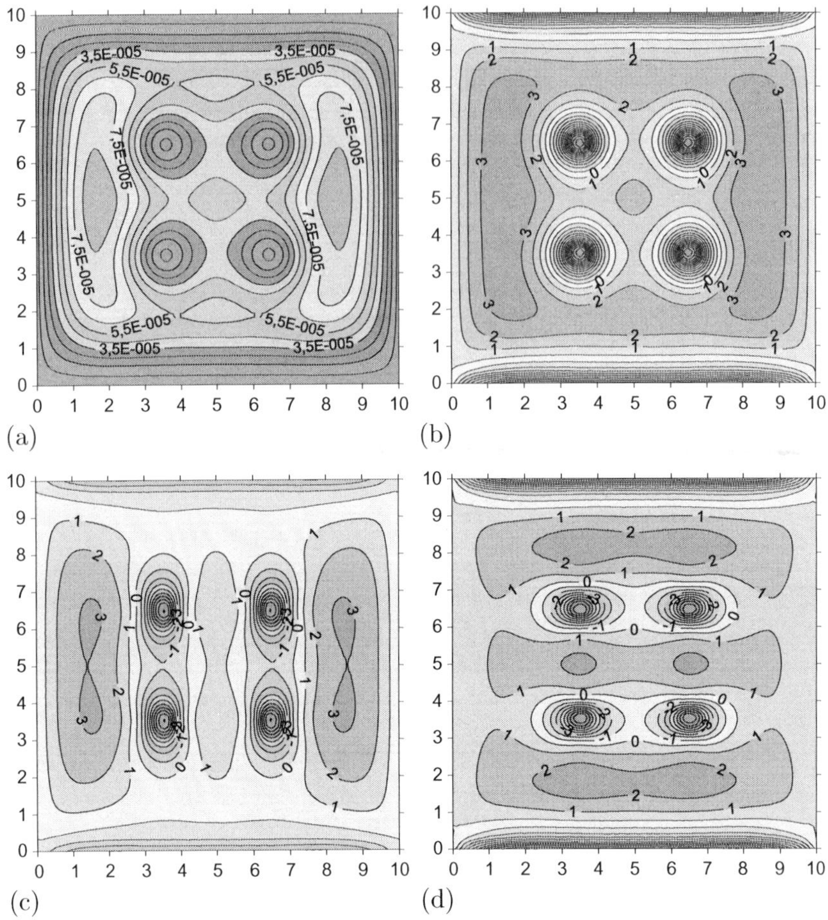

FIGURE 3.19 Contours of (a) deflection w, (b) reaction of the elastic subgrade p, (c) bending moment M_x, and (d) bending moment M_y in Example 3.4.

numerical results were obtained with $a=5$, $h=0.25$, $f=10$, $\lambda=a/(D/k)^{1/4}=3$, $s=a/\sqrt{D/G}=5$

EXAMPLE 3.5

In this example, a plate with composite shape resting on biparametric elastic foundation, $p_s=kw-G\nabla^2 w$, is analyzed. The geometry of the plate is shown in Fig. 3.20. Its boundary is clamped and is subjected to a uniform load f. The employed data are $h=0.2\text{m}$, $a=6$, $\lambda=a/(D/k)^{1/4}=15$, $s=a/\sqrt{D/G}=18$, $E=3\times 10^7\text{kN/m}^2$, and $\nu=0.3$. The solution was obtained with $N=592$ boundary elements and $M=673$ internal nodal points resulting from 1092 linear triangular elements. The perspective of the quadrant of the deflection surface is shown in Fig. 3.21. Figure 3.22 and Fig. 3.23 present the variation of the deflection w and the bending moments M_x and M_y along the semi-axes of

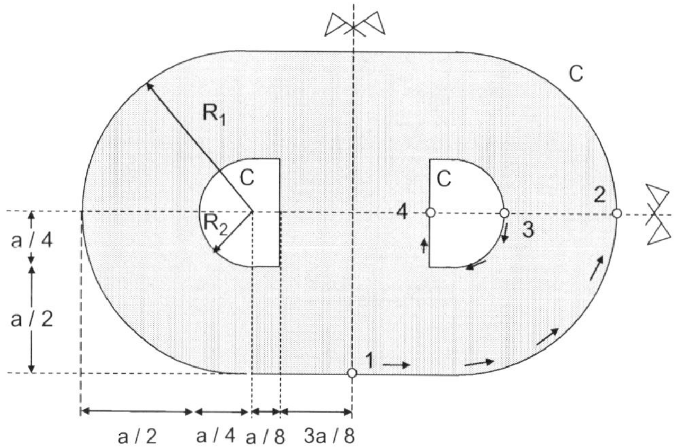

FIGURE 3.20 Geometry of the plate on biparametric elastic foundation in Example 3.5.

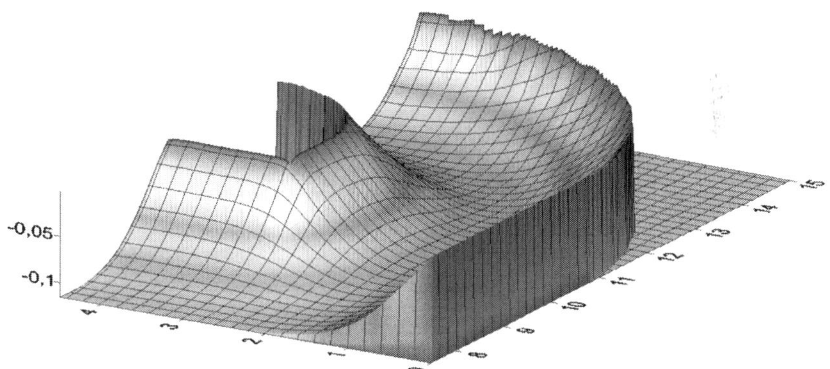

FIGURE 3.21 Quadrant of the deflection surface w of the plate in Example 3.5.

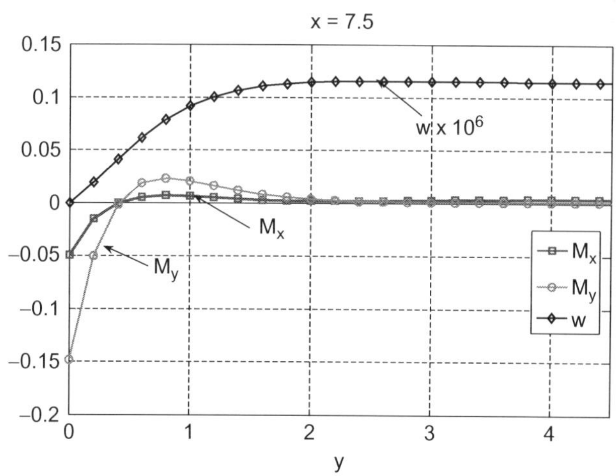

FIGURE 3.22 Variation of w, M_x and M_y along the semi-axis of symmetry $x = 7.5$ in Example 3.5.

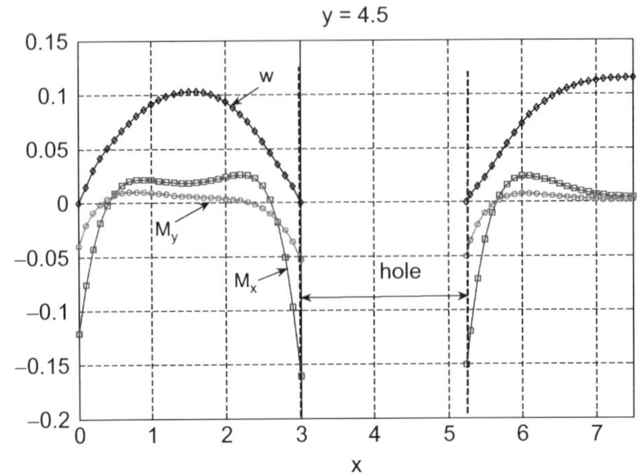

FIGURE 3.23 Variation of w, M_x and M_y along the semi-axis of symmetry $y = 4.5$ in Example 3.5.

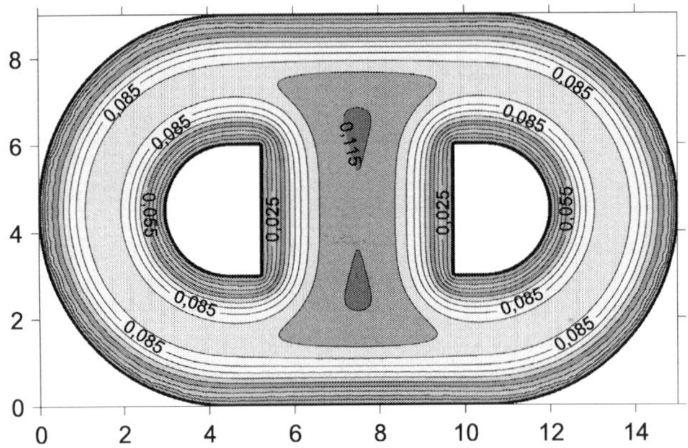

FIGURE 3.24 Contours of the deflection $w \times 10^5$ in Example 3.5.

symmetry $x = 7.5$ and $y = 4.5$. Moreover, Fig. 3.24 and Fig. 3.25 show the contour curves of the deflection surface and the reaction of the elastic foundation. Finally, Fig. 3.26 and Fig. 3.27 show the variation of the bending moment M_n and reaction V_n along the boundary, external section 1-2 and interior section 3-4, respectively.

3.5 LARGE DEFLECTIONS OF THIN PLATES

3.5.1 BEM for large deflections of plates

The BEM has proven an efficient alternative to the domain-type methods and has been used to study large deflection of plates since the 1980s. The BEM solution approaches to nonlinear analysis of plates can be classified into three groups on the basis of the used field equations. In the first group we may include the

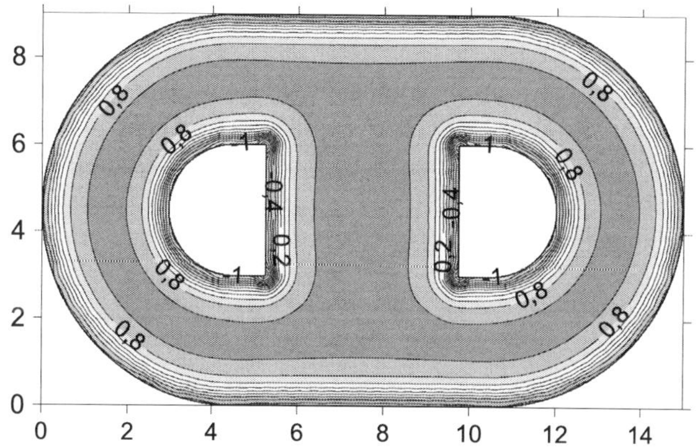

FIGURE 3.25 Contours of the reaction of the elastic subgrade p_s in Example 3.5.

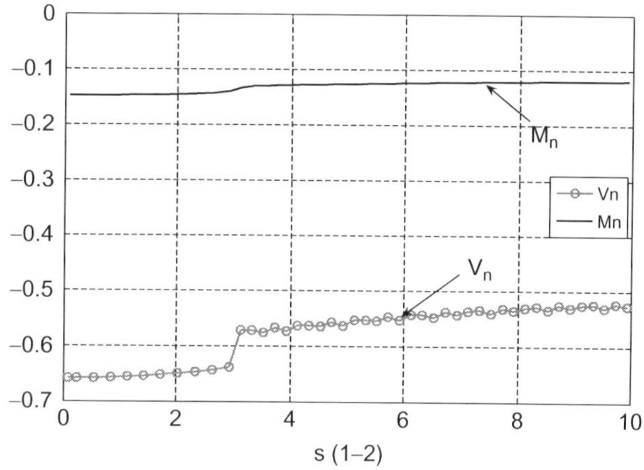

FIGURE 3.26 Variation of the bending moment M_n and reaction V_n on the external boundary (part 1-2) in Example 3.5.

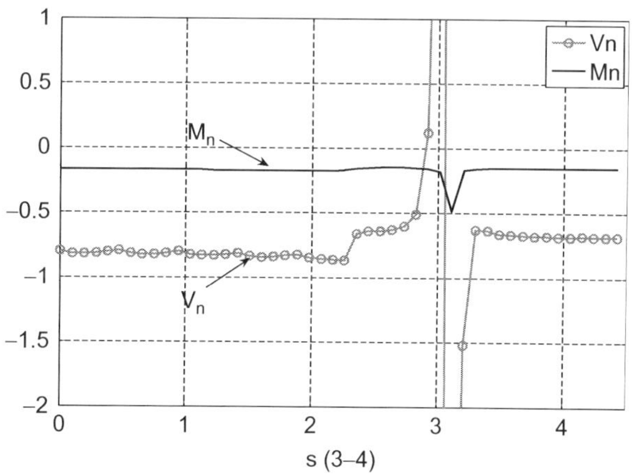

FIGURE 3.27 Variation of the bending moment M_n and reaction V_n on the interior boundary (part 3-4) in Example 3.5.

methods employing the approximate linearized uncoupled Berger's equations [39–43]. The obtained numerical results by means of this formulation are a sufficiently acceptable approximation to the more rigorous solution when inplane displacement boundary conditions are considered. In the second group we may include BEM solutions employing the von-Kármán equations in terms of the transverse deflection and the Airy's stress function. This solution is efficient only when inplane traction boundary conditions are considered. A first formulation of this approach was presented by Tanaka [44], who in a later paper [45], obtained some numerical results using an incremental form of the field equations. Another solution of these equations was presented by Nerantzaki and Katsikadelis [46] by first establishing the Green's function of the linear problem under the boundary conditions of the nonlinear one. Katsikadelis and Nerantzaki [47] and Nerantzaki [48] also presented another integral formulation that permitted the determination of the curvature tensors of the deflection surface and the stress function inside the domain by solving domain integral equations. The domain integrals were approximated using 2-D Gauss integration in domains of arbitrary shape. They analyzed several example problems and gave numerical results. The AEM has also been employed to solve these equations [49]. Elzein and Syngellakis [50] used the incremental formulation to investigate the postbuckling behavior of imperfect plates with FEM discretization for the evaluation of the domain integrals. The third group of BEM approaches employs field equations in terms of the transverse deflections and the inplane displacement components [51–54]. This formulation can treat all types of inplane boundary conditions (traction, displacement, or mixed boundary conditions). However, the nonlinear differential equations that must be solved are more complex and their numerical treatment requires special care. This formulation in its most general form is used here and the resulting equations are solved using the AEM.

3.5.2 Derivation of the equations of the nonlinear problem

The assumptions for thin plates stated in Section 2.2.2 are still valid, except the last one. Here, the displacements w are large but the strains remain still small compared with the unity. This is expressed by Eqs. (2.6), which as shown in Section 3.3.1 become

$$\overline{\varepsilon}_x = -w_{,xx}z + u_{,x} + \frac{1}{2}w_{,x}^2 \tag{3.91a}$$

$$\overline{\varepsilon}_y = -w_{,yy}z + v_{,y} + \frac{1}{2}w_{,y}^2 \tag{3.91b}$$

$$\overline{\gamma}_{xy} = -2zw_{,xy} + u_{,y} + v_{,x} + w_{,x}w_{,y} \tag{3.91c}$$

in which u, v, w are the displacements of the points of the middle point. The strain-displacement relations (3.91) express the von Kármán assumption, according to which the squares of the rotations and their products are retained in the kinematical relations. If w is small the quantities $w_{,x}^2$, $w_{,x}^2$, $w_{,x}\, w_y$ are negligible. However, with increasing w, these quantities contribute to the stretching of the middle surface of the plate and create additional membrane forces. There is no clear distinction between linear and nonlinear plate theory. Usually, nonlinear analysis is required if $w/h > 0.2$. But this criterion is not always adequate, because the shape of the plate or the boundary conditions may change this condition. A safe method is to check the values of $w_{,x}^2$, $w_{,x}^2$, $w_{,x}\, w_y$ after performing linear analysis, or even safer to analyze the plate by performing always large deflection analysis. The plate theory, based on Eqs. (3.91), is known as the *von Kármán plate theory* or *moderately large deflection theory for plates.*

For the consistency of the presentation, the equations are derived using the energy method, though the principle of virtual work could be used [55].

The total potential energy Π of a plate is the sum of the strain energy U_b due to bending, the strain energy U_m due to membrane deformation, the elastic energy U_s of the yielding support, and the potential V of the external applied loads. Thus, we have

$$\Pi = U_b + U_s + U_m + V \tag{3.92}$$

The strain energy U_b is given by Eq. (3.5) and U_s by Eq. (3.6), i.e.,

$$U_b = \frac{D}{2}\int_{\Omega}\left[w_{,xx}^2 + w_{,yy}^2 + 2\nu w_{,xx}\, w_{,yy} + 2(1-\nu)\, w_{,xy}^2\right] d\Omega \tag{3.93}$$

$$U_s = \frac{1}{2}\int_{\Gamma}\left(k_T w^2 + k_R w_{,n}^2\right) ds + \frac{1}{2}\sum_k k_c^{(k)}\, w_k^2 \tag{3.94}$$

and the potential of the external forces V is given by Eq. (3.13).

Regarding the expression of U_m we distinguish two cases:

(a) The membrane loads N_n^*, N_t^*, p_x, p_y are imposed simultaneously with the transverse load f. The strain energy U_m, in this case, is given by Eq. (2.26), namely,

$$U_m = \frac{1}{2}\int_V \left(\overline{\sigma}_x \overline{\varepsilon}_x + \overline{\sigma}_y \overline{\varepsilon}_y + \overline{\tau}_{xy} \overline{\gamma}_{xy}\right) dV \tag{3.95}$$

where

$$\overline{\varepsilon}_x = u_{,x} + \frac{1}{2} w_{,x}^2 \tag{3.96a}$$

$$\overline{\varepsilon}_y = v_{,y} + \frac{1}{2} w_{,y}^2 \tag{3.96b}$$

$$\overline{\gamma}_{xy} = u_{,y} + v_{,x} + w_{,x}\, w_{,y} \tag{3.96c}$$

$$\overline{\sigma}_x = \frac{C}{h}\left(\overline{\varepsilon}_{,x} + \nu\overline{\varepsilon}_{,y}\right) \tag{3.97a}$$

$$\overline{\sigma}_y = \frac{C}{h}\left(\overline{\varepsilon}_{,y} + \nu\overline{\varepsilon}_{,x}\right) \tag{3.97b}$$

$$\overline{\tau}_{xy} = \frac{C}{h}\frac{1-\nu}{2}\gamma_{xy} \tag{3.97c}$$

Introducing Eqs. (3.96) and (3.97) in Eq. (3.95) and integrating through the thickness, we obtain

$$\begin{aligned} U_m = \frac{C}{2}\int_\Omega \Bigg[& \left(u_{,x} + \frac{1}{2}w_{,x}^2\right)^2 + \left(v_{,y} + \frac{1}{2}w_{,y}^2\right)^2 \\ & + 2\nu\left(u_{,x} + \frac{1}{2}w_{,x}^2\right)\left(v_{,y} + \frac{1}{2}w_{,y}^2\right) + \frac{1-\nu}{2}\left(u_{,y} + v_{,x} + w_{,x}w_{,y}\right)^2\Bigg] d\Omega \end{aligned} \tag{3.98}$$

The equilibrium of the plate requires that the first variation of the total potential vanishes, $\delta\Pi = 0$. Thus, working as in Section 2.2.5 we obtain the following differential equations (the derivation procedure is left to the reader as an exercise):

(i) For the transverse displacement

$$D\nabla^4 w - \left(N_x w_{,xx} + 2N_{xy}w_{,xy} + N_y w_{,yy}\right) + p_x w_{,x} + p_y w_{,y} = f \quad \text{in } \Omega \tag{3.99}$$

with associated boundary conditions

$$Vw + N_n w_{,n} + N_t w_{,t} + k_T w = V_n^* \quad \text{or} \quad w = w^* \quad \text{on } \Gamma \tag{3.100a}$$

$$Mw - k_R w_{,n} = M_n^* \quad \text{or} \quad w_{,n} = w_{,n}^* \quad \text{on } \Gamma \tag{3.100b}$$

$$k_c^{(k)} w^{(k)} - [\![Tw]\!]_k = R_k^* \quad \text{or} \quad w^{(k)} = w_k^* \quad \text{at corner } k \tag{3.100c}$$

The operators M and V are given by Eqs. (2.45) and (2.46).

(ii) For the membrane displacements

$$\begin{aligned} \nabla^2 u + \frac{1+\nu}{1-\nu}\left(u_{,x} + v_{,y}\right)_{,x} + w_{,x}\left(\frac{2}{1-\nu}w_{,xx} + w_{,yy}\right) & \\ + \frac{1+\nu}{1-\nu}w_{,xy}w_{,y} + \frac{1}{Gh}p_x = 0 & \end{aligned} \quad \text{in } \Omega \tag{3.101a}$$

$$\nabla^2 v + \frac{1+\nu}{1-\nu}\left(u_{,x} + v_{,y}\right)_{,y} + w_{,y}\left(\frac{2}{1-\nu} w_{,yy} + w_{,xx}\right) \\ + \frac{1+\nu}{1-\nu} w_{,xy} w_{,x} + \frac{1}{Gh} p_y = 0 \qquad \text{in } \Omega \tag{3.101b}$$

with associated boundary conditions

$$N_n = N_n^* \quad \text{or} \quad u_n = u_n^* \quad \text{on } \Gamma \tag{3.102a}$$

$$N_t = N_t^* \quad \text{or} \quad u_t = u_t^* \quad \text{on } \Gamma \tag{3.102b}$$

The quantities N_x, N_y, N_{xy} represent the membrane forces, which are given by the relations

$$N_x = C\left[u_{,x} + \frac{1}{2} w_{,x}^2 + \nu\left(v_{,y} + \frac{1}{2} w_{,y}^2\right)\right] \tag{3.103a}$$

$$N_y = C\left[v_{,y} + \frac{1}{2} w_{,y}^2 + \nu\left(u_{,x} + \frac{1}{2} w_{,x}^2\right)\right] \tag{3.103b}$$

$$N_{xy} = C\frac{1-\nu}{2}\left(u_{,y} + v_{,x} + w_{,x} w_{,y}\right) \tag{3.103c}$$

It should be noted that Eqs. (3.99) and (3.101) are nonlinear and coupled. They give nonzero membrane displacements u, v even if no external membrane loads are present.

(b) The membrane loads N_n^*, N_t^*, p_x, p_y are imposed before the transverse load f, i.e., the plate is under plane prestress. In this case the membrane forces N_x^0, N_y^0, N_{xy}^0, produced by the prestress, are known and result from the solution of the linear plane stress problem under the membrane loads (N_n^*, N_t^*, p_x, p_y). The components N_x^0, N_y^0, N_{xy}^0 do not change during bending and the produced membrane strain energy given as

$$U_m^0 = \frac{1}{2h}\int_V \left(N_x^0 \bar{\varepsilon}_x^0 + N_y^0 \bar{\varepsilon}_y^0 + N_{xy}^0 \bar{\gamma}_{xy}^0\right) dV \\ + \frac{1}{h}\int_V \left(N_x^0 \bar{\varepsilon}_x + N_y^0 \bar{\varepsilon}_y + N_{xy}^0 \bar{\gamma}_{xy}\right) dV \tag{3.104}$$

where

$$\bar{\varepsilon}_x^0 = u_{,x}^0 \tag{3.105a}$$

$$\bar{\varepsilon}_y^0 = v_{,y}^0 \tag{3.105b}$$

$$\bar{\gamma}_{xy}^0 = u_{,y}^0 + v_{,x}^0 \tag{3.105c}$$

while $\bar{\varepsilon}_x, \bar{\varepsilon}_y, \bar{\gamma}_{xy}$ are given by Eqs. (3.96).

Additional membrane strain energy U_m is produced by the stretching of the middle surface due to bending. This is given by Eqs. (3.98). Thus, the total potential energy is

$$\Pi = U_b + U_s + U_m^0 + U_m + V \tag{3.106}$$

in which the potential of the external forces V is given by Eq. (3.13).

The condition $\delta\Pi = 0$ produces the following equations (the derivation procedure is left to the reader as an exercise):

(i) For the transverse displacement

$$D\nabla^4 w - \left[\left(N_x^0 + N_x\right)w_{,xx} + 2\left(N_{xy}^0 + N_{xy}\right)w_{,xy} + \left(N_y^0 + N_y\right)w_{,yy}\right)\Big] + p_x w_{,x} + p_y w_{,y} = f \quad \text{in } \Omega \tag{3.107}$$

The associated boundary conditions are given by Eqs. (3.100), where now N_n and N_t must be augmented by N_n^* and N_t^* if the boundary is free.

(ii) For the membrane displacements

$$\nabla^2 u + \frac{1+\nu}{1-\nu}\left(u_{,x} + v_{,y}\right)_{,x} + w_{,x}\left(\frac{2}{1-\nu}w_{,xx} + w_{,yy}\right) + \frac{1+\nu}{1-\nu}w_{,xy}w_{,y} = 0 \quad \text{in } \Omega \tag{3.108a}$$

$$\nabla^2 v + \frac{1+\nu}{1-\nu}\left(u_{,x} + v_{,y}\right)_{,y} + w_{,y}\left(\frac{2}{1-\nu}w_{,yy} + w_{,xx}\right) + \frac{1+\nu}{1-\nu}w_{,xy}w_{,x} = 0 \quad \text{in } \Omega \tag{3.108b}$$

with homogeneous boundary conditions

$$N_n = 0 \quad \text{or} \quad u_n = 0 \quad \text{on } \Gamma \tag{3.109a}$$

$$N_t = 0 \quad \text{or} \quad u_t = 0 \quad \text{on } \Gamma \tag{3.109b}$$

The displacements u, v, which result from the solution of the coupled nonlinear equations, represent the additional displacements at the points of the middle surface due to bending.

3.5.3 The linear problem

The linear problem can be obtained as a special case of the nonlinear one, if we set $w_{,x}^4 = w_{,y}^4 = w_{,x}^2 w_{,y}^2 \simeq 0$ in Eq. (3.98). Thus, the resulting equations for the plane stress problem do not contain nonlinear terms and, consequently, become identical to Eqs. (3.9). This implies that the membrane forces, though they contribute to the bending of the plate, are not influenced by it. The boundary conditions are the same in both problems, linear and nonlinear. Evidently, the equation for bending is independent of the equations for membrane deformation. The forces N_x, N_y, N_{xy} are established from Eqs. (3.103) with $w_{,x}^2 = w_{,y}^2 = w_{,x}w_{,y} \simeq 0$, once the displacements u, v have been computed from the linearized problem (3.101).

3.5.4 AEM solution

The analog equations of Eqs. (3.99) and (3.101) are

$$\nabla^4 w = b(\mathbf{x}) \tag{3.110a}$$

$$\nabla^2 u = b^{(1)}(\mathbf{x}) \tag{3.110b}$$

$$\nabla^2 v = b^{(2)}(\mathbf{x}) \tag{3.110c}$$

The application of the AEM as described in Section 3.3.2.1 converts Eq. (3.99) into the equation

$$D\mathbf{b} - \left(\mathbf{N}_x \mathbf{w}_{,xx} + 2\mathbf{N}_{xy}\mathbf{w}_{,xy} + \mathbf{N}_y \mathbf{w}_{,yy}\right) + \mathbf{p}_x \mathbf{w}_{,x} + \mathbf{p}_y \mathbf{w}_{,y} = \mathbf{f} \tag{3.111a}$$

and Eqs. (3.101a,b) into

$$\begin{aligned}\mathbf{b}^{(1)} + \frac{1+\nu}{1-\nu}\left(\mathbf{u}_{,x} + \mathbf{v}_{,y}\right)_{,x} + \mathbf{w}_{,x}\left(\frac{2}{1-\nu}\mathbf{w}_{,xx} + \mathbf{w}_{,yy}\right) + \frac{1+\nu}{1-\nu}\mathbf{w}_{,xy}\mathbf{w}_{,y} \\ + \frac{1}{Gh}\mathbf{p}_x = 0\end{aligned} \tag{3.111b}$$

$$\begin{aligned}\mathbf{b}^{(2)} + \frac{1+\nu}{1-\nu}\left(\mathbf{u}_{,x} + \mathbf{v}_{,y}\right)_{,y} + \mathbf{w}_{,y}\left(\frac{2}{1-\nu}\mathbf{w}_{,yy} + \mathbf{w}_{,xx}\right) + \frac{1+\nu}{1-\nu}\mathbf{w}_{,xy}\mathbf{w}_{,x} \\ + \frac{1}{Gh}\mathbf{p}_y = 0\end{aligned} \tag{3.111c}$$

Moreover, Eqs. (3.103) become

$$\mathbf{N}_x = C\left[\mathbf{u}_{,x} + \frac{1}{2}\mathbf{w}_{,x}\mathbf{w}_{,x} + \nu\left(\mathbf{v}_{,y} + \frac{1}{2}\mathbf{w}_{,y}\mathbf{w}_{,y}\right)\right] \tag{3.112a}$$

$$\mathbf{N}_y = C\left[\mathbf{v}_{,y} + \frac{1}{2}\mathbf{w}_{,y}\mathbf{w}_{,y} + \nu\left(\mathbf{u}_{,x} + \frac{1}{2}\mathbf{w}_{,x}\mathbf{w}_{,x}\right)\right] \tag{3.112b}$$

$$\mathbf{N}_{xy} = C\frac{1-\nu}{2}\left(\mathbf{u}_{,y} + \mathbf{v}_{,x} + \mathbf{w}_{,x}\mathbf{w}_{,y}\right) \tag{3.112c}$$

Referring to Eqs. (3.103), we observe that the membrane forces N_x, N_y and N_{xy}, hence N_n and N_t in the boundary conditions, include nonlinear terms. Therefore, it is necessary to modify the expressions (3.67a,b) in order to include these terms. To comply with this, Eqs. (3.71) are written as

$$\begin{bmatrix} \gamma_1 & \mathbf{0} \\ \mathbf{0} & \delta_1 \end{bmatrix} \begin{Bmatrix} \mathbf{u} \\ \mathbf{v} \end{Bmatrix} + \begin{bmatrix} \gamma_2 & \mathbf{0} \\ \mathbf{0} & \delta_2 \end{bmatrix} \begin{Bmatrix} \mathbf{t}_x \\ \mathbf{t}_y \end{Bmatrix} = \begin{Bmatrix} \gamma_3 \\ \delta_3 \end{Bmatrix} - \begin{bmatrix} \gamma_2 & \mathbf{0} \\ \mathbf{0} & \delta_2 \end{bmatrix} \begin{Bmatrix} \mathbf{t}_x^{nl} \\ \mathbf{t}_y^{nl} \end{Bmatrix} \tag{3.113}$$

where $\mathbf{t}_x^{nl}$, $\mathbf{t}_y^{nl}$ represent unknown boundary tractions, due to the nonlinear terms of Eqs. (3.103) and given as

$$\mathbf{t}_x^{nl} = \frac{C}{h}\left[\left(\frac{1}{2}\mathbf{w}_{,x}^2 + \nu\frac{1}{2}\mathbf{w}_{,y}^2\right)\mathrm{n}_x + \frac{1-\nu}{2}\mathbf{w}_{,x}\,\mathbf{w}_{,y}\,\mathbf{n}_y\right] \tag{3.114a}$$

$$\mathbf{t}_y^{nl} = \frac{C}{h}\left[\frac{1-\nu}{2}\mathbf{w}_{,x}\,\mathbf{w}_{,y}\,\mathrm{n}_x + \left(\frac{1}{2}\mathbf{w}_{,y}^2 + \frac{1}{2}\nu\mathbf{w}_{,x}^2\right)\mathbf{n}_y\right] \tag{3.114b}$$

Thus, Eq. (3.72) can be written in the following form:

$$\widetilde{\mathbf{H}}\begin{Bmatrix} \mathbf{u} \\ \mathbf{v} \end{Bmatrix} + \widetilde{\mathbf{G}}\begin{Bmatrix} \mathbf{u}_n \\ \mathbf{v}_n \end{Bmatrix} = \widetilde{\mathbf{F}} - \begin{bmatrix} \gamma_2 & \mathbf{0} \\ \mathbf{0} & \delta_2 \end{bmatrix}\begin{Bmatrix} \mathbf{t}_x^{nl} \\ \mathbf{t}_y^{nl} \end{Bmatrix} \tag{3.115}$$

Equations (3.65) and (3.115) are combined and solved to express $\mathbf{u}$, $\mathbf{v}$, $\mathbf{u}_n$, and $\mathbf{v}_n$ in terms of $\mathbf{b}^{(1)}$, $\mathbf{b}^{(2)}$, $\mathbf{t}_x^{nl}$, and $\mathbf{t}_y^{nl}$. Hence, Eqs. (3.76) become

$$\mathbf{u}_{,pq} = \mathbf{U}_{,pq}^{(1)}\mathbf{b}^{(1)} + \mathbf{U}_{,pq}^{(2)}\mathbf{b}^{(2)} + \mathbf{U}_{,pq}^{(3)}\mathbf{t}_x^{nl} + \mathbf{U}_{,pq}^{(4)}\mathbf{t}_y^{nl} + \mathbf{e}_{,pq} \tag{3.116a}$$

$$\mathbf{v}_{,pq} = \overline{\mathbf{U}}_{,pq}^{(1)}\mathbf{b}^{(1)} + \overline{\mathbf{U}}_{,pq}^{(2)}\mathbf{b}^{(2)} + \overline{\mathbf{U}}_{,pq}^{(3)}\mathbf{t}_x^{nl} + \overline{\mathbf{U}}_{,pq}^{(4)}\mathbf{t}_y^{nl} + \overline{\mathbf{e}}_{,pq} \tag{3.116b}$$

in which $\mathbf{U}_{,pq}^{(3)}$, $\mathbf{U}_{,pq}^{(4)}$, $\overline{\mathbf{U}}_{,pq}^{(3)}$, $\overline{\mathbf{U}}_{,pq}^{(4)}$ are known $M \times M$ matrices.

By virtue of Eqs. (3.46) and (3.116), Eqs. (3.111) and (3.115) yield the system of the nonlinear algebraic equations

$$D\mathbf{b} - \mathbf{F}\left(\mathbf{b}, \mathbf{b}^{(1)}, \mathbf{b}^{(2)}, \mathbf{t}_x^{nl}, \mathbf{t}_y^{nl}\right) = \mathbf{f} \tag{3.117a}$$

$$\mathbf{A}_1\mathbf{b}^{(1)} + \mathbf{B}_1\mathbf{b}^{(2)} + \mathbf{H}_1(\mathbf{b}) = \mathbf{p}_1 \tag{3.117b}$$

$$\mathbf{A}_2\mathbf{b}^{(1)} + \mathbf{B}_2\mathbf{b}^{(2)} + \mathbf{H}_2(\mathbf{b}) = \mathbf{p}_2 \tag{3.117c}$$

$$\mathbf{t}_x^{nl} = \mathbf{H}_3(\mathbf{b}) \tag{3.117d}$$

$$\mathbf{t}_y^{nl} = \mathbf{H}_4(\mathbf{b}) \tag{3.117e}$$

where

$\mathbf{A}_i$, $\mathbf{B}_i$ $(i=1,2)$ are $M \times M$ known matrices.

$\mathbf{F}\left(\mathbf{b}, \mathbf{b}^{(1)}, \mathbf{b}^{(2)}, \mathbf{t}_x^{nl}, \mathbf{t}_y^{nl}\right)$ is a vector nonlinearly depending on the components of $\mathbf{b}$, $\mathbf{b}^{(1)}$, $\mathbf{b}^{(2)}$, $\mathbf{t}_x^{nl}$, $\mathbf{t}_y^{nl}$.

$\mathbf{H}_i$ $(i=1, 2, 3, 4)$ are vectors nonlinearly depending on the components of $\mathbf{b}$.

$\mathbf{f}, \mathbf{p}_1, \mathbf{p}_2$ are known load vectors.

Equations (3.117) are solved to give $\mathbf{b}$, $\mathbf{b}^{(1)}, \mathbf{b}^{(2)}, \mathbf{t}_x^{nl}, \mathbf{t}_y^{nl}$. It should be noted that Eqs. (3.117b,c,d,e) are linear with respect to $\mathbf{b}^{(1)}, \mathbf{b}^{(2)}, \mathbf{t}_x^{nl}, \mathbf{t}_y^{nl}$. This fact facilitates the solution of the nonlinear system [56].

Subsequently, the displacements and their derivatives are computed from Eqs. (3.46) and (3.116). The stress resultants are computed using Eqs. (2.19), (2.21), and (3.112). The AEM either by converting the domain integrals to boundary line integrals or with domain discretization as presented herein has been employed for the solution of many nonlinear plate problems [49,56–58].

3.5.5 Examples

EXAMPLE 3.6

In this example the nonlinear response of the plate of Fig. 3.28 is studied. The plate is loaded by a uniform load f. The boundary conditions are shown in the figure. The employed data are $E=3\times10^7\,\mathrm{kN/m^2}$, $\nu=0.3$, and $h=0.05\,\mathrm{m}$. The results were obtained with $N=410$ boundary elements and $M=244$ domain nodal points resulting from 409 linear triangular elements. Fig. 3.29 shows the variation of the displacements w and u at point $A(7.42,0)$ versus the load f. Fig. 3.30 presents the variation of the quantities w, u, M_x and M_y along the axis $y=0$ for $f=10\,\mathrm{kN/m^2}$ as compared with those of a FEM solution using 16614 quadrilateral elements. Finally, the contours of w, M_x, M_{xy}, and M_y are shown in Fig. 3.31 and Fig. 3.32.

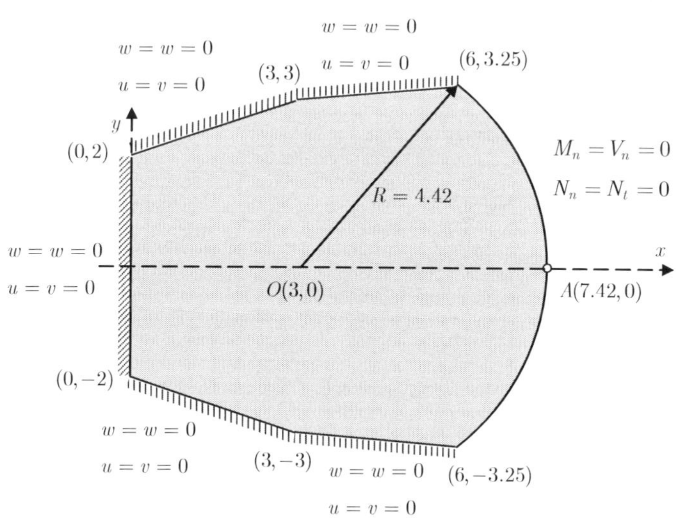

FIGURE 3.28 Plate of composite shape in Example 3.6.

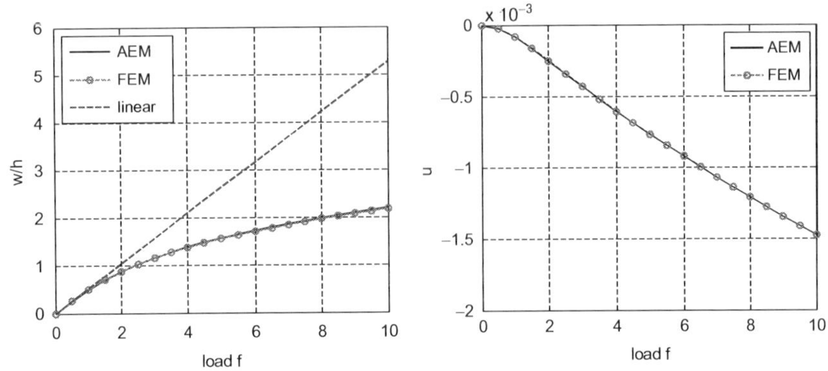

FIGURE 3.29 Variation of the displacements w and u at point $A(7.42,0)$ versus the load f in Example 3.6.

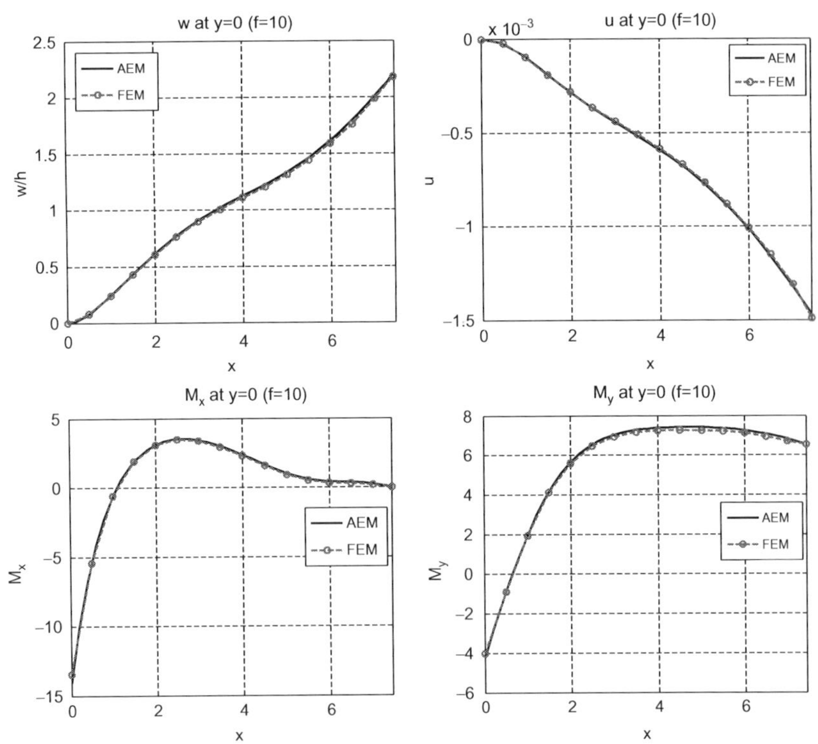

FIGURE 3.30 Variation of the displacements w, u and bending moments M_x, M_y along the axis $y=0$ in Example 3.6.

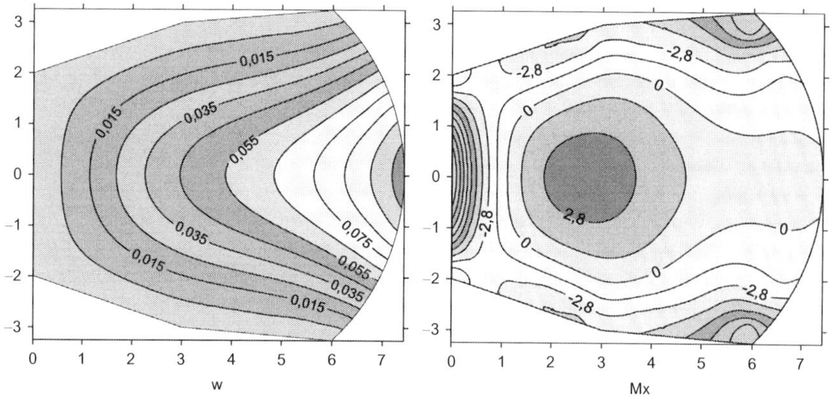

FIGURE 3.31 Contours of deflection w and bending moment M_x in Example 3.6.

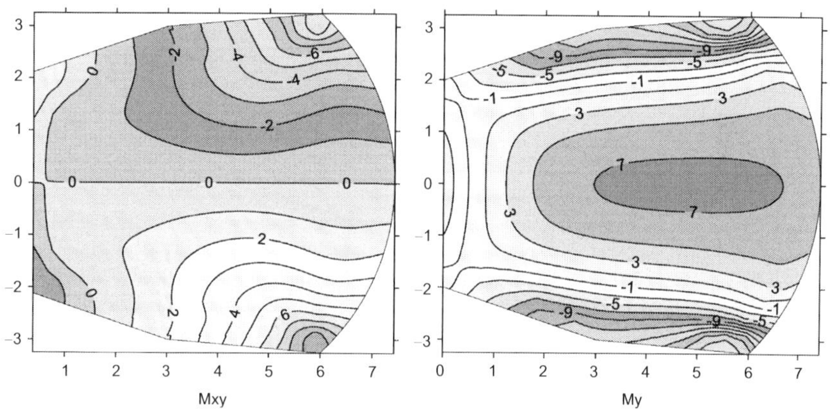

FIGURE 3.32 Contours of the twisting moment M_{xy} and bending moment M_y in Example 3.6.

EXAMPLE 3.7

In this example the postbuckling response of the square plate shown in Fig. 3.33 is investigated. All edges are simply supported and movable in the plane of the plate. The employed data are $E = 3 \times 10^7 \text{kN/m}^2$, $\nu = 0.3$. The results were obtained with $N = 200$ boundary elements and $M = 137$ domain nodal points resulting from 216 linear triangular elements. First the linear problem was solved, which gave the critical buckling load $P_{cr}^1 = 4332.6\text{kN}$ (exact value $P_{cr}^1 = 4333.3kN$). Figure 3.34 shows the load-deflection curve for the transverse displacement at the central point of the plate as compared with the analytical result [59] for different values of the initial deflection w_0, when the first linear buckling mode is taken as initial deflection surface; w_0 is the value of the mode shape at the center of the plate. Moreover, Fig. 3.35 presents the load-displacement curve for the membrane displacement v at point (2.5, 0). It is worth noting how the curvature of this curve increases with decreasing w_0 in the vicinity of the bifurcation point.

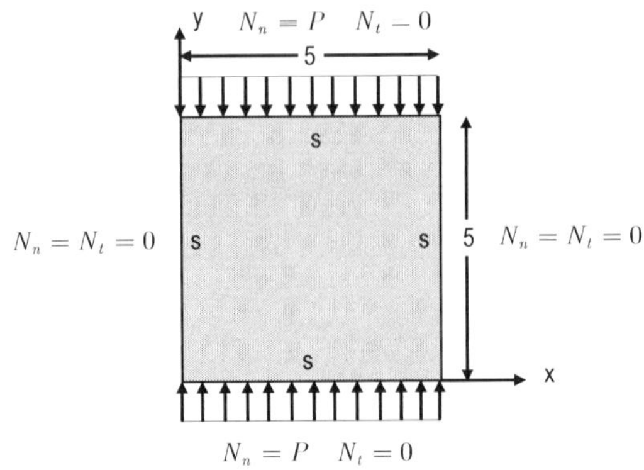

FIGURE 3.33 Square plate with simply supported movable edges in Example 3.7.

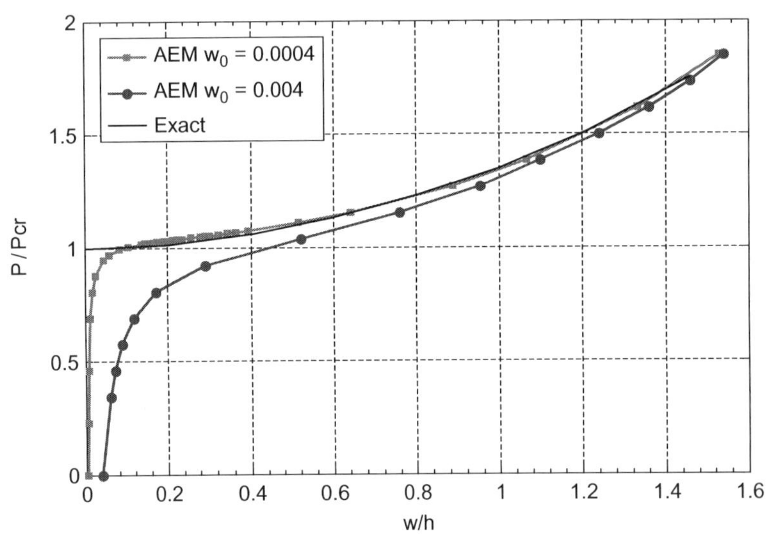

FIGURE 3.34 Load-displacement curve for the deflection at the center of the plate for various values of w_0 in Example 3.7.

EXAMPLE 3.8

In this example the postbuckling response of the circular plate of Fig. 3.36 is studied. The boundary is clamped and movable. The employed data are $E = 3\times 10^7 \mathrm{kN/m^2}$, $\nu = 0.3$, $h = 0.0625\mathrm{m}$. The results were obtained with $N = 200$ boundary elements and $M = 300$ domain nodal points resulting from 516 linear triangular elements. Fig. 3.37 shows the variation of the central deflection w/h versus the load parameter $\lambda = P/P_{cr}$ as compared with other available solutions [60,61]. The computed linear buckling load is $P_{cr} = 1103.15\mathrm{kN}$ (exact $P_{cr} = 1094.01\mathrm{kN}$ [59]).

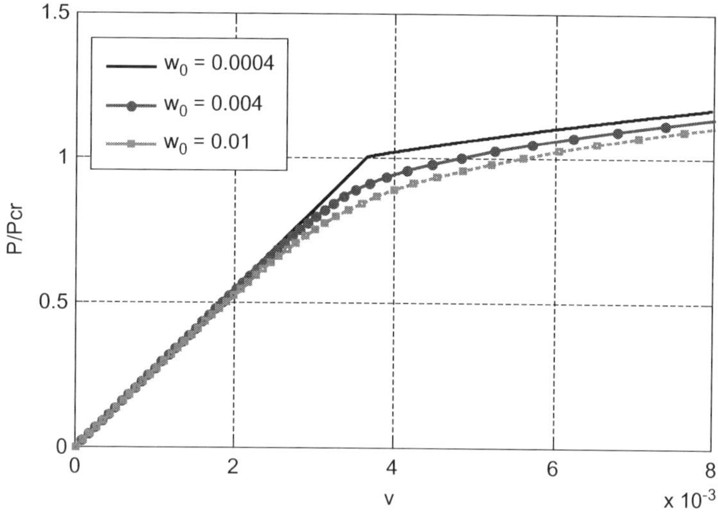

FIGURE 3.35 Load-displacement curve for the membrane displacement v at point $(2.5, 0)$ for various values of w_0 in Example 3.7.

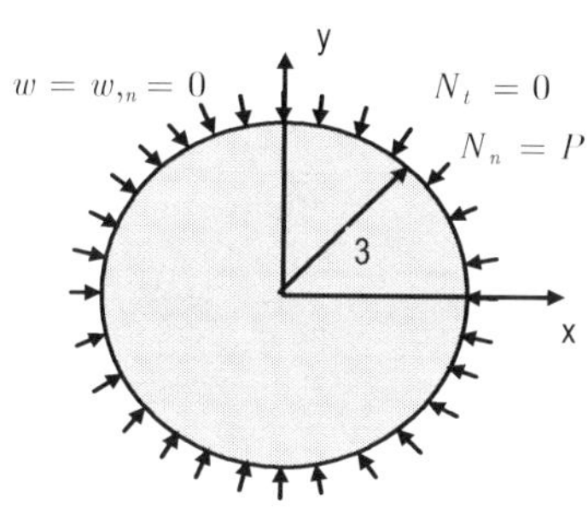

FIGURE 3.36 Circular plate with clamped movable boundary in Example 3.8.

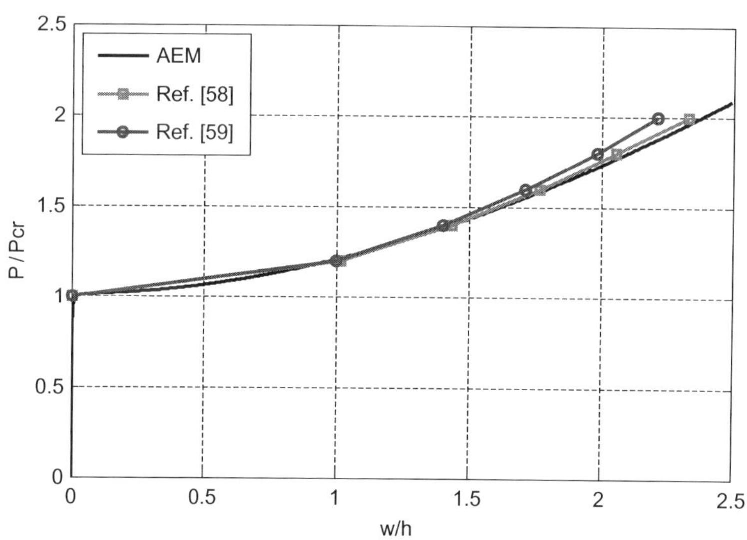

FIGURE 3.37 Load-deflection curve at the center of the circular plate in Example 3.8.

3.6 PLATES WITH VARIABLE THICKNESS

3.6.1 Derivation of the differential equation

To derive the differential equation for the plate with variable thickness, we assume that there is no "abrupt" variation in the thickness $h(x, y)$. The consequence of this is that the expressions (2.19) for the bending and twisting moments derived for plates with constant thickness also apply with sufficient accuracy to this case. Then

$$M_x = -D\left(w_{,xx} + \nu w_{,yy}\right) \tag{3.118a}$$

$$M_y = -D\left(w_{,yy} + \nu w_{,xx}\right) \tag{3.118b}$$

$$M_{xy} = -M_{yx} = D(1-\nu)w_{,xy} \tag{3.118c}$$

where now the bending stiffness of the plate is position dependent, i.e., $D = D(x, y)$.

Application of Eqs. (2.20) yields the shear forces

$$Q_x = -D\left(\nabla^2 w\right)_{,x} + \ln(D)_{,x} M_x - \ln(D)_{,y} M_{xy} \tag{3.119a}$$

$$Q_y = -D\left(\nabla^2 w\right)_{,y} + \ln(D)_{,y} M_y - \ln(D)_{,x} M_{xy} \tag{3.119b}$$

The differential equation is readily obtained by considering the equilibrium of the plate element in the transverse direction. Thus referring to Fig. 2.6 we obtain

$$\frac{\partial Q_x}{\partial x} + \frac{\partial Q_y}{\partial y} = -f(x, y) \tag{3.120}$$

which by virtue of Eqs. (2.20) becomes

$$\frac{\partial^2 M_x}{\partial x^2} + \frac{\partial^2 M_{yx}}{\partial x \partial y} - \frac{\partial^2 M_{xy}}{\partial x \partial y} + \frac{\partial^2 M_y}{\partial^2 y} = -f(x, y) \tag{3.121}$$

Finally, introducing Eqs. (3.118a,b,c) into Eqs. (3.121) yields the differential equation for the plate with variable thickness

$$\begin{aligned} &D\nabla^4 w + 2D_{,x}\nabla^2 w_{,x} + 2D_{,y}\nabla^2 w_{,y} + \nabla^2 D \nabla^2 w \\ &\quad -(1-\nu)\left(D_{,xx}w_{,yy} - 2D_{,xy}w_{,xy} + D_{,yy}w_{,xx}\right) = f \end{aligned} \quad \text{in } \Omega \tag{3.122}$$

The associated boundary conditions result from Eqs. (2.40) by considering that the bending stiffness is variable. Thus, we have

$$\hat{V}w + k_T w = V_n^* \quad \text{or} \quad w = w^* \quad \text{on } \Gamma \tag{3.123a}$$

$$Mw - k_R w_{,n} = M_n^* \quad \text{or} \quad w_{,n} = w_{,n}^* \quad \text{on } \Gamma \tag{3.123b}$$

$$k_c^{(k)} w_k - [\![Tw]\!]_k = R_k^* \quad \text{or} \quad w_k = w_k^* \quad \text{at corner } k \tag{3.123c}$$

where the operators $\hat{V}$, M, T are defined as

$$\hat{V} = V - 2\frac{1}{D}\frac{\partial D}{\partial s}T + \frac{1}{D}\frac{\partial D}{\partial n}M \tag{3.124a}$$

$$M = -D\left[\nabla^2 - (1-\nu)\left(\frac{\partial^2}{\partial s^2} + \kappa\frac{\partial}{\partial n}\right)\right] \tag{3.124b}$$

$$T = D(1-\nu)\left(\frac{\partial^2}{\partial s \partial n} - \kappa\frac{\partial}{\partial s}\right) \tag{3.124c}$$

The operator V is defined by Eq. (2.46), i.e., it produces the effective shear force of the plate with constant thickness. Note that the operators M, T are identical with those of the plate with constant thickness.

The concept of the *thin plate* is characterized by the ratio h/a, where h is the plate thickness and a the smallest of the other two dimensions(*) so that $(h/a)^2 \ll 1$. On the other hand, the concept of the *abrupt* thickness variation is not clearly defined and mathematically expressed so that the validity of Eq. (3.122) can be checked. This means that the solution, which is obtained from this equation may not satisfy the Kirchhoff assumptions for thin plates. This difficulty becomes pronounced in plate optimization problems, such as the determination of the thickness variation to maximize the buckling load or the plate stiffness, while keeping the volume of the plate constant. Thus, the use of Eq. (3.122) may lead to solutions that are not unacceptable, if they are obtained without imposing a constraint on the thickness variation to ensure the validity of the thin plate theory. Certain investigators [62,63] imposed the constraint of small slopes $h_{,x}$, $h_{,y}$ for thickness variation law, but it proves to be insufficient. Katsikadelis and Babouskos [64] derived the equation for the thin plate with variable thickness starting from the three-dimensional theory of elasticity and concluded with the following constraint that ensures the validity of thin plate theory:

$$\begin{aligned} g(x,y) &= \frac{\nu}{24(1-\nu)}\frac{|\nabla^2(h^3)|}{h} \\ &= \frac{\nu}{8(1-\nu)}\left|2h_{,x}^2 + 2h_{,y}^2 + hh_{,xx} + hh_{,yy}\right| < \varepsilon \end{aligned} \tag{3.125}$$

where ε is a small positive number, e.g., $\varepsilon = 0.001$. We observe that relation (3.125) contains not only the first derivatives of the thickness function but also its second ones. We can say that the constraint (3.125) quantifies the term "abrupt variation" and should be imposed in addition to $(h/a)^2 \ll 1$. This constraint has been employed in several plate thickness optimization problems and has performed well.

(*)The value $h/a < 0.1$ is usually acceptable so that a plate is considered thin.

The formulation of the direct BEM for the plate with variable thickness requires the knowledge of the fundamental solution of Eq. (3.122), but it cannot be established. Therefore, recourse to BEM formulations using the known fundamental solution of the biharmonic equation was inevitable to develop BEM solutions for plates with variable thickness. The first publication using this formulation was by Sapountzakis and Katsikadelis [65]. Later, Chaves et al. [66] presented another formulation for the problem. The AEM has been successfully applied to the problem at hand [67]. Moreover, the AEM has been employed to solve more difficult problems for plates with variable thickness such as buckling problems [68], large deflections of plates [69], plates on biparametric elastic foundation [70] as well as plate thickness optimization problems to maximize the buckling load or to minimize the deflection [64].

3.6.2 AEM solution

Application of the AEM converts Eq. (3.122) into the following algebraic equation:

$$\begin{aligned}&\mathbf{Db}+2\mathbf{D}_{,x}\nabla^2\mathbf{w}_{,x}+2\mathbf{D}_{,y}\nabla^2\mathbf{w}_{,y}+\nabla^2\mathbf{D}\nabla^2\mathbf{w}\\&-(1-\nu)\left(\mathbf{D}_{,xx}\mathbf{w}_{,yy}-2\mathbf{D}_{,xy}\mathbf{w}_{,xy}+\mathbf{D}_{,yy}\mathbf{w}_{,xx}\right)=\mathbf{f}\end{aligned}\quad \text{in } \Omega \tag{3.126}$$

The boundary conditions (3.123) can be written in the form

$$\alpha_1 w + a_2[Vw - 2\ln(D)_{,s}Tw + \ln(D)_{,n}Mw] = a_3 \quad \text{on } \Gamma \tag{3.127a}$$

$$\beta_1 w_{,n} + \beta_2 Mw = \beta_3 \quad \text{on } \Gamma \tag{3.127b}$$

$$c_{1k}w_k + c_{2k}R_k = c_{3k} \quad \text{at corner } k \tag{3.127c}$$

Apparently, for appropriate values of a_i, β_i, c_{ik} all admissible boundary conditions are obtained. Tw appears as a new boundary quantity, but it can be expressed in terms of boundary quantities that can be specified. A simple method is to use Eq. (2.52c), which is written as

$$Tw = D(1-\nu)[(w_{,n})_{,s} - \kappa w_{,s}] \tag{3.128}$$

Approximation of the tangential derivatives $w_{,s}$ and $(w_{,n})_{,s}$ with finite differences, as in Eqs. (3.31), permits the expression of Tw in terms of the nodal values of w and $w_{,n}$. This technique is adopted here. The next step is to apply the boundary conditions at the boundary nodal points and to express the actual boundary quantities in terms of those of the analog equation (3.17), which, we recall, represents the equation of a plate of constant thickness with $D=1$. Thus, the application of the boundary conditions (3.127) at the boundary nodal points and corner points yields

$$\widetilde{\mathbf{H}}\begin{Bmatrix}\mathbf{w}\\ \mathbf{w}_c\\ \mathbf{w}_n\end{Bmatrix}+\widetilde{\mathbf{G}}\begin{Bmatrix}\mathbf{V}\\ \mathbf{R}\\ \mathbf{M}\end{Bmatrix}=\widetilde{\mathbf{F}} \tag{3.129}$$

where

$$\widetilde{\mathbf{H}}=\begin{bmatrix}\boldsymbol{\alpha}_1+2\boldsymbol{\alpha}_2(1-\nu)\mathbf{D}_{,s}\,\kappa\mathbf{d} & \mathbf{0} & -2\boldsymbol{\alpha}_2(1-\nu)\mathbf{D}_{,s}\,\mathbf{d}\\ \mathbf{0} & \mathbf{c}_1 & \mathbf{0}\\ \mathbf{0} & \mathbf{0} & \boldsymbol{\beta}_1\end{bmatrix} \tag{3.130a}$$

$$\widetilde{\mathbf{G}}=\begin{bmatrix}\boldsymbol{\alpha}_2\mathbf{D} & \mathbf{0} & \boldsymbol{\alpha}_2\mathbf{D}_{,n}\\ \mathbf{0} & \mathbf{c}_2\mathbf{D} & \mathbf{0}\\ \mathbf{0} & \mathbf{0} & \boldsymbol{\beta}_2\mathbf{D}\end{bmatrix},\quad \widetilde{\mathbf{F}}=\begin{Bmatrix}\boldsymbol{\alpha}_3\\ \mathbf{c}_3\\ \boldsymbol{\beta}_3\end{Bmatrix} \tag{3.130b,c}$$

in which $\mathbf{D}$, $\mathbf{D}_{,s}$ and $\mathbf{D}_{,n}$ are known $N\times N$ diagonal matrices including the values of $D(x,y)$, $D_{,s}(x,y)$, and $D_{,n}(x,y)$ at the boundary nodal points. The matrices $\boldsymbol{\alpha}_i$, $\boldsymbol{\beta}_i$, $\mathbf{c}_i$ and $\mathbf{d}$ have been defined in Eqs. (3.31). Eqs. (3.129) are combined with Eqs. (3.28) and solved to give the boundary quantities in terms of the fictitious load $\mathbf{b}$. Then Eq. (3.46) is used to express the derivatives $\mathbf{w}_{,x}$, $\mathbf{w}_{,y}$, $\mathbf{w}_{,xx}$, $\mathbf{w}_{,yy}$, $\mathbf{w}_{,xy}$ in terms of $\mathbf{b}$, which are then inserted into Eq. (3.126). This yields the linear algebraic equation, which is solved to give the fictitious load $\mathbf{b}$. Once $\mathbf{b}$ has been computed the displacements and their derivatives at the interior nodal points are computed from Eqs. (3.46). The stress resultants are computed using Eqs. (3.118) and (3.119).

3.6.3 Plate with variable thickness under combined transverse and membrane load

The differential equation is readily derived if the influence of the membrane force is included in Eq. (3.122). Thus, referring to Eq. (3.14) we have

$$\begin{aligned}&D\nabla^4 w+2D_{,x}\nabla^2 w_{,x}+2D_{,y}\nabla^2 w_{,y}+\nabla^2 D\nabla^2 w\\ &\quad-(1-\nu)\left(D_{,xx}w_{,yy}-2D_{,xy}w_{,xy}+D_{,yy}w_{,xx}\right)\qquad \text{in } \Omega\\ &\quad-\left(N_x w_{,xx}+2N_{xy}w_{,xy}+N_y w_{,yy}\right)+p_x w_{,x}+p_y w_{,y}=f\end{aligned} \tag{3.131}$$

with associated boundary conditions

$$\hat{V}w+N_n w_{,n}+N_t w_{,t}+k_T w=V_n^* \quad\text{or}\quad w=w^* \quad \text{on } \Gamma \tag{3.132a}$$

$$Mw-k_R w_{,n}=M_n^* \quad\text{or}\quad w_{,n}=w_n^* \quad \text{on } \Gamma \tag{3.132b}$$

$$k_c^{(k)} w_k-[\![Tw]\!]_k=R_k^* \quad\text{or}\quad w_k=w_k^* \quad \text{at corner } k \tag{3.132c}$$

The membrane forces are obtained from the solution of the plane stress problem for a thin plane body with variable thickness. The governing equations of this problem can be readily obtained by considering the equilibrium of the plate element.

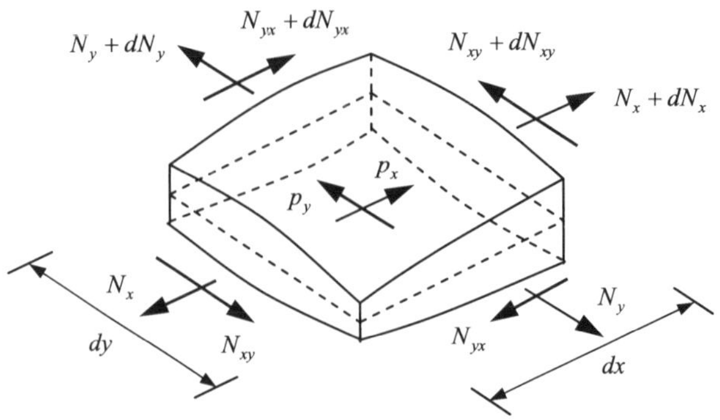

FIGURE 3.38 Plate element with variable thickness under membrane forces.

Thus, referring to Fig. 3.38 we have

$$N_{x,x}+N_{yx,y}+p_x=0 \tag{3.133a}$$

$$N_{xy,x}+N_{y,y}+p_y=0 \tag{3.133b}$$

Introducing Eqs. (3.8) in Eqs. (3.133) and taking into account that the membrane stiffness is variable, $C=C(x,y)=Eh(x,y)/(1-\nu^2)$, we obtain the following Navier-type equations for the plane stress problem:

$$\begin{aligned}&\mu h\nabla^2 u+(\lambda+\mu)h\left(u_{,xx}+v_{,xy}\right)\\&+h_{,x}\left[\lambda\left(u_{,x}+v_{,y}\right)+2\mu u_{,x}\right]+h_{,y}\mu\left(u_{,y}+v_{,x}\right)+\frac{p_x}{Gh}=0\end{aligned} \quad \text{in } \Omega \tag{3.134a}$$

$$\begin{aligned}&\mu h\nabla^2 v+(\lambda+\mu)h\left(u_{,xy}+v_{,yy}\right)\\&+h_{,x}\mu\left(u_{,y}+v_{,x}\right)+h_{,y}\left[\lambda\left(u_{,x}+v_{,y}\right)+2\mu v_{,y}\right]+\frac{p_y}{Gh}=0\end{aligned} \quad \text{in } \Omega \tag{3.134b}$$

with the boundary conditions

$$u_n=u_n^* \quad \text{or} \quad N_n=N_n^* \quad \text{on } \Gamma \tag{3.135a}$$

$$u_t=u_t^* \quad \text{or} \quad N_t=N_t^* \quad \text{on } \Gamma \tag{3.135b}$$

where $\lambda=E\nu/(1-\nu^2)$ and $\mu=E/2(1+\nu)$ are the Lamé constants.

Both boundary value problems are solved using the AEM as described in Sections 3.3.2 and 3.6.2. The plane stress problem is first solved to evaluate the membrane stress resultants N_x, N_{xy}, N_y.

3.6.4 Large deflections of plates with variable thickness

The differential equation for bending is of the same form with (3.131). The only difference is that the membrane forces N_x, N_y, N_{xy} are now given by (3.103) and are unknown. Thus, we have the following two boundary value problems:

(i) For the transverse displacement

$$\begin{aligned} &D\nabla^4 w + 2D_{,x}\nabla^2 w_{,x} + 2D_{,y}\nabla^2 w_{,y} + \nabla^2 D\nabla^2 w \\ &\quad -(1-\nu)\left(D_{,xx}w_{,yy} - 2D_{,xy}w_{,xy} + D_{,yy}w_{,xx}\right) \\ &\quad -\left(N_x w_{,xx} + 2N_{xy}w_{,xy} + N_y w_{,yy}\right) + p_x w_{,x} + p_y w_{,y} = f \end{aligned} \quad \text{in } \Omega \quad (3.136)$$

with boundary conditions

$$\hat{V}w + N_n w_{,n} + N_t w_{,t} + k_T w = V_n^* \quad \text{or} \quad w = w^* \quad \text{on } \Gamma \qquad (3.137\text{a})$$

$$Mw - k_R w_{,n} = M_n^* \quad \text{or} \quad w_{,n} = w_{,n}^* \quad \text{on } \Gamma \qquad (3.137\text{b})$$

$$k_c^{(k)} w^{(k)} - [\![Tw]\!]_k = R_k^* \quad \text{or} \quad w^{(k)} = w_k^* \quad \text{at corner k} \qquad (3.137\text{c})$$

(ii) For the membrane displacements
Introducing Eqs. (3.103) in Eqs. (3.133) we obtain the equations for the plane stress problem [71]:

$$\begin{aligned} &\left\{C\left[\left(u_{,x} + \frac{1}{2}w_{,x}^2\right) + \nu\left(v_{,y} + \frac{1}{2}w_{,y}^2\right)\right]\right\}_{,x} \\ &\quad + \left\{C\frac{(1-\nu)}{2}\left(u_{,y} + v_{,x} + w_{,x}w_{,y}\right)\right\}_{,y} = 0 \end{aligned} \quad \text{in } \Omega \qquad (3.138\text{a})$$

$$\begin{aligned} &\left\{C\frac{(1-\nu)}{2}\left(u_{,y} + v_{,x} + w_{,x}w_{,y}\right)\right\}_{,x} \\ &\quad + \left\{C\left[\left(v_{,y} + \frac{1}{2}w_{,y}^2\right) + \nu\left(u_{,x} + \frac{1}{2}w_{,x}^2\right)\right]\right\}_{,y} = 0 \end{aligned} \quad \text{in } \Omega \qquad (3.138\text{b})$$

with boundary conditions

$$u_n = u_n^* \quad \text{or} \quad N_n = N_n^* \quad \text{on } \Gamma \qquad (3.139\text{a})$$

$$u_t = u_t^* \quad \text{or} \quad N_t = N_t^* \quad \text{on } \Gamma \qquad (3.139\text{b})$$

The differentiations in Eqs. (3.138) have not been performed for the simplicity of the expressions. Eqs. (3.136) and (3.138) are nonlinear and coupled. The application of the AEM results in the system of nonlinear equations for the evaluation of the fictitious loads $b(\mathbf{x})$, $b^{(1)}(\mathbf{x})$, $b^{(2)}(\mathbf{x})$ (see Section 3.5.4).

3.6.5 Examples

EXAMPLE 3.9

In this example, a square simply supported plate with side length $a = 5\text{m}$ $(0 \le x,\ y \le 5)$, linear thickness variation law $h = 0.005(x + y) + 0.10$ and elastic constants $E = 3 \times 10^7\,\text{kN/m}^2$, $\nu = 0.3$ is analyzed. The plate is subjected to the

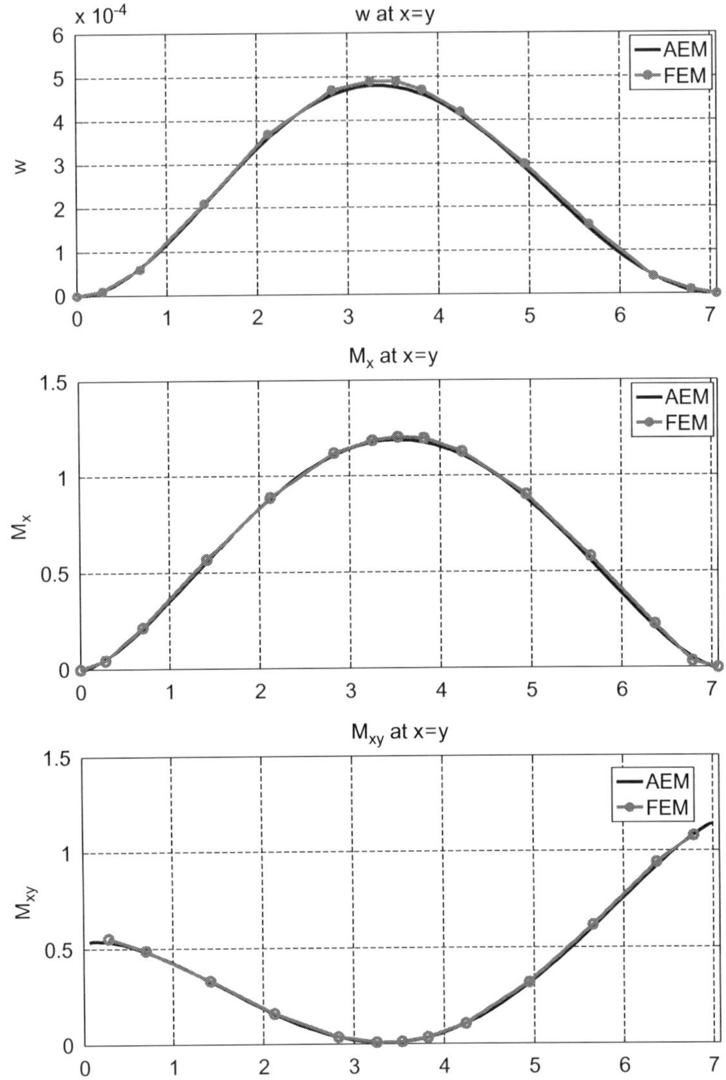

FIGURE 3.39 Variation of w, M_x and M_{xy} along the line $x = y$ in Example 3.9.

uniform transverse load $f = 1\text{kN/m}^2$. The results were obtained with $N = 200$ boundary elements and $M = 181$ domain nodal points resulting from 300 linear triangular elements. Figure 3.39 shows the variation of w, M_x and M_{xy} along the line $x = y$ as compared with a FEM solution obtained with 2500 square elements.

EXAMPLE 3.10

The buckling of the rectangular plate ($0 \le x \le a$, $-b/2 \le y \le b/2$) shown in Fig. 3.40 is investigated under the uniform compressive load N_n^* acting along the edges $x = 0, a$. All edges are simply supported and movable in the plane of the plate. The thickness varies according to the exponential law

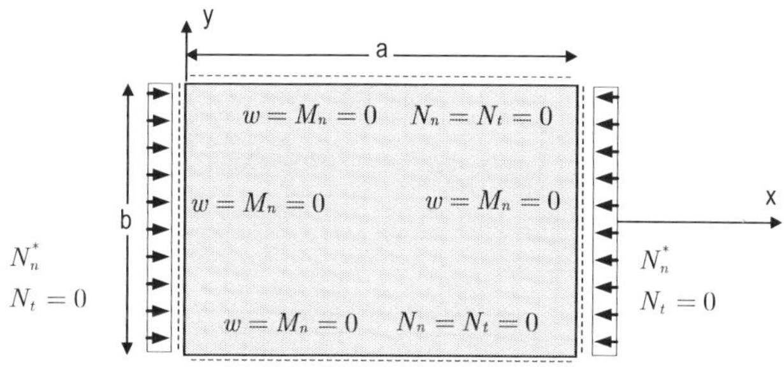

FIGURE 3.40 Rectangular plate under membrane load in Example 3.10.

$$h(x) = h_0 \exp\left(\frac{x}{a} \ln\frac{h_a}{h_0}\right) \tag{3.140}$$

The results were obtained with $N = 250$ boundary elements and $M = 237$ domain nodal points resulting from 404 linear triangular elements. Table 3.6 gives the values of the buckling parameter $k_M = N_n^* b^2/\pi^2 D_M$, where $D_M = Eh_M^3/12(1-\nu^2)$ and $h_M = h(a/2)$, for various values of the aspect ratio a/b and thickness ratio h_a/h_0. Figure 3.41 shows the first buckling mode for $a/b = 1$ and $a/b = 2$ with $h_a/h_0 = 2.0$ in both cases. It is worth noting how the ratio a/b affects the shape of the buckling mode.

EXAMPLE 3.11

In this example the following two problems are solved:

1. The minimization of the compliance $C = \int_\Omega wf\,d\Omega$ for a given load (or the maximization of the bending stiffness) of the simply supported square plate of Fig. 3.42a.
2. The maximization of the buckling load of the square clamped plate with inplane movable edges of Fig. 3.42b.

TABLE 3.6 Values of the buckling parameter k_M for the plate in Example 3.10 ($\nu = 1/3$). Upper row: Ref [72]; Middle row: Ref. [73]; Lower row: AEM

	h_a/h_0				
a/b	1.125	1.250	1.500	1.750	2.00
0.5	6.450	7.339	6.670	6.995	7.455
	6.232	6.187	6.049	5.885	5.722
	6.275	6.220	6.052	5.845	5.623
1.00	4.073	4.000	3.950	3.843	3.686
	3.972	3.901	3.700	3.476	3.262
	3.992	3.918	3.709	3.476	3.253
1.50	4.184	3.949	3.541	3.206	2.933
	4.153	3.884	3.407	3.024	2.715
	4.213	3.934	3.440	3.043	2.722
2.00	3.897	3.674	3.224	2.850	2.552
	3.866	3.616	3.122	2.720	2.401
	3.935	3.679	3.169	2.752	2.420

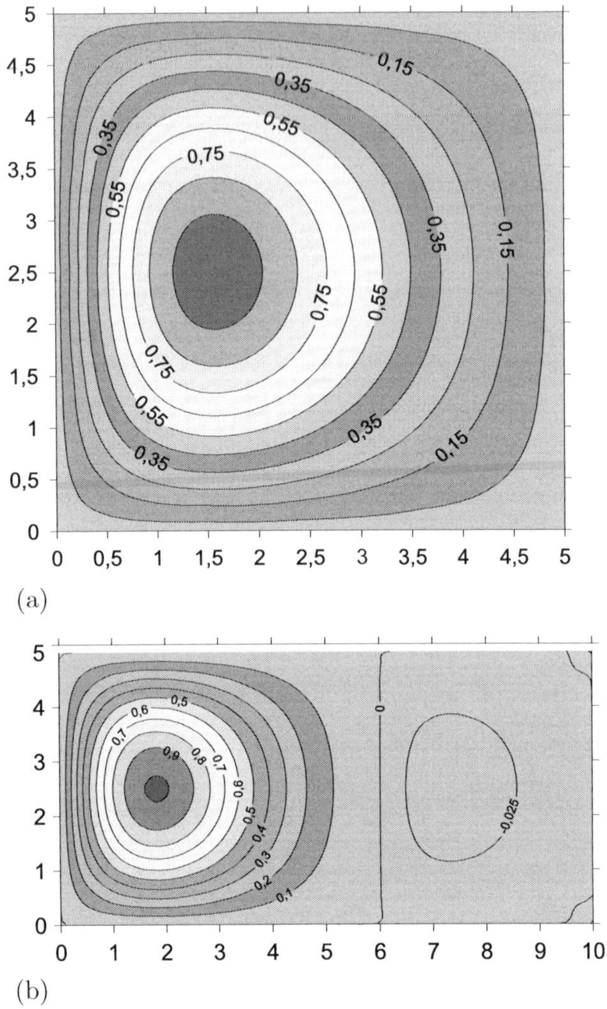

FIGURE 3.41 First buckling mode for (a) $a/b = 1$ and (b) $a/b = 2$ of the plate in Example 3.10 ($h_a/h_0 = 2.0$).

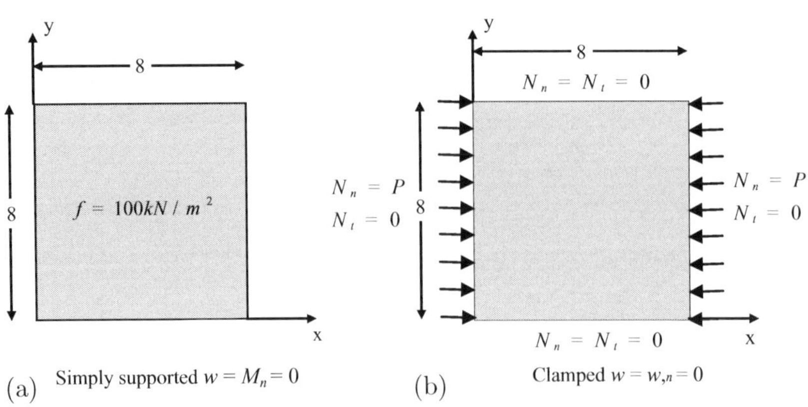

FIGURE 3.42 Square plate in Example 3.11. (a) For minimum compliance and (b) for maximum buckling load.

In both cases the volume of the plate remains constant. The employed data are $E = 3 \times 10^7\,\mathrm{kN/m^2}$ and $\nu = 0.3$. The plate has an initial thickness $h = 0.1$ with allowable thickness bounds $0.02 < h < 0.2$ and constraint parameter $\varepsilon = 0.0002$ (see Eq. (3.125)). The results were obtained with $N = 200$ boundary elements and $M = 185$ domain nodal points resulting from 304 linear triangular elements. Figure 3.43 shows the optimum thickness distribution for minimum compliance when $f = 100\mathrm{kN/m^2}$, while Fig. 3.44 shows the optimum thickness distribution for maximum buckling load. The initial compliance is $0.160\mathrm{kNm}$ and the optimum $0.103\mathrm{kNm}$, i.e., a 35.6% reduction was achieved. The corresponding values of the buckling load are $4404.1\ \mathrm{kN/m}$ (initial value) and $7144.2\ \mathrm{kN/m}$ optimum value, i.e., an increase of 62%.

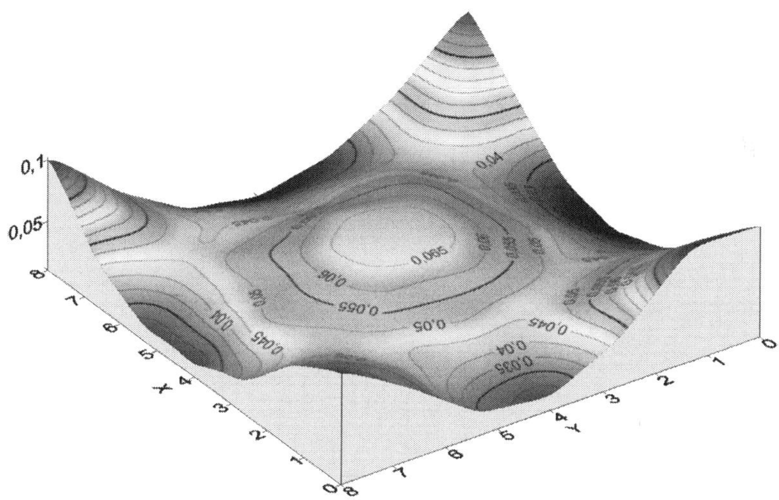

FIGURE. 3.43 Plate thickness distribution for minimum compliance in Example 3.11.

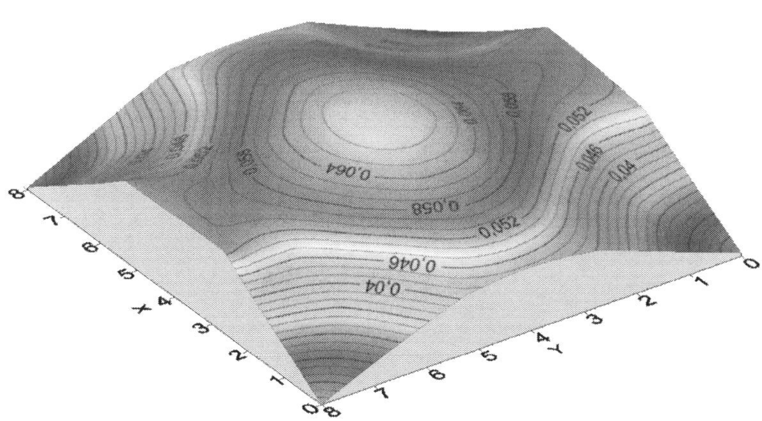

FIGURE 3.44 Plate thickness distribution for maximum buckling load in Example 3.11.

EXAMPLE 3.12

The optimum thickness distribution of the rectangular plate of Fig. 3.45 is studied for maximum stiffness [64]. The plate is subjected to a transverse uniform load $f=1\ \mathrm{kN/m^2}$ and is simply supported along the edges with $x=0$ and $x=10$, while the other two edges are free. Moreover, there are two point supports at points A(5,1) and B(5,3). The employed elastic constants are $E=21\times10^6\ \mathrm{kN/m^2}$ and $\nu=0.3$. The results were obtained using $N=560$ boundary elements and $M=268$ domain nodal points resulting from 426 linear triangular elements. The initial thickness was $h_0=0.2$ m with allowable thickness bounds $0.1<h<0.3$ and constraint parameter $\varepsilon=5\times10^{-3}$ (see Eq. 3.125). The initial compliance is 5.36×10^{-3} kNm. Figure 3.46 presents the optimum thickness distribution for maximum stiffness. This distribution results in a minimum compliance of 3.98×10^{-3}, which corresponds to an increase of 26% in stiffness. It is observed that the thickness approaches the upper bound in the central region and approaches the lower bound in the region of the supported edges.

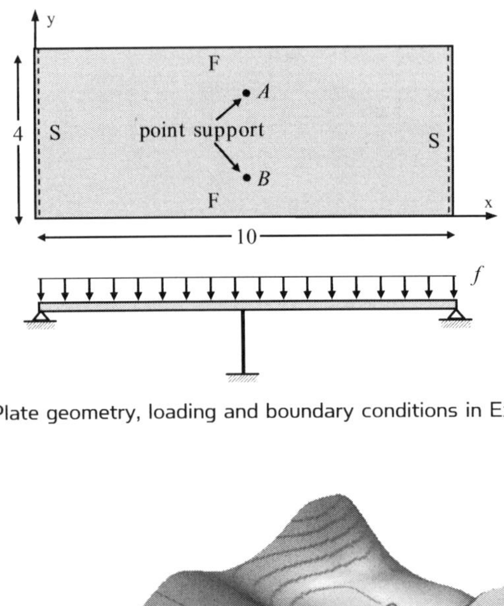

FIGURE 3.45 Plate geometry, loading and boundary conditions in Example 3.12.

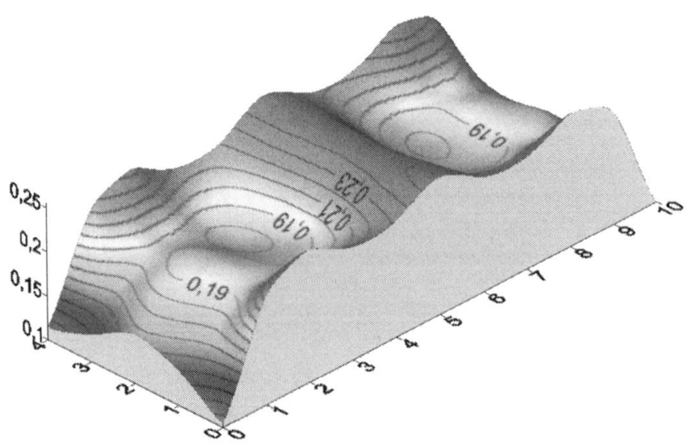

FIGURE 3.46 Optimum thickness distribution for maximum stiffness in Example 3.12.

EXAMPLE 3.13

In this example the nonlinear response of a circular plate with radius a and axisymmetric thickness variation law $h = h_0 \exp\left[\beta(r/a)^2\right], 0 \le r \le a$, under a uniform load f is studied. The central deflection of the plate is presented in Table 3.7 as compared with that obtained from a one-dimensional solution using the AEM [69] and two approximate solutions [74,75]. Two types of boundary conditions were considered: (i) $w(a)=0,\ w_{,r}(a)=0,\ u_r(a)=0$ (fixed with immovable boundary) and (ii) $w(a)=0\ M_r(a)=0,\ N_r(a)=0$ (simply supported with movable boundary).

3.7 THICK PLATES

3.7.1 Two-dimensional plate theories

As stated in Section 2.2.1, the plate is a cylindrical or prismatic elastic body whose thickness (height) is small compared with the dimensions of its base. The external load may be applied as tractions on the upper and lower surface or as a body force. The produced deformation and stress are evidently three-dimensional. A number of plate theories have been developed. All of them attempt to describe the three-dimensional state of deformation and stress of the three-dimensional prismatic body using two-dimensional quantities, which are defined at the middle surface of the plate. Hence, all theories are, by their nature, approximate. In general, the two-dimensional plate theories can be categorized as two types: (1) the classical plate theory, in which the transverse shear deformation effects are neglected, also known as Kirchhoff's plate theory,

TABLE 3.7 Deflection w/h_0 at the center of the circular plate in Example 3.13

β		Clamped immovable boundary			Simply supported movable boundary		
		$\bar{f}=fa^4/Eh_0^4$			$\bar{f}=fa^4/Eh_0^4$		
		1	3	5	5	7	10
−0.2	AEM	0.2228	0.5728	0.8126	1.990	2.317	2.668
	AEM [69]	0.2209	0.5641	0.7973	1.951	2.284	2.668
	Ref. [75]	0.2195	0.5564	0.7847			
0	AEM	0.1680	0.4594	0.6815	1.809	2.123	2.476
	AEM [69]	0.1680	0.4588	0.6796	1.800	2.121	2.492
	Ref. [74]	0.1680	0.4588	0.6808	1.800	2.123	2.495
0.2	AEM	0.1258	0.3585	0.5537	1.616	1.916	2.261
	AEM [69]	0.1265	0.3608	0.5576	1.650	1.961	2.319
	Ref. [75]	0.1268	0.3634	0.5648			
0.4	AEM	0.0942	0.2751	0.4380	1.442	1.724	2.050
	AEM [69]	0.0950	0.2777	0.4430	1.499	1.800	2.148
	Ref. [75]	0.0966	0.2838	0.4555			

already discussed in Section 2.2, and (2) shear deformation plate theories, also known as thick plate theories. There are several possibilities to derive plate theories. The usual approach to derive plate theories is to expand the displacements in power series through the plate thickness. The number of the terms in the truncated series, which approximates the displacement field, defines the order of the plate theory [76]. Usually, the truncation of the series terms is supplemented with a priori assumptions based on engineering intuition. The need to develop higher-order plate theories resulted from the observation that the Kirchhoff theory gives displacements, which become considerably smaller than the actual ones with the increase of the thickness. This evidently occurs, because the Kirchhoff plate theory neglects the transverse shear deformation.

Among the first plate theories that coped with this problem adequately were the thick plate theories of Reissner [77] and Mindlin [78]. Both theories, though they consider the shear deformation effect, continue to adopt linear variation of the strains through the thickness, which implies that plane cross-sections remain plane after deformation. This is an inconsistency of the theory, since the shear effect deforms the cross-section. For this reason correction coefficients are introduced to overcome the error due to this contradiction.

Consistent plate theories are presented systematically by Kienzler [79]. The complexity of the differential equations increases with increasing order of the plate theory. The BEM has been applied to thick plates using different formulations. There is extensive literature on the subject, to which the interested reader can refer to [80–82]. Nevertheless, the problem of establishing the fundamental solution and the complexity of the kernels of the boundary integral equations dominate these formulations. Therefore, the use of D/BEMs seems to offer efficient solution approaches. The AEM overcomes any difficulties and simplifies the solution highly as it reduces the thick plate equations to two-dimensional Poisson's equations, which are readily treated by the direct BEM. Here, the AEM is applied to solve the equations of thick plates resulting from the Mindlin plate theory [83]. From the description of the solution procedure, it becomes obvious that the AEM can be employed to analyze shear deformable plates of order higher than the first.

3.7.2 Mindlin theory

The Mindlin theory is a first-order shear deformation plate theory, which gives good results for moderately thick plates, i.e., when $h/a < 0.2$ with a being the smallest of the other two dimensions. The displacement field is expressed by the relations

$$u(x, y, z) = z\phi_x(x, y), \tag{3.141a}$$

$$v(x, y, z) = z\phi_y(x, y), \tag{3.141b}$$

$$w(x, y, z) = w(x, y), \tag{3.141c}$$

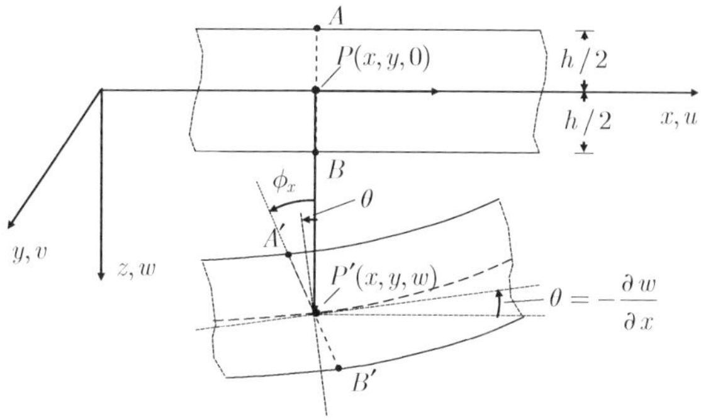

FIGURE 3.47 Section of the plate by the xz plane before and after deformation in the Mindlin theory.

where u, v, w are the displacement components in the x, y, z axes, respectively, while the functions ϕ_x, $-\phi_y$ represent the rotations about the y and x axes, respectively (see Fig. 3.47).

The strain components result from Eqs. (3.141) as

$$\bar{\varepsilon}_x = z\frac{\partial \phi_x}{\partial x} \tag{3.142a}$$

$$\bar{\varepsilon}_y = z\frac{\partial \phi_y}{\partial y} \tag{3.142b}$$

$$\bar{\varepsilon}_z = 0 \tag{3.142c}$$

$$\bar{\gamma}_{xy} = z\left(\frac{\partial \phi_x}{\partial y} + \frac{\partial \phi_y}{\partial x}\right) \tag{3.142d}$$

$$\bar{\gamma}_{xz} = \phi_x + \frac{\partial w}{\partial x} \tag{3.142e}$$

$$\bar{\gamma}_{yz} = \phi_y + \frac{\partial w}{\partial y} \tag{3.142f}$$

Introducing Eqs. (3.142) in the constitutive equations of the plane stress problem (see [2], Section 7.2.2) we obtain

$$\bar{\sigma}_x = \frac{E}{1-\nu^2} z\left(\frac{\partial \phi_x}{\partial x} + \nu\frac{\partial \phi_y}{\partial y}\right) \tag{3.143a}$$

$$\bar{\sigma}_y = \frac{E}{1-\nu^2} z\left(\frac{\partial \phi_y}{\partial y} + \nu\frac{\partial \phi_x}{\partial x}\right) \tag{3.143b}$$

$$\bar{\sigma}_z = 0 \tag{3.143c}$$

$$\bar{\tau}_{xy} = Gz\left(\frac{\partial \phi_x}{\partial y} + \frac{\partial \phi_y}{\partial x}\right) \tag{3.143d}$$

$$\overline{\tau}_{xz} = G\left(\phi_x + \frac{\partial w}{\partial x}\right) \tag{3.143e}$$

$$\overline{\tau}_{yz} = G\left(\phi_y + \frac{\partial w}{\partial y}\right) \tag{3.143f}$$

Equations (3.142e) and (3.143f) express an inconsistency of the Mindlin theory. Obviously, we observe that Eqs. (3.143e,f) give constant distribution of the shear stresses through the thickness. This is in contradiction with the parabolic variation in thin plate theory. Moreover, the condition that the shear stresses must vanish at the upper and lower surfaces is also violated.

A remedy to cope with this inconsistency is to modify the transverse shear stiffnesses of the plate by introducing a *shear correction factor* K_s, which should multiply the shear stresses in Eqs. (3.143e,f). The factor K_s is computed such that the strain energy due to the transverse shear stresses in the first-order shear plate theory equals the strain energy due to the transverse shear stresses predicted by the three-dimensional elasticity theory or its equivalent [76]. For $\nu = 0.3$ it is $K_s = 0.86$. A more accurate estimation is obtained from the relation $K_s = 20(1+\nu)/(24+25\nu+\nu^2)$, which for $\nu = 0.3$ gives $K_s = 0.823$. The latter value is closer to the second Mindlin's estimation $K_s = \pi^2/12 = 0.822$, which is based on the velocity of the shear wave.

The constitutive plate equations result from Eqs. (2.15) using the expressions (3.143) and the correction factor for the transverse shear forces. Thus, we have

$$M_x = D\left(\frac{\partial \phi_x}{\partial x} + \nu\frac{\partial \phi_y}{\partial y}\right) \tag{3.144a}$$

$$M_y = D\left(\frac{\partial \phi_y}{\partial y} + \nu\frac{\partial \phi_x}{\partial x}\right) \tag{3.144b}$$

$$M_{xy} = -M_{yx} = -\frac{D(1-\nu)}{2}\left(\frac{\partial \phi_x}{\partial y} + \frac{\partial \phi_y}{\partial x}\right) \tag{3.144c}$$

$$Q_x = K_s Gh\left(\phi_x + \frac{\partial w}{\partial x}\right) \tag{3.144d}$$

$$Q_y = K_s Gh\left(\phi_y + \frac{\partial w}{\partial y}\right) \tag{3.144e}$$

Note that the minus sign in Eq. (3.144c) is due to the definition of M_{xy} adopted in this book (see Fig. 2.4 and Fig. 2.5).

The equilibrium of the plate requires that the first variation of the total potential, $\delta\Pi$, vanishes:

$$\delta\Pi = \delta U + \delta V = 0 \tag{3.145}$$

where U is strain energy due to bending and transverse shear deformation given by

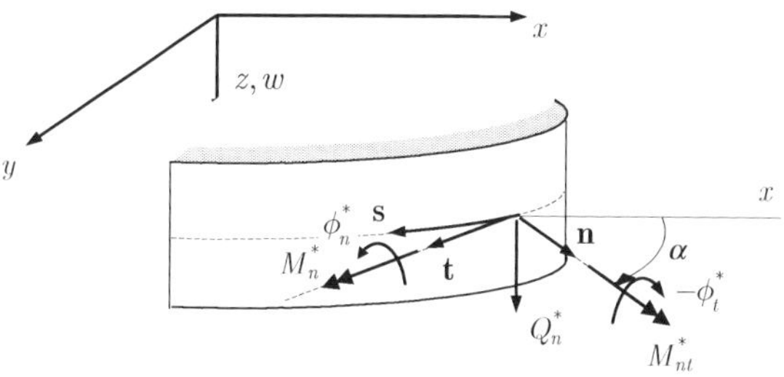

FIGURE 3.48 Boundary forces and displacements.

$$U=\frac{1}{2}\int_V\left(\overline{\sigma}_x\overline{\varepsilon}_x+\overline{\sigma}_y\overline{\varepsilon}_y+\overline{\tau}_{xy}\overline{\gamma}_{xy}+\overline{\tau}_{xz}\overline{\gamma}_{xz}+\overline{\tau}_{yz}\overline{\gamma}_{yz}\right)dV \tag{3.146a}$$

and V is the potential of the external forces due to the lateral load $f(x, y)$ and the boundary forces M_n^*, M_{nt}^*, Q_n^* given by

$$V=-\int_\Omega fwd\Omega-\int_\Gamma\left(Q_n^* w+M_n^*\phi_n-M_{nt}^*\phi_t\right)ds \tag{3.146b}$$

where $\phi_n, -\phi_t$ represent the rotations about the t and n axes, respectively, on the boundary surface with normal vector $\mathbf{n}(n_x, n_y)$, Fig. 3.48.

Introducing Eqs. (3.146) in (3.145), using Eqs. (3.142) and (3.143), taking into account that(*)

$$\begin{Bmatrix}-\phi_y\\ \phi_x\end{Bmatrix}=\begin{bmatrix}n_x & -n_y\\ n_y & n_x\end{bmatrix}\begin{Bmatrix}-\phi_t\\ \phi_n\end{Bmatrix} \tag{3.147}$$

and

$$Q_n=Q_x n_x+Q_y n_y \tag{3.148}$$

and working as in Section 2.2.5 we obtain

$$\begin{aligned}&-\int_\Omega\left[\left(\frac{\partial M_x}{\partial x}-\frac{\partial M_{xy}}{\partial y}-Q_x\right)\delta\phi_x+\left(\frac{\partial M_{yx}}{\partial x}+\frac{\partial M_y}{\partial y}-Q_y\right)\delta\phi_y\right.\\&\qquad\left.+\left(\frac{\partial Q_x}{\partial x}+\frac{\partial Q_y}{\partial y}+f_y\right)\delta w\right]d\Omega\\&\quad+\int_\Gamma\left[(M_n-M_n^*)\delta\phi_n+(M_{nt}-M_{nt}^*)\delta\phi_t+(Q_n-Q_n^*)\delta w\right]ds=0\end{aligned} \tag{3.149}$$

(*)Eq. (3.147) expresses the transformation of the rotation vector from the axes (n, t) to (x, y). This is permitted, because the small rotations transform as vectors.

which yields the equilibrium equations

$$\frac{\partial M_x}{\partial x} - \frac{\partial M_{xy}}{\partial y} - Q_x = 0 \tag{3.150a}$$

$$\frac{\partial M_{yx}}{\partial x} + \frac{\partial M_y}{\partial y} - Q_y = 0 \tag{3.150b}$$

$$\frac{\partial Q_x}{\partial x} + \frac{\partial Q_y}{\partial y} + f = 0 \tag{3.150c}$$

with the boundary conditions

$$Q_n = Q_n^* \quad \text{or} \quad w = w^* \tag{3.151a}$$

$$M_n = M_n^* \quad \text{or} \quad \phi_n = \phi_n^* \tag{3.151b}$$

$$M_{nt} = M_{nt}^* \quad \text{or} \quad \phi_t = \phi_t^* \tag{3.151c}$$

in which

$$M_n = D\left(\frac{\partial \phi_n}{\partial n} + \nu \frac{\partial \phi_t}{\partial t}\right) \tag{3.152a}$$

$$M_{nt} = -\frac{D(1-\nu)}{2}\left(\frac{\partial \phi_n}{\partial t} + \frac{\partial \phi_t}{\partial n}\right). \tag{3.152b}$$

$$Q_n = K_s Gh\left(\phi_n + \frac{\partial w}{\partial n}\right) \tag{3.152c}$$

In deriving Eqs. (3.150) the integrals $\int_{-h/2}^{h/2} \begin{Bmatrix} \tau_{xz} \\ \tau_{yz} \end{Bmatrix} dz$ have been replaced with

$$\begin{Bmatrix} Q_x \\ Q_y \end{Bmatrix} = K_s \int_{-h/2}^{h/2} \begin{Bmatrix} \tau_{xz} \\ \tau_{yz} \end{Bmatrix} dz \tag{3.153}$$

which represent the corrected transverse shear forces.

Finally, using Eqs. (3.144) in Eqs. (3.150) we obtain the equilibrium equations of the plate in terms of the functions (w, ϕ_x, ϕ_y)

$$-K_s Gh\left(\nabla^2 w + \frac{\partial \phi_x}{\partial x} + \frac{\partial \phi_y}{\partial y}\right) = f(x, y), \tag{3.154a}$$

$$-\frac{D}{2}(1-\nu)\nabla^2 \phi_x - \frac{D}{2}(1+\nu)\left(\frac{\partial^2 \phi_x}{\partial x^2} + \frac{\partial^2 \phi_y}{\partial x \partial y}\right) + K_s Gh\left(\frac{\partial w}{\partial x} + \phi_x\right) = 0 \tag{3.154b}$$

$$-\frac{D}{2}(1-\nu)\nabla^2\phi_y-\frac{D}{2}(1+\nu)\left(\frac{\partial^2\phi_x}{\partial x\partial y}+\frac{\partial^2\phi_y}{\partial y^2}\right) \\ +K_sGh\left(\frac{\partial w}{\partial y}+\phi_y\right)=0 \tag{3.154c}$$

with the boundary conditions (3.151).

We observe that the boundary conditions in this case are three. Apparently, their implementation requires that one element of each of the pairs (w^*, Q_n^*), (ϕ_n^*, M_n^*), (ϕ_t^*, M_{nt}^*) be specified in order that the mathematical problem is well posed. Thus, we may distinguish the following typical boundary conditions for an edge:

Clamped: $w=0,\ \phi_t=0,\ \phi_n=0$

Simply supported of type I: $w=0,\ M_{nt}=0,\ M_n=0$

Simply supported of type II: $w=0,\ \phi_t=0,\ M_n=0$

Free: $Q_n=0,\ M_{nt}=0,\ M_n=0$

Note that there are two types of simply supported edge in the first-order shear deformation plate theory. The boundary conditions may be nonhomogeneous if the quantities w^*, $Q_n^*\ldots$ are nonzero.

The boundary conditions (3.151) can be written in the following general form, which may include also elastic supports:

$$\alpha_1 w+\alpha_2 Q_n=\alpha_3, \tag{3.155a}$$

$$\beta_1\phi_n+\beta_2 M_n=\beta_3, \tag{3.155b}$$

$$\gamma_1\phi_t+\gamma_2 M_{nt}=\gamma_3, \tag{3.155c}$$

Apparently, the typical boundary conditions result for appropriate values of $\alpha_i, \beta_i, \gamma_i$.

3.7.3 AEM solution

Equations (3.154) are of the second order. Hence, the Laplace operator can give three uncoupled analog equations:

$$\nabla^2 w=b^{(1)}(\mathbf{x}) \tag{3.156a}$$

$$\nabla^2\phi_x=b^{(2)}(\mathbf{x}) \tag{3.156b}$$

$$\nabla^2\phi_y=b^{(3)}(\mathbf{x}) \tag{3.156c}$$

where $b^{(i)}(\mathbf{x})(i=1,2,3)$ are the three fictitious sources, unknown in the first instance.

Equations (3.156) can be readily treated by the BEM. Their solution is given in integral form (see [2], Section 3.4):

$$\varepsilon w(\mathbf{x})=\int_\Omega v^* b^{(1)}\,d\Omega-\int_\Gamma (v^* w_{,n}-v^*_{,n}w)\,ds \quad \mathbf{x}\in\Omega\cup\Gamma \tag{3.157a}$$

$$\varepsilon\phi_x(\mathbf{x}) = \int_\Omega v^* b^{(2)}\, d\Omega - \int_\Gamma \left(v^*\phi_{x,n} - v^*_{,n}\phi_x\right) ds \quad \mathbf{x}\in\Omega\cup\Gamma \tag{3.157b}$$

$$\varepsilon\phi_y(\mathbf{x}) = \int_\Omega v^* b^{(3)}\, d\Omega - \int_\Gamma \left(v^*\phi_{y,n} - v^*_{,n}\phi_y\right) ds \quad \mathbf{x}\in\Omega\cup\Gamma \tag{3.157c}$$

where $v^* = \ln r/2\pi$ is the fundamental solution of Eqs. (3.156), $r = |\mathbf{y}-\mathbf{x}|$ $\mathbf{x}\in\Omega\cup\Gamma$ and $\mathbf{y}\in\Gamma$; ε is the free term coefficient, which takes the values $\varepsilon=1$ if $\mathbf{x}\in\Omega$, $\varepsilon=1/2$ if $\mathbf{x}\in\Gamma$ (for smooth boundary), and $\varepsilon=0$ if $\mathbf{x}\notin\Omega\cup\Gamma$.

Using the BEM with N constant boundary elements and linear triangular domain elements, Eqs. (3.157) give the following system of algebraic equations for the boundary quantities w, $w_{,n}$, ϕ_x, $\phi_{x,n}$, ϕ_y, $\phi_{y,n}$ at the N boundary nodal points and fictitious sources $b^{(k)}$ at the M domain nodal points:

$$\mathbf{H}\begin{Bmatrix}\mathbf{w}\\ \boldsymbol{\phi}_x\\ \boldsymbol{\phi}_y\end{Bmatrix} = \mathbf{G}\begin{Bmatrix}\mathbf{w}_{,n}\\ \boldsymbol{\phi}_{x,n}\\ \boldsymbol{\phi}_{y,n}\end{Bmatrix} + \mathbf{D}\begin{Bmatrix}\mathbf{b}^{(1)}\\ \mathbf{b}^{(2)}\\ \mathbf{b}^{(3)}\end{Bmatrix} \tag{3.158}$$

where

$$\mathbf{H} = \begin{bmatrix} -\left[v^*_n\right]^{ij} + \frac{1}{2}\delta_{ij} & \mathbf{0} & \mathbf{0} \\ \mathbf{0} & -\left[v^*_n\right]^{ij} + \frac{1}{2}\delta_{ij} & \\ \mathbf{0} & \mathbf{0} & -\left[v^*_n\right]^{ij} + \frac{1}{2}\delta_{ij} \end{bmatrix} \tag{3.159a}$$

$$\mathbf{G} = \begin{bmatrix} -\left[v^*\right]^{ij} & \mathbf{0} & \mathbf{0} \\ \mathbf{0} & -\left[v^*\right]^{ij} & \mathbf{0} \\ \mathbf{0} & \mathbf{0} & -\left[v^*\right]^{ij} \end{bmatrix} \tag{3.159b}$$

$$\mathbf{D} = \begin{bmatrix} \left[D^{ik}\right] & \mathbf{0} & \mathbf{0} \\ \mathbf{0} & \left[D^{ik}\right] & \mathbf{0} \\ \mathbf{0} & \mathbf{0} & \left[D^{ik}\right] \end{bmatrix} \tag{3.159c}$$

The matrices $[v_n]^{ij}$, $\left[v^*_n\right]^{ij}$, and $\left[D^{ik}\right]$ are defined in Section 3.3.2.3.

Applying the boundary conditions (3.155) at the N boundary nodal points gives

$$\alpha_1\mathbf{w} + \alpha_2 K_s Gh(\boldsymbol{\phi}_n + \mathbf{w}_{,n}) = \alpha_3 \tag{3.160a}$$

$$\beta_1\boldsymbol{\phi}_n + \beta_2 D(\boldsymbol{\phi}_{n,n} + \nu\boldsymbol{\phi}_{t,t}) = \beta_3 \tag{3.160b}$$

$$\gamma_1\boldsymbol{\phi}_t - \gamma_2\frac{D(1-\nu)}{2}(\boldsymbol{\phi}_{n,t} + \boldsymbol{\phi}_{t,n}) = \gamma_3, \tag{3.160c}$$

The boundary quantities ϕ_n, ϕ_t, $\phi_{n,n}$, $\phi_{t,n}$, $\phi_{n,t}$ and $\phi_{t,t}$ need to be expressed in terms of ϕ_x, ϕ_y, $\phi_{x,n}$, $\phi_{y,n}$ This is achieved as follows:

From Eqs. (3.147) we have

$$\phi_n = \mathbf{n}_x \phi_x + \mathbf{n}_y \phi_y \tag{3.161a}$$

$$\phi_t = -\mathbf{n}_y \phi_x + \mathbf{n}_x \phi_y \tag{3.161b}$$

Moreover, differentiation of Eq. (3.61a) with respect to n gives

$$\phi_{n,n} = \mathbf{n}_x \phi_{x,n} + \mathbf{n}_y \phi_{y,n} \tag{3.162a}$$

The tangential derivatives $\phi_{n,t}$ and $\phi_{t,t}$ at the boundary nodal points can be expressed in terms of the values of ϕ_n and ϕ_t with finite differences using the matrix $\mathbf{d}$ defined in Section 3.3.2.1. Thus, we have

$$\begin{aligned}\phi_{n,t} &= \mathbf{d}\phi_n \\ &= \mathbf{d}(\mathbf{n}_x \phi_x + \mathbf{n}_y \phi_y)\end{aligned} \tag{3.163a}$$

$$\begin{aligned}\phi_{t,t} &= \mathbf{d}\phi_t \\ &= \mathbf{d}(-\mathbf{n}_y \phi_x + \mathbf{n}_x \phi_y)\end{aligned} \tag{3.163b}$$

Equations (3.160) by virtue of Eqs. (3.161) and (3.163) become

$$\overline{\mathbf{H}} \begin{Bmatrix} \mathbf{w} \\ \phi_x \\ \phi_y \end{Bmatrix} + \overline{\mathbf{G}} \begin{Bmatrix} \mathbf{w}_{,n} \\ \phi_{x,n} \\ \phi_{y,n} \end{Bmatrix} = \overline{\mathbf{F}} \tag{3.164}$$

where

$$\overline{\mathbf{H}} = \begin{bmatrix} \alpha_1 & K_s G h \alpha_2 \mathbf{n}_x & K_s G h \alpha_2 \mathbf{n}_y \\ \mathbf{0} & \beta_1 \mathbf{n}_x - D\nu\beta_1 \mathbf{d}\mathbf{n}_y & \beta_1 \mathbf{n}_y - D\nu\beta_2 \mathbf{d}\mathbf{n}_x \\ \mathbf{0} & -\left(\gamma_1 \mathbf{n}_y + \gamma_2 \dfrac{D(1-\nu)}{2} \mathbf{d}\mathbf{n}_x\right) & \gamma_1 \mathbf{n}_x + \gamma_2 \dfrac{D(1-\nu)}{2} \mathbf{d}\mathbf{n}_y \end{bmatrix} \tag{3.165a}$$

$$\overline{\mathbf{G}} = \begin{bmatrix} K_s G h \alpha_2 & \mathbf{0} & \mathbf{0} \\ \mathbf{0} & D\beta_2 \mathbf{n}_x & D\beta_2 \mathbf{n}_y \\ \mathbf{0} & \gamma_2 \dfrac{D(1-\nu)}{2} \mathbf{n}_y & -\gamma_2 \dfrac{D(1-\nu)}{2} \mathbf{n}_x \end{bmatrix} \tag{3.165b}$$

$$\overline{\mathbf{F}} = \begin{Bmatrix} \alpha_3 \\ \beta_3 \\ \gamma_3 \end{Bmatrix} \tag{3.165c}$$

Equations (3.158) and (3.164) constitute a system of $6N$ algebraic equations, which are solved to express the vectors $\mathbf{w}$, $\mathbf{w}_{,n}$, ϕ_x, $\phi_{x,n}$, ϕ_y, $\phi_{y,n}$ in terms of the vectors of the fictitious sources $\mathbf{b}^{(i)} (i = 1, 2, 3)$. The vectors $\mathbf{w}, \phi_x, \phi_y$ and their derivatives at the M domain nodal points are also expressed in terms of the vectors of fictitious sources using Eqs. (3.75) appropriately. Hence,

$$\mathbf{w}_{,pq} = \mathbf{D}_{,pq}\mathbf{b}^{(1)} - [v^*]_{,pq}\mathbf{w}_{,n} + [v_n^*]_{,pq}\mathbf{w} \tag{3.166a}$$

$$\boldsymbol{\phi}_{x,pq} = \mathbf{D}_{,pq}\mathbf{b}^{(2)} - [v^*]_{,pq}\boldsymbol{\phi}_{x,n} + [v_n^*]_{,pq}\boldsymbol{\phi}_x \tag{3.166b}$$

$$\boldsymbol{\phi}_{y,pq} = \mathbf{D}_{,pq}\mathbf{b}^{(3)} - [v^*]_{,pq}\boldsymbol{\phi}_{y,n} + [v_n^*]_{,pq}\boldsymbol{\phi}_y \tag{3.166c}$$

for $p,\ q=0,\ x,\ y$.

Then, substituting the quantities $\mathbf{w}$, $\mathbf{w}_{,n}$, $\boldsymbol{\phi}_x$, $\boldsymbol{\phi}_{x,n}$, $\boldsymbol{\phi}_y$, $\boldsymbol{\phi}_{y,n}$ in Eqs. (3.166) gives the derivatives in terms of the fictitious sources:

$$\mathbf{w}_{,pq} = \mathbf{R}_{,pq}^{(1)}\mathbf{b}^{(1)} + \mathbf{R}_{,pq}^{(2)}\mathbf{b}^{(2)} + \mathbf{R}_{,pq}^{(3)}\mathbf{b}^{(3)} + \mathbf{r}_{,pq} \tag{3.167a}$$

$$\boldsymbol{\phi}_{x,pq} = \mathbf{S}_{,pq}^{(1)}\mathbf{b}^{(1)} + \mathbf{S}_{,pq}^{(2)}\mathbf{b}^{(2)} + \mathbf{S}_{,pq}^{(3)}\mathbf{b}^{(3)} + \mathbf{s}_{,pq} \tag{3.167b}$$

$$\boldsymbol{\phi}_{y,pq} = \mathbf{T}_{,pq}^{(1)}\mathbf{b}^{(1)} + \mathbf{T}_{,pq}^{(2)}\mathbf{b}^{(2)} + \mathbf{T}_{,pq}^{(3)}\mathbf{b}^{(3)} + \mathbf{t}_{,pq} \tag{3.167c}$$

where $\mathbf{R}_{,pq}^{(i)}$, $\mathbf{S}_{,pq}^{(i)}$, $\mathbf{T}_{,pq}^{(i)}$ $(i=1,2,3)$ are known $M\times M$ matrices and $\mathbf{r}_{,pq}$, $\mathbf{s}_{,pq}$, $\mathbf{t}_{,pq}$ are known $M\times 1$ vectors, whose expressions are not given here as they can be readily formulated and computed in the computer program for the numerical solution.

We are now ready to evaluate the fictitious sources $\mathbf{b}^{(i)}$. For this purpose Eqs. (3.154) are applied at the M domain nodal points. This yields

$$-K_s Gh\left(\mathbf{b}^{(1)} + \boldsymbol{\phi}_{x,x} + \boldsymbol{\phi}_{y,y}\right) = \mathbf{f} \tag{3.168a}$$

$$-\frac{D}{2}(1-\nu)\mathbf{b}^{(2)} - \frac{D}{2}(1+\nu)\left(\boldsymbol{\phi}_{x,xx} + \boldsymbol{\phi}_{y,xy}\right) + K_s Gh\left(\mathbf{w}_{,x} + \boldsymbol{\phi}_x\right) = 0 \tag{3.168b}$$

$$-\frac{D}{2}(1-\nu)\mathbf{b}^{(3)} - \frac{D}{2}(1+\nu)\left(\boldsymbol{\phi}_{x,xy} + \boldsymbol{\phi}_{y,yy}\right) + K_s Gh\left(\mathbf{w}_{,y} + \boldsymbol{\phi}_y\right) = 0 \tag{3.168c}$$

Substituting the derivatives from Eqs. (3.167) in Eqs. (3.168), we obtain the linear system of $3M$ equations whose solution gives the fictitious sources $\mathbf{b}^{(i)}(i=1,2,3)$. Subsequently, these fictitious sources are used in Eqs. (3.167) to evaluate $\mathbf{w}_{,pq}$, $\boldsymbol{\phi}_{x,pq}$, and $\boldsymbol{\phi}_{y,pq}$. The stress resultants at the interior points are computed using Eqs. (3.144) and the boundary forces using Eqs. (3.152).

3.7.4 Examples

EXAMPLE 3.14

The developed solution procedure is applied to analyze a circular plate of radius a under a uniform load f. The examined boundary conditions are (i) simply supported edge of type I and (ii) clamped edge. The results were obtained with $N=200$ constant boundary elements and $M=200$ internal points randomly distributed in the domain. The deflection $\overline{w}=wD/fa^4$ and the bending

TABLE 3.8 Deflection $\overline{w}$ and bending moments $\overline{M}_{rr}$, $\overline{M}_{\theta\theta}$ at the center of the simply supported thick circular plate ($\nu=0.30$) in Example 3.14

h/a	$\overline{w}$		$\overline{M}_{rr}=\overline{M}_{\theta\theta}$	
	Analytic [76]	AEM	Analytic [76]	AEM
0.20	6.65591e-02	6.68705e-02	2.06250e-01	2.07133e-01
0.25	3.49011e-02	3.50025e-02	2.06250e-01	2.06868e-01
0.30	2.07794e-02	2.08196e-02	2.06250e-01	2.06676e-01
0.35	1.35187e-02	1.35369e-02	2.06250e-01	2.06577e-01
0.40	9.39131e-03	9.40034e-03	2.06250e-01	2.06510e-01

TABLE 3.9 Deflection $\overline{w}$ and bending moments $\overline{M}_{rr}$, $\overline{M}_{\theta\theta}$ at the center of the center of the clamped thick circular plate ($\nu=0.30$) in Example 3.14

h/a	$\overline{w}$		$\overline{M}_{rr}=\overline{M}_{\theta\theta}$	
	Analytic [76]	AEM	Analytic [76]	AEM
0.20	1.84821e-02	1.84868e-02	8.12500e-02	8.12744e-02
0.25	1.02857e-02	1.02859e-02	8.12500e-02	8.12635e-02
0.30	6.53439e-03	6.53366e-03	8.12500e-02	8.12571e-02
0.35	4.54810e-03	4.54722e-03	8.12500e-02	8.12529e-02
0.40	3.38170e-03	3.38087e-03	8.12500e-02	8.12500e-02

moments $\overline{M}_{rr}=M_{rr}/fa^2$, $\overline{M}_{\theta\theta}=M_{\theta\theta}/fa^2$ at the center of the plate for various values of ratio h/a are given in Table 3.8 (simply supported of type I) and Table 3.9 (clamped). The obtained results are compared with those obtained from an analytical solution [76]. The results validate the efficiency and accuracy of the presented computational method.

3.8 ANISOTROPIC PLATES

3.8.1 BEM for anisotropic plates

Anisotropic plates are widely used in modern engineering structures. Multilayered composite plates are an example of anisotropic plates whose design demands reliable modeling and numerical treatment. The direct BEM formulation could offer an efficient computational tool for the analysis of anisotropic plates. However, we come across practically insurmountable difficulties, because the fundamental solution is too complicated and its analytical and numerical treatment requires special care. This is reflected in the literature; the number of publications in which the BEM is applied to anisotropic plates is notably very few. Nevertheless, in the last 25 years important advances in the application of the BEM techniques to anisotropic plates have been reported, in which both Kirchhoff and shear deformable plate theories have been considered. Shi and Bezine [84] presented a BEM formulation for thin plates using a Rayleigh-Green type reciprocal identity for governing operator and the fundamental solutions proposed by Wu and Altiero [85]. Rajamohan and Raamachandran [86] proposed a formulation in which the singularities were avoided by placing source points outside the domain. Shear deformable plates have also been analyzed by the BEM [87,88].

In the direct BEM formulation, the problems we have to cope with are the treatment of singular and hypersingular boundary integrals [89] and the evaluation of the domain integrals [90]. Special boundary integral equation techniques have also been developed for polygonal anisotropic plates [91]. The treatment of plates of arbitrary curvilinear boundary is not discussed. All solved examples are limited to plates of simple geometry with rectilinear boundaries, mainly rectangular. Therefore, the reported BEM formulations do not offer an efficient computational tool for analyzing anisotropic plates of arbitrary shape and boundary conditions. The AEM overcomes all these difficulties, since the problem of the anisotropic plate is reduced to that of an isotropic plate for Kirchhoff's theory or to three potential problems for shear deformation plate theory. Moreover, the AEM is preferred here, because it can be readily employed to solve nonlinear static and dynamic problems for anisotropic plates. This approach is presented in the following subsections.

3.8.2 Thin anisotropic plates

The theory of thin anisotropic plates is an extension of the Kirchhoff plate theory to plates of anisotropic material. Thus, the displacement field is given by Eqs. (2.8) and the strain-displacement relations by Eqs. (2.9) with $\varepsilon_x = \varepsilon_y = \gamma_{xy} = 0$.

We consider a single-layer plate consisting of an anisotropic material, which has a plane of elastic symmetry (*monoclinic* material) located at the middle surface of the plate (xy plane). The constitutive equations of the material are given as [92]

$$\begin{Bmatrix} \overline{\sigma}_x \\ \overline{\sigma}_y \\ \overline{\tau}_{xy} \end{Bmatrix} = \begin{bmatrix} C_{11} & C_{12} & C_{16} \\ C_{12} & C_{22} & C_{26} \\ C_{16} & C_{26} & C_{66} \end{bmatrix} \begin{Bmatrix} \overline{\varepsilon}_x \\ \overline{\varepsilon}_y \\ \overline{\gamma}_{xy} \end{Bmatrix} \tag{3.169}$$

which by virtue of Eqs. (2.9) are written as

$$\begin{Bmatrix} \overline{\sigma}_x \\ \overline{\sigma}_y \\ \overline{\tau}_{xy} \end{Bmatrix} = \begin{bmatrix} C_{11} & C_{12} & C_{16} \\ C_{12} & C_{22} & C_{26} \\ C_{16} & C_{26} & C_{66} \end{bmatrix} \begin{Bmatrix} -zw_{,xx} \\ -zw_{,yy} \\ -2zw_{,xy} \end{Bmatrix} \tag{3.170}$$

with C_{ij} being the elastic constants of the anisotropic material.

The constitutive plate equations result from Eqs. (2.15) using the expressions (3.170) for $(\overline{\sigma}_x, \overline{\sigma}_y, \overline{\tau}_{xy})$. Thus, we have

$$\begin{Bmatrix} M_x \\ M_y \\ M_{yx} \end{Bmatrix} = \int_{-h/2}^{h/2} \begin{Bmatrix} \overline{\sigma}_x \\ \overline{\sigma}_y \\ \overline{\tau}_{xy} \end{Bmatrix} z dz = - \begin{bmatrix} D_{11} & D_{12} & 2D_{16} \\ D_{12} & D_{22} & 2D_{26} \\ D_{16} & D_{26} & 2D_{66} \end{bmatrix} \begin{Bmatrix} w_{,xx} \\ w_{,yy} \\ w_{,xy} \end{Bmatrix} \tag{3.171a}$$

$$M_{xy} = -M_{yx} \tag{3.171b}$$

where

$$D_{ij} = \frac{C_{ij} h^3}{12} \tag{3.172}$$

are the *bending and twisting stiffnesses* of the anisotropic plate.

The shear forces Q_x, Q_y in Kirchhoff's theory result from the equilibrium of the moments acting on a plate element about x and y axes (see Section 2.2.4), namely,

$$Q_x = \frac{\partial M_x}{\partial x} + \frac{\partial M_{yx}}{\partial y} \tag{3.173a}$$

$$Q_y = \frac{\partial M_{yx}}{\partial x} + \frac{\partial M_y}{\partial y} \tag{3.173b}$$

The differential equation and the boundary conditions are obtained from the minimization of the total potential Π, i.e., from the condition

$$\delta\Pi = \delta U_b + \delta U_s + \delta V = 0 \tag{3.174}$$

where δU_b, δU_s, and δV are the variations of the bending strain energy, the elastic energy of the yielding support, and the potential of the external forces, respectively, and are expressed as

$$\begin{aligned} \delta U_b &= \frac{1}{2}\delta \int_V \left(\overline{\sigma}_x \overline{\varepsilon}_x + \overline{\sigma}_y \overline{\varepsilon}_y + \overline{\tau}_{xy} \overline{\gamma}_{xy}\right) dV \\ &= \int_V \left(\overline{\sigma}_x \delta\overline{\varepsilon}_x + \overline{\sigma}_y \delta\overline{\varepsilon}_y + \overline{\tau}_{xy} \delta\overline{\gamma}_{xy}\right) dV \end{aligned} \tag{3.175}$$

$$\delta U_s = \int_\Gamma \left(k_T w \delta w + k_R w_{,n} \delta w_{,n}\right) ds + \sum_k k_c^{(k)} w_k \delta w_k \tag{3.176}$$

$$\delta V = -\int_\Omega f \delta w d\Omega - \int_\Gamma \left(V_n^* \delta w - M_n^* \delta w_{,n}\right) ds - \sum_k R_k^* \delta w_k \tag{3.177}$$

where M_n^*, V_n^*, R_n^* are the prescribed boundary forces (Fig. 2.2). Equation (3.175) expresses the principle of virtual work for the strain energy. Its proof is left as an exercise to the reader.

Substituting the expressions for $(\overline{\sigma}_x, \overline{\sigma}_y, \overline{\tau}_{xy})$ from Eq. (3.170) and for $(\overline{\varepsilon}_x, \overline{\varepsilon}_y, \overline{\gamma}_{xy})$ from Eqs. (2.9) in Eq. (3.175) and integrating through the thickness, we obtain

$$\delta U_b = -\int_\Omega \left(M_x \delta w_{,xx} + M_y \delta w_{,yy} - 2M_{xy} \delta w_{,xy}\right) dxdy \tag{3.178}$$

Further, introducing Eqs. (3.176), (3.177), and (3.178) in Eq. (3.174) yields

$$\begin{aligned} \delta\Pi &= \int_\Omega \left(-M_x \delta w_{,xx} + 2M_{xy} \delta w_{,xy} - M_y \delta w_{,yy}\right) d\Omega \\ &+ \int_\Gamma \left(k_T w \delta w + k_R w_{,n} \delta w_{,n}\right) ds - \int_\Omega f \delta w d\Omega \\ &- \int_\Gamma \left(V_n^* \delta w - M_n^* \delta w_{,n}\right) ds + \sum_k \left[k_c^{(k)} w_k - R_k^*\right] \delta w_k = 0 \end{aligned} \tag{3.179}$$

Equation (3.179) expresses the principle of virtual work for the anisotropic plate in bending. Apparently, it has the same form with Eq. (2.34), but the expressions for the moments are different.

Using now integration by parts to eliminate the derivatives from the virtual quantities and working as for Eq. (2.34), Eq. (3.179) gives

$$\begin{aligned} \delta\Pi = &\int_\Omega \left(-M_{x,xx} + 2M_{xy,xy} - M_{y,yy} - f\right) \delta w d\Omega \\ &+ \int_\Gamma \left[\left(-M_n + k_R w_{,n} + M_n^*\right) \delta w_{,n} + \left(Q_n - M_{nt,s} + kTw - V_n^*\right) \delta w\right] ds \\ &+ \sum_k \left[k_c^{(k)} w_k - [\![M_{nt}]\!]_k - R_k^*\right] \delta w_k = 0 \end{aligned} \tag{3.180}$$

The expressions for M_n, M_{nt}, Q_n are given by Eqs. (2.23a,c,d).

On the basis of the calculus of variations, Eq. (3.180) yields the equilibrium equation

$$M_{x,xx} - 2M_{xy,xy} + M_{y,yy} = -f \quad \text{in } \Omega \tag{3.181}$$

and the boundary conditions

$$V_n + k_T w = V_n^* \quad \text{or} \quad w = w^* \quad \text{on } \Gamma \tag{3.182a}$$

$$M_n - k_R w_{,n} = M_n^* \quad \text{or} \quad w_{,n} = w_{,n}^* \quad \text{on } \Gamma \tag{3.182b}$$

$$k_c^{(k)} w_k - R_k = R_k^* \quad \text{or} \quad w_k = w_k^* \quad \text{at corner } k \tag{3.182c}$$

in which

$$V_n = Q_n - M_{nt,s} \tag{3.183a}$$

$$R_k = [\![M_{nt}]\!]_k \tag{3.183b}$$

Substituting the expressions (3.171) for M_x, M_y, M_{xy} in Eq. (3.181), we obtain the differential equation of the anisotropic plate in terms of the deflection

$$D_{11}\frac{\partial^4 w}{\partial x^4} + 4D_{16}\frac{\partial^4 w}{\partial x^3 \partial y} + 2(D_{12} + 2D_{66})\frac{\partial^4 w}{\partial x^2 \partial y^2} + 4D_{26}\frac{\partial^4 w}{\partial x \partial y^3} + D_{22}\frac{\partial^4 w}{\partial y^4} = f \tag{3.184}$$

3.8.3 AEM for the thin anisotropic plate

Equation (3.184) is of the fourth order. Consequently, Eq. (3.17) can be used as the analog equation in this case, too. Therefore, Eq. (3.28), which relates the boundary quantities to the fictitious load, also applies here, namely,

$$\mathbf{H}\begin{Bmatrix} \mathbf{w} \\ \mathbf{w}_c \\ \mathbf{w}_n \end{Bmatrix} = \mathbf{G}\begin{Bmatrix} \mathbf{V} \\ \mathbf{R} \\ \mathbf{M} \end{Bmatrix} + \mathbf{A}\mathbf{b} + \frac{1}{D}\mathbf{F} \tag{3.185}$$

The additional $2N + N_c$ equations that relate the boundary quantities involved in Eq. (3.185) will result from the application of the boundary conditions at the boundary nodal points. This gives

$$\begin{bmatrix} \boldsymbol{\alpha}_1 & \mathbf{0} & \mathbf{0} \\ \mathbf{0} & \mathbf{c}_1 & \mathbf{0} \\ \mathbf{0} & \mathbf{0} & \boldsymbol{\beta}_1 \end{bmatrix}\begin{Bmatrix} \mathbf{w} \\ \mathbf{w}_c \\ \mathbf{w}_n \end{Bmatrix} + \begin{bmatrix} \boldsymbol{\alpha}_2 & \mathbf{0} & \mathbf{0} \\ \mathbf{0} & \mathbf{c}_2 & \mathbf{0} \\ \mathbf{0} & \mathbf{0} & \boldsymbol{\beta}_2 \end{bmatrix}\begin{Bmatrix} \mathbf{V}_n \\ \mathbf{R}_n \\ \mathbf{M}_n \end{Bmatrix} = \begin{Bmatrix} \boldsymbol{\alpha}_3 \\ \mathbf{c}_3 \\ \boldsymbol{\beta}_3 \end{Bmatrix} \tag{3.186}$$

The $N \times N$ diagonal matrices $\boldsymbol{\alpha}_i$, $\boldsymbol{\beta}_i$, the $N_c \times N_c$ diagonal matrix c_i ($i = 1, 2$) and the vectors $\boldsymbol{\alpha}_3$, $\mathbf{c}_3$, $\boldsymbol{\beta}_3$ express the same quantities as in Eq. (3.31). In as much as the quantities $\mathbf{M}_n$, $\mathbf{V}_n$, $\mathbf{R}_n$ refer to the anisotropic plate, they should be expressed in terms of $\mathbf{M}$, $\mathbf{V}$, $\mathbf{R}$. This is achieved as follows.

We write Eqs. (2.23a,b,c) in matrix form

$$\begin{Bmatrix} M_n \\ M_t \\ M_{nt} \end{Bmatrix} = \mathbf{A}\begin{Bmatrix} M_x \\ M_y \\ M_{yx} \end{Bmatrix} \tag{3.187}$$

where

$$\mathbf{A} = \begin{bmatrix} \cos^2 a & \sin^2 a & 2\sin a \cos a \\ \sin^2 a & \cos^2 a & -2\sin a \cos a \\ \sin a \cos a & -\sin a \cos a & -\cos^2 a + \sin^2 a \end{bmatrix} \tag{3.188}$$

Equation (3.187) by virtue of Eqs. (3.171a) becomes

$$\begin{Bmatrix} M_n \\ M_t \\ M_{nt} \end{Bmatrix} = -\mathbf{AD} \begin{Bmatrix} w_{,xx} \\ w_{,yy} \\ w_{,xy} \end{Bmatrix} \tag{3.189}$$

with **D** being the plate stiffness matrix, namely

$$\mathbf{D} = \begin{bmatrix} D_{11} & D_{12} & 2D_{16} \\ D_{12} & D_{22} & 2D_{26} \\ D_{16} & D_{26} & 2D_{66} \end{bmatrix} \tag{3.190}$$

Taking now into account the transformation of the second derivatives due to rotation of the axes

$$\begin{Bmatrix} w_{,xx} \\ w_{,yy} \\ w_{,xy} \end{Bmatrix} = \mathbf{B}^{-1} \begin{Bmatrix} w_{,nn} \\ w_{,tt} \\ w_{,nt} \end{Bmatrix} \tag{3.191}$$

in which

$$\mathbf{B} = \begin{bmatrix} \cos^2 a & \sin^2 a & 2\sin a \cos a \\ \sin^2 a & \cos^2 a & -2\sin a \cos a \\ -\sin a \cos a & \sin a \cos a & \cos^2 a - \sin^2 a \end{bmatrix} \tag{3.192}$$

we obtain finally

$$\begin{Bmatrix} M_n \\ M_t \\ M_{nt} \end{Bmatrix} = \begin{bmatrix} e_{11} & e_{12} & e_{13} \\ e_{21} & e_{22} & e_{23} \\ e_{31} & e_{32} & e_{33} \end{bmatrix} \begin{Bmatrix} w_{,nn} \\ w_{,tt} \\ w_{,nt} \end{Bmatrix} \tag{3.193}$$

where

$$\begin{bmatrix} e_{11} & e_{12} & e_{13} \\ e_{21} & e_{22} & e_{23} \\ e_{31} & e_{32} & e_{33} \end{bmatrix} = -\mathbf{ADB}^{-1} \tag{3.194}$$

After appropriate manipulations, Eq. (3.193) gives

$$\begin{aligned} M_n &= (e_{11} w_{,nn} + e_{12} w_{,tt} + e_{13} w_{,nt}) \\ &= \left[-e_{11} \overline{M}_n + (e_{12} - e_{11}\nu) w_{,tt} + e_{13} w_{,nt}\right] \end{aligned} \tag{3.195a}$$

$$\begin{aligned} M_{nt} &= e_{31} w_{,nn} + e_{32} w_{,tt} + e_{33} w_{,nt} \\ &= -e_{31} \overline{M}_n + (e_{32} - e_{31}\nu) w_{,tt} + \frac{e_{33}}{1-\nu} \overline{M}_{nt} \end{aligned} \tag{3.195b}$$

$$\begin{aligned}M_{nt,s} &= -e_{31}\overline{M}_{n,s} - e_{31,s}\overline{M}_n + (e_{32} - e_{31}\nu)w_{,tts} + \frac{e_{33}}{1-\nu}\overline{M}_{nt,s} \\ &+ (e_{32,s} - e_{31,s}\nu)w_{,tt} + e_{33,s}w_{,tt}\end{aligned} \tag{3.195c}$$

$$\begin{aligned}Q_n &= \frac{\partial M_n}{\partial n} - \frac{\partial M_{nt}}{\partial t} \\ &= -e_{11}\overline{Q}_n + (e_{12} - e_{11} - e_{33})w_{,ttn} - [(e_{13} - e_{31})\overline{M}_{n,t} \\ &-[(e_{13} - e_{31})\nu + e_{32}]w_{,tt}\end{aligned} \tag{3.195d}$$

$$\begin{aligned}V_n &= Q_n - M_{nt,s} \\ &= -e_{11}\overline{V}_n + (e_{12} - e_{11} - e_{33})w_{,ntt} - (e_{13} - e_{31})\overline{M}_{n,t} \\ &+ e_{31}\overline{M}_{n,s} - e_{31,s}\overline{M}_n - (e_{32} - e_{31}\nu)w_{,tts} - [e_{11}(1-\nu) + e_{33}]w_{,nts} \\ &+ (e_{13} - e_{31} + e_{32})w_{,ttt} + (e_{31,s}\nu - e_{32,s})w_{,tt} - e_{33,s}w_{,nt}\end{aligned} \tag{3.195e}$$

$$R_k = [\![M_{nt}]\!]_k = -[\![e_{31}\overline{M}_n]\!]_k + [\![(e_{32} - e_{31}\nu)w_{,tt}]\!]_k + \left[\!\!\left[\frac{e_{33}}{1-\nu}\overline{M}_{nt}\right]\!\!\right]_k \tag{3.195f}$$

The overbar here designates the boundary quantities of the analog equation.

Using finite differences, the derivatives $w_{,ntt}$, $w_{,ttt}$ and $w_{,tts}$ are expressed in terms of the boundary nodal values of w and $w_{,n}$ (see Section A.4 in Appendix A). Thus, all quantities in Eq. (3.186) are expressed in terms of the boundary quantities of the analog equation. Equations (3.185) and (3.186) are combined to express the boundary quantities in terms of the fictitious load **b**. Then they are inserted into Eq. (3.46), which is now extended to include the fourth-order derivatives, namely,

$$\mathbf{w}_{,pqrs} = \mathbf{S}_{,pqrs}\mathbf{b} + \mathbf{c} \tag{3.196}$$

where p, q, r, s represent 0, x, y.

Finally, application of Eq. (3.184) at the M domain nodal points results in the following algebraic equation:

$$\mathbf{D}_{11}\mathbf{w}_{,xxxx} + 4\mathbf{D}_{16}\mathbf{w}_{,xxxy} + 2(\mathbf{D}_{12} + 2\mathbf{D}_{66})\mathbf{w}_{,xxyy} + 4\mathbf{D}_{26}\mathbf{w}_{,xyyy} + \mathbf{D}_{22}\mathbf{w}_{,yyyy} = \mathbf{f} \tag{3.197}$$

Substituting the derivatives from Eq. (3.196) in Eq. (3.197) gives the linear algebraic equation for the evaluation of the fictitious load **b**.

3.8.4 Examples

EXAMPLE 3.15

A three-layer cross-ply $(0°/90°/0°)$ square anisotropic plate subjected to a uniform load f_0 is analyzed. The computed nondimensional central deflections $\overline{w} = w(a/2, a/2)\dfrac{E_2h^3}{a^4f_0}\times 10^2$ and in-plane normal stresses at levels $z = h/2$ and $z = z/6$, i.e., $\overline{\sigma}_x = \sigma_x(a/2, a/2, h/2)\dfrac{h^2}{a^2f_0}\times 10$ and $\overline{\sigma}_y = \sigma_y(a/2, a/2, h/6)\dfrac{h^2}{a^2f_0}\times 10$, for various boundary conditions are given in

TABLE 3.10 Nondimensional central deflections ($\overline{w}$) and in-plane normal stresses ($\overline{\sigma}_x, \overline{\sigma}_y$) in symmetric cross-ply ($0°/90°/0°$) anisotropic square plates under a uniform transverse load f_0 for various boundary conditions in Example 3.15. Upper value: Computed; Lower value: Analytical (Levy solution) [93]

Boundary conditions	Nodal points N/M	$\overline{w}=w\dfrac{E_2h^3}{a^4f_0}$ $\times 10^2$	$\overline{\sigma}_x=\sigma_x\dfrac{h^2}{a^2f_0}$ $\times 10$ $(z=h/2)$	$\overline{\sigma}_y=\sigma_y\dfrac{h^2}{a^2f_0}$ $\times 10$ $(z=h/6)$
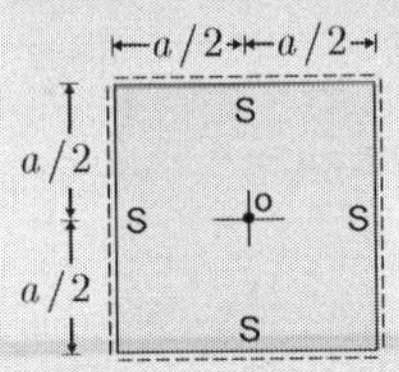	400/133	1.192	7.174	1.929
	400/245	1.203	7.235	1.930
	Analytic	1.206	7.251	1.938
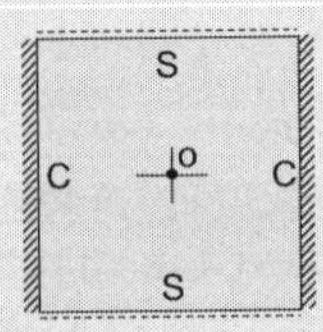	400/133	0.278	2.770	0.206
	400/245	0.280	2.784	0.194
	Analytic	0.280	2.787	0.185
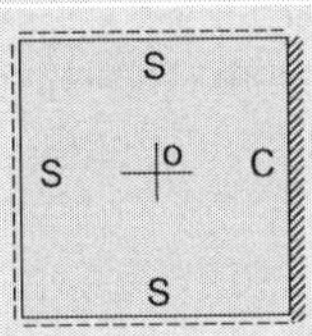	400/133	0.540	4.048	0.663
	400/245	0.543	4.072	0.657
	Analytic	0.544	4.082	0.651
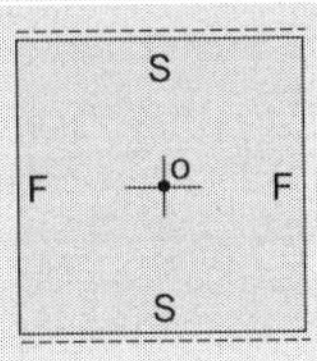	600/133	10.902	0.153	21.566
	600/245	10.932	0.180	21.611
	Analytic	10.920	0.195	21.597
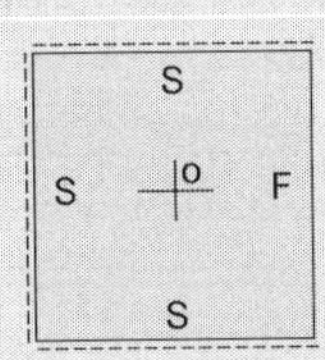	600/133	5.967	3.724	11.581
	600/245	5.986	3.770	11.605
	Analytic	5.992	3.778	11.621
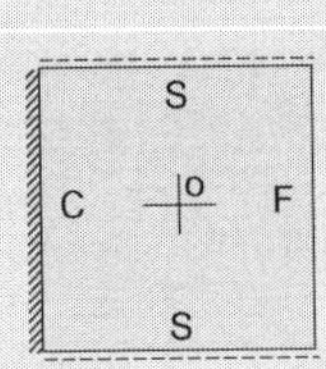	600/133	2.369	1.697	4.311
	600/245	2.373	1.684	4.312
	Analytic	2.376	1.685	4.313

Table 3.10 as compared with those obtained by Levy's solution [93]. The employed data are $E_1 = 19.2\,\text{MPa}$, $E_2 = 1.56\,\text{MPa}$, $G_{12} = 0.82\,\text{MPa}$, and $\nu_{12} = 0.24$. The stiffness matrix of the laminated plate is computed as follows.

The constitutive matrix **C** in Eq. (3.169) for the orthotropic lamina is expressed in terms of the engineering constants [93]:

$$\mathbf{C} = \begin{bmatrix} \dfrac{E_1}{1-\nu_{12}\nu_{21}} & \dfrac{\nu_{12}E_2}{1-\nu_{12}\nu_{21}} & 0 \\ \dfrac{\nu_{12}E_2}{1-\nu_{12}\nu_{21}} & \dfrac{E_2}{1-\nu_{12}\nu_{21}} & 0 \\ 0 & 0 & G_{12} \end{bmatrix}, \quad \nu_{21} = \nu_{12}E_2/E_1 \tag{3.198}$$

and it is transformed from the material axes (x', y') in the working axes (x, y) using the relation

$$\overline{\mathbf{C}} = \mathbf{TCT}^T \tag{3.199}$$

where

$$T = \begin{bmatrix} \cos^2\theta & \sin^2\theta & -2\cos\theta\sin\theta \\ \cos^2\theta & \cos^2\theta & 2\cos\theta\sin\theta \\ \cos\theta\sin\theta & -\cos\theta\sin\theta & \cos^2\theta - \sin^2\theta \end{bmatrix}, \quad \theta = \measuredangle(x', x) \tag{3.200}$$

Thus, we obtain the coefficients matrices in the working axes (x, y):

Layer 1: $-h/2 \le z \le -h/6$, $(\theta = 0)$

$$\overline{\mathbf{C}}_1 = \begin{bmatrix} 19290278.50 & 376160.43 & 0 \\ 376160.43 & 1567335.13 & 0 \\ 0 & 0 & 820000.00 \end{bmatrix} \tag{3.201}$$

Layer 2: $-h/6 \le z \le h/6$, $(\theta = 90°)$

$$\overline{\mathbf{C}}_2 = \begin{bmatrix} 1567335.13 & 376160.43 & 0 \\ 376160.43 & 19290278.50 & 0 \\ 0 & 0 & 820000.00 \end{bmatrix} \tag{3.202}$$

Layer 3: $h/6 \le z \le h/2$, $(\theta = 0°)$

$$\overline{\mathbf{C}}_3 = \overline{\mathbf{C}}_1 \tag{3.203}$$

The bending stiffnesses result from (3.171a) if it is applied for the three-layer cross-ply plate, namely,

$$\begin{Bmatrix} M_x \\ M_y \\ M_{yx} \end{Bmatrix} = \sum_{k=1}^{3} \int_{z_k}^{z_{k+1}} \begin{Bmatrix} \overline{\sigma}_x \\ \overline{\sigma}_y \\ \overline{\tau}_{xy} \end{Bmatrix} z dz = -\begin{bmatrix} D_{11} & D_{12} & 2D_{16} \\ D_{12} & D_{22} & 2D_{26} \\ D_{16} & D_{26} & 2D_{66} \end{bmatrix} \begin{Bmatrix} w_{,xx} \\ w_{,yy} \\ w_{,xy} \end{Bmatrix} \tag{3.204}$$

in which $z_1=-h/2$, $z_2=-h/6$, $z_3=h/6$, $z_4=h/2$ and $k=1, 2, 3$. Thus, we obtain

$$\begin{bmatrix} D_{11} & D_{12} & 2D_{16} \\ D_{12} & D_{22} & 2D_{26} \\ D_{16} & D_{26} & 2D_{66} \end{bmatrix} = h^3 \begin{bmatrix} 1552822.766 & 31346.702 & 0 \\ 31346.702 & 185311.703 & 0 \\ 0 & 0 & 136666.667 \end{bmatrix} \tag{3.205}$$

EXAMPLE 3.16

A three-layer ($-45°/0°/90°$) clamped elliptic anisotropic plate having thickness h and semi-axes a, b is analyzed when subjected to a uniform load f_0. The layers are identical and have thickness $h/3$ and material constants $E_1=19.2\,\mathrm{GPa}$, $E_2=1.56\,\mathrm{GPa}$, $G_{12}=0.82\,\mathrm{GPa}$, $\nu_{12}=0.24$. The problem admits an exact solution [94]:

$$w=\frac{f_0 a^4}{64D'}\left(1-\frac{x^2}{a^2}-\frac{y^2}{b^2}\right)^2 \tag{3.206}$$

where

$$D'=\frac{1}{8}\left[3D_{11}+2(D_{12}+2D_{66})c^2+3D_{22}c^4\right], \quad c=\frac{a}{b} \tag{3.207}$$

The stiffness matrix of the plate obtained as in previous example is

$$\begin{bmatrix} D_{11} & D_{12} & 2D_{16} \\ D_{12} & D_{22} & 2D_{26} \\ D_{16} & D_{26} & 2D_{66} \end{bmatrix} = \begin{bmatrix} 80.37 & 43.22 & 76.80 \\ 43.22 & 222.15 & 76.80 \\ 38.40 & 38.40 & 102.43 \end{bmatrix} \tag{3.208}$$

The numerical results were obtained with $a=3.0\,m$, $b=2.0\,m$, $h=0.06\,m$, $f_0=5\mathrm{kN/m^2}$ using $N=200$ boundary elements and $M=229$ domain nodal points resulting from 370 linear triangular elements (Fig. 3.49). The computed deflections and stress resultants are shown in Figs. 3.50 through 3.54. Apparently, the obtained results are in excellent agreement with those obtained from the analytical solution.

3.9 THICK ANISOTROPIC PLATES

We consider first-order shear deformable anisotropic plates with Mindlin's assumption for the shear forces. The constitutive equations of the material for $(\overline{\sigma}_x, \overline{\sigma}_y, \overline{\tau}_{xy})$ are given by (3.169), while for (τ_{xz}, τ_{yz}) by the equations [93]:

$$\begin{Bmatrix} \tau_{xz} \\ \tau_{yz} \end{Bmatrix} = \begin{bmatrix} C_{55} & C_{45} \\ C_{45} & C_{44} \end{bmatrix} \begin{Bmatrix} \gamma_{xz} \\ \gamma_{yz} \end{Bmatrix} \tag{3.209}$$

Equations (3.169) and (3.209) with the help of Eqs. (3.142) become

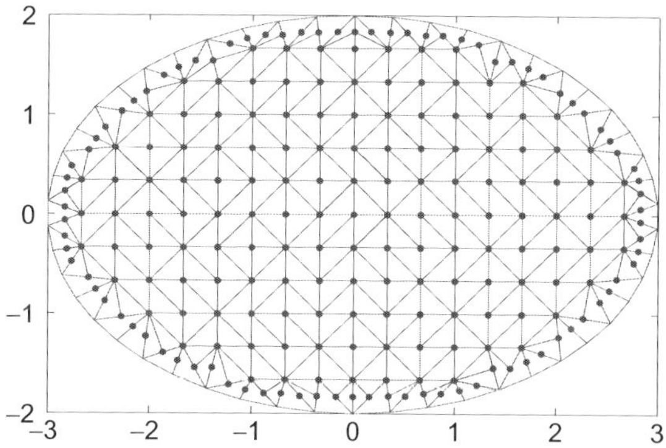

FIGURE 3.49 Domain discretization.

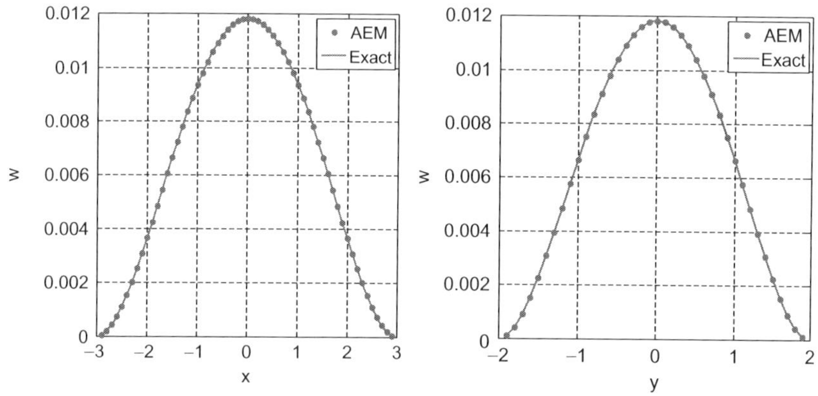

FIGURE 3.50 Deflections w along the axes $y=0$ and $x=0$ in Example 3.15.

$$\begin{Bmatrix} \overline{\sigma}_x \\ \overline{\sigma}_y \\ \overline{\tau}_{xy} \end{Bmatrix} = \begin{bmatrix} C_{11} & C_{12} & C_{16} \\ C_{12} & C_{22} & C_{26} \\ C_{16} & C_{26} & C_{66} \end{bmatrix} \begin{Bmatrix} z\phi_{x,x} \\ z\phi_{y,y} \\ z(\phi_{x,y}+\phi_{y,x}) \end{Bmatrix} \tag{3.210a}$$

$$\begin{Bmatrix} \tau_{xz} \\ \tau_{yz} \end{Bmatrix} = \begin{bmatrix} C_{55} & C_{45} \\ C_{45} & C_{44} \end{bmatrix} \begin{Bmatrix} \phi_x + w_{,x} \\ \phi_y + w_{,y} \end{Bmatrix} \tag{3.210b}$$

The constitutive plate equations result from Eqs. (2.15) using the expressions (3.210) for the stresses and including the correction factor for the transverse shear forces. Thus, we obtain

$$\begin{Bmatrix} M_x \\ M_y \\ M_{yx} \end{Bmatrix} = \int_{-h/2}^{h/2} \begin{Bmatrix} \overline{\sigma}_x \\ \overline{\sigma}_y \\ \overline{\tau}_{xy} \end{Bmatrix} z dz = \begin{bmatrix} D_{11} & D_{12} & D_{16} \\ D_{12} & D_{22} & D_{26} \\ D_{16} & D_{26} & D_{66} \end{bmatrix} \begin{Bmatrix} \phi_{x,x} \\ \phi_{y,y} \\ \phi_{x,y}+\phi_{y,x} \end{Bmatrix} \tag{3.211a}$$

$$M_{xy} = -M_{yx} \tag{3.211b}$$

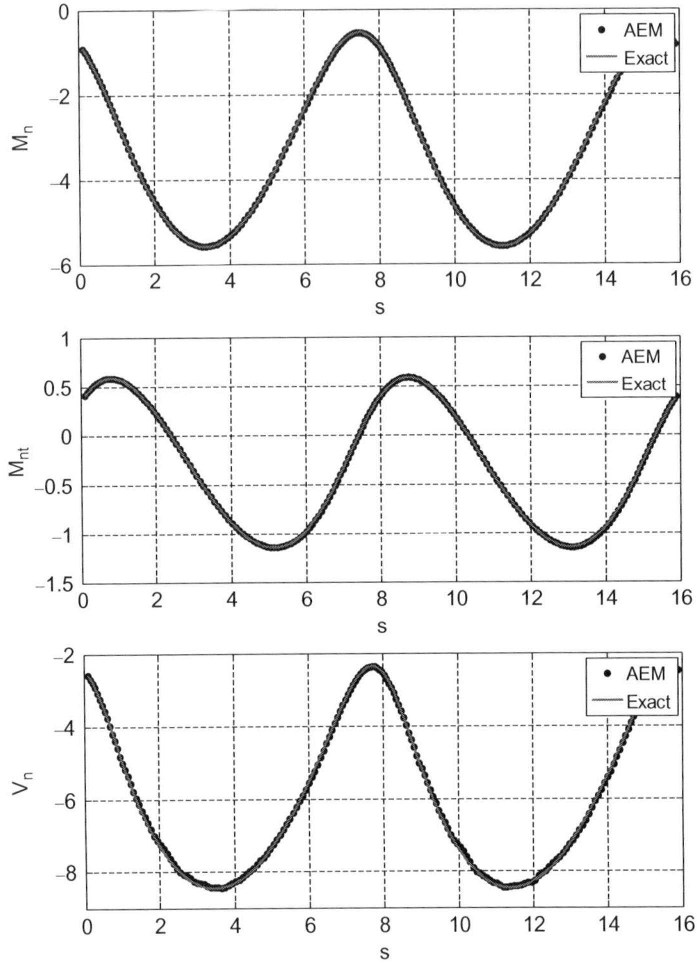

FIGURE 3.51 Bending moment M_n, twisting moment M_{nt} and reaction V_n along the boundary in Example 3.15.

$$\begin{Bmatrix} Q_x \\ Q_y \end{Bmatrix} = K_s \int_{-h/2}^{h/2} \begin{Bmatrix} \tau_{xz} \\ \tau_{yz} \end{Bmatrix} dz = \begin{bmatrix} A_{55} & A_{45} \\ A_{45} & A_{44} \end{bmatrix} \begin{Bmatrix} \phi_x + w_{,x} \\ \phi_y + w_{,y} \end{Bmatrix} \tag{3.211c}$$

in which

$$D_{ij} = \frac{C_{ij} h^3}{12}, \quad A_{ij} = K_s h C_{ij} \tag{3.212}$$

Finally, using Eqs. (3.211) in Eqs. (3.150), we obtain the equilibrium equations of the plate in terms of the functions (w, ϕ_x, ϕ_y)

$$\begin{aligned} A_{55}(\phi_{x,x} + w_{,xx}) + A_{45}(\phi_{x,y} + \phi_{y,x} + 2w_{,xy}) \\ + A_{44}(\phi_{y,y} + w_{,yy}) = -f \end{aligned} \tag{3.213a}$$

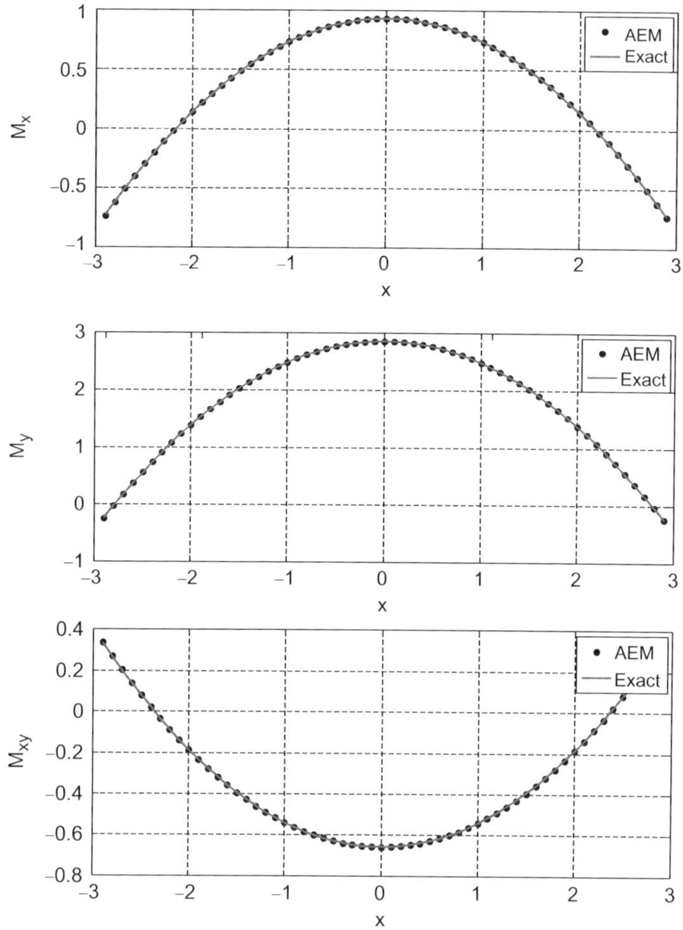

FIGURE 3.52 Bending moments M_x, M_y and twisting moment M_{xy} along the axis $y = 0$ in Example 3.15.

$$\begin{aligned} D_{11}\phi_{x,xx} + 2D_{16}\phi_{x,xy} + D_{66}\phi_{x,yy} + D_{16}\phi_{y,xx} + (D_{12} + D_{66})\phi_{y,xy} \\ + D_{26}\phi_{y,yy} - A_{55}(\phi_x + w_{,x}) - A_{45}(\phi_y + w_{,y}) = 0 \end{aligned} \tag{3.213b}$$

$$\begin{aligned} D_{66}\phi_{y,xx} + 2D_{26}\phi_{y,xy} + D_{22}\phi_{y,yy} + D_{16}\phi_{x,xx} + (D_{12} + D_{66})\phi_{x,xy} \\ + D_{26}\phi_{x,yy} - A_{45}(\phi_x + w_{,x}) - A_{44}(\phi_y + w_{,y}) = 0 \end{aligned} \tag{3.213c}$$

The associated boundary conditions are given by Eqs. (3.155).

3.9.1 AEM for the thick anisotropic plate

The boundary value problem described by Eqs. (3.213) and (3.155) is solved using the AEM. The procedure is the same as in Section (3.7.3). Hence, Eq. (3.158) applies to this case, too, which we rewrite here for convenience:

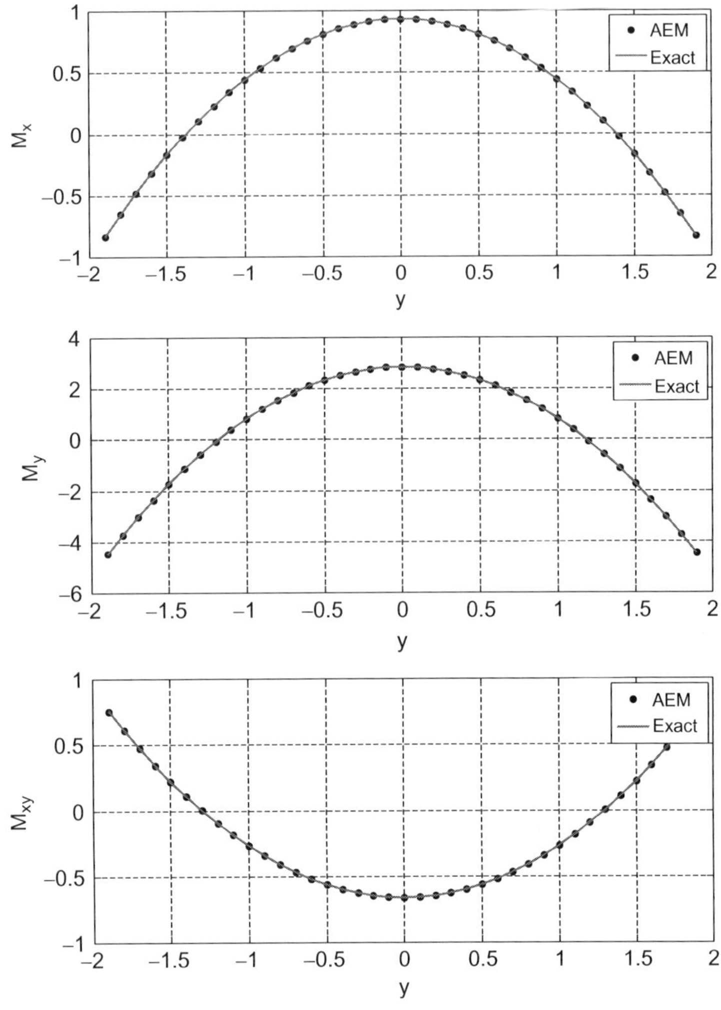

FIGURE 3.53 Bending moments M_x, M_y and twisting moment M_{xy} along the axis $x = 0$ in Example 3.15.

$$\mathbf{H}\begin{Bmatrix} \mathbf{w} \\ \phi_x \\ \phi_y \end{Bmatrix} = \mathbf{G}\begin{Bmatrix} \mathbf{w}_{,n} \\ \phi_{x,n} \\ \phi_{y,n} \end{Bmatrix} + \mathbf{D}\begin{Bmatrix} \mathbf{b}^{(1)} \\ \mathbf{b}^{(2)} \\ \mathbf{b}^{(3)} \end{Bmatrix} \tag{3.214}$$

Application of the boundary conditions (3.155) at the boundary nodal points yields

$$\begin{bmatrix} \alpha_1 & \mathbf{0} & \mathbf{0} \\ \mathbf{0} & \beta_1 & \mathbf{0} \\ \mathbf{0} & \mathbf{0} & \gamma_1 \end{bmatrix}\begin{Bmatrix} \mathbf{w} \\ \phi_n \\ \phi_t \end{Bmatrix} + \begin{bmatrix} \alpha_2 & \mathbf{0} & \mathbf{0} \\ \mathbf{0} & \beta_2 & \mathbf{0} \\ \mathbf{0} & \mathbf{0} & \gamma_2 \end{bmatrix}\begin{Bmatrix} \mathbf{Q}_n \\ \mathbf{M}_n \\ \mathbf{M}_{nt} \end{Bmatrix} = \begin{Bmatrix} \alpha_3 \\ \beta_3 \\ \gamma_3 \end{Bmatrix} \tag{3.215}$$

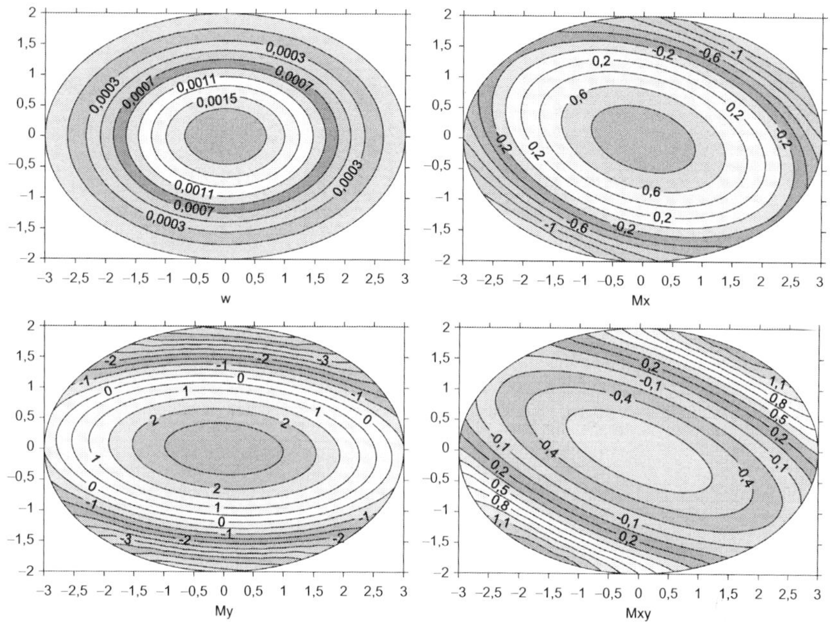

FIGURE 3.54 Contours of the deflection w, bending moments M_x, M_y and twisting moment M_{xy} in Example 3.15.

In order to combine Eqs. (3.214) with (3.215) the boundary quantities ϕ_n, ϕ_t, $\mathbf{Q}_n$, $\mathbf{M}_n$, $\mathbf{M}_{nt}$ must be first expressed in terms of $\mathbf{w}$, ϕ_x, ϕ_y, $\mathbf{w}_{,n}$, $\phi_{x,n}$, $\phi_{y,n}$. The rotations ϕ_x, ϕ_y are readily expressed in terms of ϕ_n, ϕ_t using Eqs. (3.161). Then starting from the transformation equation (3.187) and using Eqs. (3.211a) and (3.147), we obtain

$$\begin{Bmatrix} M_n \\ M_t \\ M_{nt} \end{Bmatrix} = \begin{bmatrix} E_{11} & E_{12} & E_{13} & E_{14} \\ E_{21} & E_{22} & E_{23} & E_{24} \\ E_{21} & E_{32} & E_{33} & E_{34} \end{bmatrix} \begin{Bmatrix} \phi_{x,n} \\ \phi_{x,t} \\ \phi_{y,n} \\ \phi_{y,t} \end{Bmatrix} \tag{3.216}$$

where

$$\begin{bmatrix} E_{11} & E_{12} & E_{13} & E_{14} \\ E_{21} & E_{22} & E_{23} & E_{24} \\ E_{21} & E_{32} & E_{33} & E_{34} \end{bmatrix} = \mathbf{AD} \begin{bmatrix} 1 & 0 & 0 & 0 \\ 0 & 0 & 0 & 1 \\ 0 & 1 & 1 & 0 \end{bmatrix} \begin{bmatrix} \mathbf{R}^T & \mathbf{0} \\ \mathbf{0} & \mathbf{R}^T \end{bmatrix} \tag{3.217}$$

in which

$$\mathbf{D} = \begin{bmatrix} D_{11} & D_{12} & D_{16} \\ D_{12} & D_{22} & D_{26} \\ D_{16} & D_{26} & D_{66} \end{bmatrix} \quad \text{and} \quad \mathbf{R} = \begin{bmatrix} \cos a & \sin a \\ -\sin a & \cos a \end{bmatrix} \tag{3.218a,b}$$

Further, the relation

$$Q_n = [\cos a \;\; \sin a]\begin{Bmatrix} Q_x \\ Q_y \end{Bmatrix} \tag{3.219a,b}$$

by virtue of Eq. (3.211c) yields

$$Q_n = [Z_{11} \;\; Z_{12} \;\; Z_{13} \;\; Z_{14}]\begin{Bmatrix} \phi_x \\ \phi_y \\ w_{,n} \\ w_{,t} \end{Bmatrix} \tag{3.220}$$

where

$$[Z_{11} \;\; Z_{12} \;\; Z_{13} \;\; Z_{14}] = [\cos a \;\; \sin a]\begin{bmatrix} A_{55} & A_{45} \\ A_{45} & A_{44} \end{bmatrix}\begin{bmatrix} 1 & 0 & 1 & 0 \\ 0 & 1 & 0 & 1 \end{bmatrix}\begin{bmatrix} \mathbf{R}^T & \mathbf{0} \\ \mathbf{0} & \mathbf{R}^T \end{bmatrix} \tag{3.221}$$

On the basis of Eqs. (3.161) and (3.221), Eqs. (3.215) become

$$\overline{\mathbf{H}}\begin{Bmatrix} \mathbf{w} \\ \boldsymbol{\phi}_x \\ \boldsymbol{\phi}_y \end{Bmatrix} + \overline{\mathbf{G}}\begin{Bmatrix} \mathbf{w}_{,n} \\ \boldsymbol{\phi}_{x,n} \\ \boldsymbol{\phi}_{y,n} \end{Bmatrix} = \overline{\mathbf{F}} \tag{3.222}$$

where

$$\overline{\mathbf{H}} = \begin{bmatrix} \boldsymbol{\alpha}_1 + \boldsymbol{\alpha}_2\mathbf{Z}_{14}\mathbf{d} & \boldsymbol{\alpha}_2\mathbf{Z}_{11} & \boldsymbol{\alpha}_2\mathbf{Z}_{12} \\ \mathbf{0} & \boldsymbol{\beta}_1\mathbf{n}_x + \boldsymbol{\beta}_2\mathbf{E}_{12}\mathbf{d} & \boldsymbol{\beta}_1\mathbf{n}_y + \boldsymbol{\beta}_2\mathbf{E}_{14}\mathbf{d} \\ \mathbf{0} & -\boldsymbol{\gamma}_1\mathbf{n}_y + \boldsymbol{\gamma}_2\mathbf{E}_{32}\mathbf{d} & \boldsymbol{\gamma}_1\mathbf{n}_x + \boldsymbol{\gamma}_2\mathbf{E}_{34}\mathbf{d} \end{bmatrix} \tag{3.223a}$$

$$\overline{\mathbf{G}} = \begin{bmatrix} \boldsymbol{\alpha}_2\mathbf{Z}_{13} & \mathbf{0} & \mathbf{0} \\ \mathbf{0} & \boldsymbol{\beta}_2\mathbf{E}_{11} & \boldsymbol{\beta}_2\mathbf{E}_{13} \\ \mathbf{0} & \boldsymbol{\gamma}_2\mathbf{E}_{31} & \boldsymbol{\gamma}_2\mathbf{E}_{33} \end{bmatrix} \tag{3.223b}$$

$$\overline{\mathbf{F}} = \begin{Bmatrix} \boldsymbol{\alpha}_3 \\ \boldsymbol{\beta}_3 \\ \boldsymbol{\gamma}_3 \end{Bmatrix} \tag{3.223c}$$

where $\mathbf{n}_x$, $\mathbf{n}_y$ $\boldsymbol{\alpha}_i$, $\boldsymbol{\beta}_i$, $\boldsymbol{\gamma}_i$, $\mathbf{d}$, $\mathbf{E}_{ij}$, $\mathbf{Z}_{ij}$ are $N \times N$ diagonal matrices and $\boldsymbol{\alpha}_3$, $\boldsymbol{\beta}_3$, $\boldsymbol{\gamma}_3$ are N×1 vectors; $n_x = \cos a$, $n_y = \sin a$ are the direction cosines of the normal vector to the boundary.

Equations (3.214) and (3.222) constitute a system of $6N$ algebraic equations, which are solved to express the vectors $\mathbf{w}$, $\mathbf{w}_{,n}$, $\boldsymbol{\phi}_x$, $\boldsymbol{\phi}_{x,n}$, $\boldsymbol{\phi}_y$, $\boldsymbol{\phi}_{y,n}$ in terms of the vectors of the fictitious sources $\mathbf{b}^{(i)}(i = 1, 2, 3)$. The vectors $\mathbf{w}$, $\boldsymbol{\phi}_x$, $\boldsymbol{\phi}_y$ and their derivatives at the M domain nodal points are also expressed in terms of the vectors of fictitious sources using Eqs. (3.167).

We are now ready to evaluate the fictitious sources $\mathbf{b}^{(i)}$. For this purpose Eqs. (3.213) are applied at the M domain nodal points. This yields

$$\begin{aligned} A_{55}\left(\phi_{x,x}+\mathbf{w}_{,xx}\right)+A_{45}\left(\phi_{x,y}+\phi_{y,x}+\mathbf{2w}_{,xy}\right) \\ +A_{44}\left(\phi_{y,y}+\mathbf{w}_{,yy}\right)=-f \end{aligned} \tag{3.224a}$$

$$\begin{aligned} D_{11}\phi_{x,xx}+2D_{16}\phi_{x,xy}+D_{66}\phi_{x,yy}+D_{16}\phi_{y,xx}+(D_{12}+D_{66})\phi_{y,xy} \\ +D_{26}\phi_{y,yy}-A_{55}(\phi_x+\mathbf{w}_{,x})-A_{45}\left(\phi_y+\mathbf{w}_{,y}\right)=0 \end{aligned} \tag{3.224b}$$

$$\begin{aligned} D_{66}\phi_{y,xx}+2D_{26}\phi_{y,xy}+D_{22}\phi_{y,yy}+D_{16}\phi_{x,xx}+(D_{12}+D_{66})\phi_{x,xy} \\ +D_{26}\phi_{x,yy}-A_{45}(\phi_x+\mathbf{w}_{,x})-A_{44}\left(\phi_y+\mathbf{w}_{,y}\right)=0 \end{aligned} \tag{3.224c}$$

Substituting the derivatives from Eqs. (3.167) in Eqs. (3.224), we obtain the linear system of $3M$ equations whose solution gives the fictitious sources $\mathbf{b}^{(i)}$, $i=1, 2, 3$. Subsequently, these fictitious sources are used in Eqs. (3.167) to evaluate $\mathbf{w}_{,pq}$, $\phi_{,x,pq}$, and $\phi_{,y,pq}$. The stress resultants at the interior points are computed using Eqs. (3.211) and the boundary forces using Eqs. (3.216) and (3.220).

3.10 REFERENCES

[1] J.T. Katsikadelis, The analog equation method. A powerful BEM-based solution technique for solving linear and nonlinear engineering problems, in: C.A. Brebbia (Ed.), Boundary Element Method XVI, Proc. 16th International Boundary Element Method Conference (BEM XVI), July 12–15, Computational Mechanics Publications, Southampton, UK, 1994, pp. 167–182.

[2] J.T. Katsikadelis, Boundary Elements, Theory and Applications, Elsevier, Oxford, 2002.

[3] G. Gospodinov, D. Ljutskanov, The boundary element method applied to plates, Appl. Math. Model. 6 (1982) 237–244.

[4] W. Chen, Z.J. Shen, L.J. Shen, G.W. Yuan, General solutions and fundamental solutions of varied orders to the vibrational thin, the Berger and the Winkler plates, Eng. Anal. Bound. Elem. 29 (2005) 699–702.

[5] G.D. Manolis, D.E. Beskos, M.F. Pineros, Beam and plate stability by boundary elements, Comput. Struct. 22 (6) (1986) 917–923.

[6] M.R. Spiegel, Real Variables; Lebesgue Measure and Integration With Applications to Fourier Series, Scaum's Outline Series, McGraw-Hill, New York, 1969.

[7] M.A. Goldberg, C.S. Chen, The Dual Reciprocity Method and Radial Basis Functions, Elsevier, Oxford, 2000.

[8] E.J. Kansa, Y.C. Hon, Circumventing the ill-conditioning problem with multiquadric radial basis functions: applications to elliptic partial differential equations, Comput. Math. Appl. 39 (2005) 123–137.

[9] J.T. Katsikadelis, A generalized Ritz method for partial differential equations in domains of arbitrary geometry using global shape functions, Eng. Anal. Bound. Elem. 32 (5) (2008) 353–367.

[10] J.T. Katsikadelis, N. Babouskos, The post-buckling analysis of plates. A BEM based meshless variational solution, Facta Universitatis, Series Mechanics, Automatic Control and Robotics 6 (2007) 113–118.

[11] N. Nick Babouskos, J.T. Katsikadelis, Flutter instability of damped plates under combined conservative and nonconservative loads, Arch. Appl. Mech. 79 (2009) 541–556.

[12] S. Timoshenko, S. Woinowsky-Krieger, Theory of Plates and Shells, second ed., McGraw-Hill Book Company, New York, 1959.

[13] S. Bharatha, M. Levinson, A theory of elastic foundations, Arch. Ration. Mech. Anal. 74 (3) (1980) 249–266.
[14] E. Winkler, Die Lehre Von der Elasticität und Festigkeit, Dominicus, Prague, 1867.
[15] A.D. Kerr, Elastic and viscoelastic foundation models, J. Appl. Mech. 31 (1965) 491–498.
[16] A.D. Kerr, On the formal development of elastic foundation models, Ingenieur Archiv 54 (6) (1984) 455–464.
[17] A.P.S. Selvadurai, Elastic Analysis of Soil-foundation Interaction, Elsevier, New York, 1979.
[18] M.M. Filonenko-Borodich, Some approximate theories of the elastic foundation, Scientific Notes of the Moscow State University, Mechanics 46 (1940) 3–18 (in Russian).
[19] M. Hetenyi, Beams on Elastic Foundation, The University of Michigan Press, Ann Arbor, Michigan, 1946.
[20] P.L. Pasternak, On the New Method of Analysis of an Elastic Foundation by means of Two Foundation Constants, State Publications on Construction Literature, Architecture, Moscow, 1954 (in Russian).
[21] E. Reissner, A note on deflections of plates on viscoelastic foundation, J. Appl. Mech., Transactions ASME 25 (1958) 144–145.
[22] V.Z. Vlasov, N.N. Leontiev, Beams, Plates and Shells on Elastic Foundations, Israel Program for Scientific Translations, Jerusalem, Israel, 1966.
[23] J. Ratzersdorfer, Die Knickfestigkeitvon Stäben und Stabwerken, Springer-Verlag, New York, 1936.
[24] J.T. Katsikadelis, A.E. Armenakas, Analysis of clamped plates on elastic foundation by the boundary integral equation method, J. Appl. Mech., Transactions ASME 51 (1984) 547–580.
[25] J.T. Katsikadelis, A.E. Armenakas, Plates on elastic foundation by the BIE method, ASCE, J. Eng. Mech. 110 (7) (1984) 1086–1105.
[26] J.A. Costa Jr., C.A. Brebbia, The boundary element method applied to plates on elastic foundations, Eng. Anal. 2 (4) (1985) 174–183.
[27] J.T. Katsikadelis, L. Kallivokas, Clamped plates on Pasternak-type elastic foundation by the boundary element method, J. Appl. Mech., Transactions ASME 53 (4) (1986) 909–917.
[28] J.T. Katsikadelis, L. Kallivokas, Plates on biparametric elastic foundation by BDIE method, ASCE J. Eng. Mech. 114 (5) (1988) 847–875.
[29] J. Balas, V. Sladek, J. Sladek, The boundary integral equation method for plates resting on a two parameter foundation, ZAMM 64 (1984) 137–146.
[30] J. Puttonen, P. Varpasuo, Boundary element analysis of a plate on elastic foundations, Int. J. Numer. Meth. Eng. 23 (1986) 287–303.
[31] G. Bezine, A New boundary element method for bending of plates on elastic foundations, Int. J. Solid. Struct. 24 (6) (1988) 557–565.
[32] N. Kamiya, Y. Sawaki, The plate bending analysis by the dual reciprocity boundary elements, Eng. Anal. Bound. Elem. 5 (1) (1988) 36–40.
[33] J.T. Katsikadelis, E.J. Sapountzakis, A New method for the analysis of plates on elastic foundation, in: Proc. of the First Hellenic Conference on Geotechnical Engineering, Athens, Greece, vol. 2, 1988, pp. 197–202.
[34] N.A. Silva, W.S. Venturini, Dual reciprocity process applied to solve bending plates on elastic foundations, in: C.A. Brebbia (Ed.), Boundary Element X, Stress Analysis, Springer-Verlag, New York, 1988, pp. 95–105.
[35] S. de Leon, F. Paris, Analysis of thin plates on elastic foundations with boundary element method, Eng. Anal. Bound. Elem. 6 (4) (1989) 190–196.
[36] E.J. Sapountzakis, J.T. Katsikadelis, Unilaterally supported plates on elastic foundation by the boundary element method, J. Appl. Mech., Transactions ASME 59 (3) (1992) 580–586.
[37] S. Chucheepsakul, B. Chinnaboon, An alternative domain/ boundary element technique for analyzing plates on two-parameter elastic foundations, Eng. Anal. Bound. Elem. 26 (2002) 547–555.

[38] B. Chinnaboon, J.T. Katsikadelis, S. Chucheepsakul, A BEM-based meshless method for plates on biparametric elastic foundation with internal supports, Comput. Meth. Appl. Mech. Eng. 196 (33–34) (2007) 3165–3177.
[39] N. Kamiya, Y. Sawaki, An integral equation approach to finite deflection of elastic plates, Int. J. Nonlin. Mech. 17 (1982) 187–194.
[40] N. Kamiya, Y. Sawaki, Y. Nakamura, A. Fukui, An approximate finite deflection of a heated elastic plate by the boundary element method, Appl. Math. Model. 6 (1982) 23–27.
[41] N. Kamiya, Y. Sawaki, Y. Nakamura, Postbuckling analysis by the boundary element method, Eng. Anal. 1 (1984) 40–44.
[42] J. Sladek, V. Sladek, The BIE analysis of the Berger equation, Inginieur Archiv 53 (1983) 385–397.
[43] Y. Sawaki, K. Tachinchi, N. Kamiya, Finite deflection analysis of plates by dual reciprocity boundary elements, in: C.A. Brebbia, J.T. Connor (Eds.), in: Advances in Boundary Elements, vol. 3, Springer-Verlag, Berlin, 1989, pp. 269–284.
[44] M. Tanaka, Integral equation approach to small and large displacements of thin elastic plates, in: C.A. Brebbia (Ed.), in: Boundary Element Methods in Engineering, Springer-Verlag, Berlin, 1982, pp. 526–539.
[45] M. Tanaka, Large deflection analysis of thin elastic plates, in: D.K. Banerjee, S. Mukherjee (Eds.), in: Developments in Boundary Element Methods, vol. 3, Elsevier Applied Science Publishers Ltd., London and New York, 1984, pp. 115–136
[46] M. Nerantzaki, J.T. Katsikadelis, A Green's function method for large deflection analysis of plates, Acta Mech. 25 (1988) 211–225.
[47] J.T. Katsikadelis, M.S. Nerantzaki, Large deflections of thin plates by the boundary element method, in: C.A. Brebbia (Ed.), Boundary Elements X, vol. 3, Berlin, Springer-Verlag, 1988, pp. 435–456.
[48] M.S. Nerantzaki, Non-linear Analysis of Plates by the Boundary Element Method, Doctoral Dissertation, National Technical, University, Athens, Greece, 1992.
[49] J.T. Katsikadelis, M.S. Nerantzaki, Non-linear analysis of plates by the analog equation method, Comput. Mech. 14 (2) (1994) 154–164.
[50] A. Elzein, S. Syngellakis, High-order elements for the BEM stability analysis of imperfect plates, in: C.A. Brebbia, J.J. Connor (Eds.), in: Advances in Boundary Elements, 3, Springer-Verlag, Berlin, 1989, pp. 269–284.
[51] N. Kamiya, Y. Sawaki, An integral equation approach to finite deflection of elastic plates, Int. J. Nonlin. Mech. 17 (1982) 187–194.
[52] P.E. O'Donoghue, S.N. Atluri, Field/boundary element approach to the large deflection of thin flat plates, Comput. Struct. 27 (3) (1987) 427–435.
[53] S.N. Atluri, D.S. Pipkins, Large deformation analysis of plates and shells, in: D. Beskos (Ed.), Boundary Element Analysis of Plates and Shells, Springer-Verlag, Berlin, 1991, pp. 141–166.
[54] N. Kamiya, Structural non-linear analysis by boundary element methods, in: C.A. Brebbia (Ed.), Boundary Element, X, vol. 3, Springer-Verlag, Berlin, 1988, pp. 17–27.
[55] K. Washizu, Variational Methods in Elasticity and Plasticity, Pergamon Press, Oxford, 1975.
[56] N. Babouskos, Analysis and Optimization of Elastic and Viscoelastic Plates by the BEM, Doctoral Dissertation, National Technical University, Athens, 2011.
[57] J.T. Katsikadelis, N. Babouskos, The post-buckling analysis of plates. A BEM based meshless variational solution, Facta Universitatis, Series Mechanics, Automatic Control and Robotics 6 (2007) 113–118.
[58] J.T. Katsikadelis, N. Babouskos, A BEM based meshless variational method for solving linear and nonlinear plate problems, in: Proc. of First Serbian (26th YU) Congress on Theoretical and Applied Mechanics, Kopaonik, Serbia, April 10–13, 2007, pp. 463–474.
[59] P.S. Bulson, The Stability of Flat Plates, Chatoo & Windus, London, 1970.
[60] Q. Qin, Y. Huang, BEM of post-buckling analysis of thin plates, Appl. Math. Model. 14 (1990) 544–548.

[61] J.M.T. Thompson, G.W. Hunt, A General Theory of Elastic Stability, Wiley, London, 1973.
[62] M.P. Bendsee, On obtaining a solution to optimization problems for solid, elastic plates by restriction of the design space, Mech. Base. Des. Struct. Mach. 11 (4) (1983) 501–521.
[63] F. Niordson, Optimal design of elastic plates with a constraint on the slope of the thickness function, Int. J. Solid. Struct. 19 (2) (1983) 141–151.
[64] J.T. Katsikadelis, N.G. Babouskos, Stiffness and buckling optimization of thin plates with BEM, Arch. Appl. Mech. 82 (2012) 1403–1422.
[65] E.J. Sapountzakis, J.T. Katsikadelis, Boundary element solution for plates of variable thickness, ASCE, J. Eng. Mech. 117 (6) (1991) 1241–1256.
[66] E.W.V. Chaves, G.R. Fernandes, W.S. Venturini, Plate bending boundary element formulation considering variable thickness, Eng. Anal. Bound. Elem. 23 (5–6) (1991) 405–418.
[67] M.S. Nerantzaki, J.T. Katsikadelis, Analysis of plates with variable thickness. An analog equation solution, in: M.H. Aliabadi (Ed.), Plate Bending Analysis with Boundary Elements, Computational Mechanics Publications, Southampton, 1998, pp. 275–307 Chapter 9.
[68] M.S. Nerantzaki, J.T. Katsikadelis, Buckling of plates with variable thickness - an analog equation solution, Eng. Anal. Bound. Elem. 18 (2) (1996) 149–154.
[69] M.S. Nerantzaki, J.T. Katsikadelis, Large deflections of axisymmetric circular plates with variable thickness, Int. J. Comput. Civ. Struct. Eng. 1 (5) (2003) 75–83.
[70] J.T. Katsikadelis, A.J. Yiotis, The BEM for plates of variable thickness on nonlinear biparametric elastic foundation. An analog equation solution, J. Eng. Math. 46 (3–4) (2003) 313–330.
[71] M. Nerantzaki, J.T. Katsikadelis, The BEM for large deflection analysis of plates with variable thickness, in: R. Callego, M.H. Aliabadi (Eds.), Proc. of the 4th International Conference on Boundary Element Techniques BeTEq IV, Granada July 15–17, 2003, pp. 353–358, Granada, Spain.
[72] W.H. Wittrik, C.H. Ellen, Buckling of tapered rectangular plates in compression, Aeronaut. Q. 13 (1962) 238–291.
[73] L.E. Harik, X. Liu, R. Ekambaram, Elastic stability of plates with varying rigidities, Comput. Struct. 38 (1991) 161–168.
[74] C.Y. Chia, Nonlinear Analysis of Plates, McGraw-Hill, New York, 1980.
[75] Y. Jianqiao, Large deflection analysis of axisymmetric circular plates with variable thickness by the boundary element method, Appl. Math. Model. 55 (1991) 325–328.
[76] C.M. Wang, J.N. Reddy, K.H. Lee, Shear Deformable Beams and Plates, Elsevier Science, Oxford, 2000.
[77] E. Reissner, The effect of transverse shear deformation on the bending of elastic plates, ASME J. Appl. Mech. 12 (1945) 69–77.
[78] R.D. Mindlin, Influence of rotary inertia and shear on flexural motions of isotropic, elastic plates, ASME J. Appl. Mech. 18 (1951) 31–38.
[79] R. Kienzler, On consistent plate theories, Arch. Appl. Mech. 72 (2002) 229–247.
[80] M.H. Aliabadi (Ed.), Plate Bending Analysis with Boundary Elements, Computational Mechanics Publications, Southampton, 1998.
[81] Y.F. Rashed, Boundary Element Formulations for Thick Plates, in: Topics in Engineering, vol. 35, WIT Press, Southampton and Boston, 1999.
[82] M.H. Aliabadi, The Boundary Element Method, John Wiley & Sons, London, 2002 Volume 2.
[83] B. Chinnaboon, S. Chucheepsakul, J.T. Katsikadelis, A BEM based domain meshless method for the analysis of Mindlin plates with general boundary conditions, Comput. Meth. Appl. Mech. Eng. (2010).
[84] G. Shi, G. Bezine, A general boundary integral formulation for the anisotropic plate bending problems, J. Compos. Mater. 22 (1988) 694–716.
[85] B.C. Wu, N.J. Altiero, A new numerical method for the analysis of anisotropic thin plate bending problems, Comput. Meth. Appl. Mech. Eng. 25 (1981) 343–353.

[86] C. Rajamohan, J. Raamachandran, Bending of anisotropic plates by charge simulation method, Adv. Eng. Software 30 (1999) 369–373.
[87] J. Wang, K. Schweizerhof, The fundamental solution of moderately thick laminated anisotropic shallow shells, Int. J. Eng. Sci. 33 (1995) 995–1004.
[88] J. Wang, K. Schweizerhof, Study on free vibration of moderately thick orthotropic laminated shallow shells by boundary-domain elements, Appl. Math. Model. 20 (1996) 579–584.
[89] W.P. Paiva, P. Sollero, E.L. Albuquerque, Treatment of hypersingularities in boundary element anisotropic plate bending problems, Lat. Am. J. Solid. Struct. 1 (2003) 49–73.
[90] E.L. Albuquerque, P. Sollero, W.S. Venturini, M.H. Aliabadi, Boundary element analysis of anisotropic Kirchhoff plates, Int. J. Solid. Struct. 43 (2006) 4029–4046.
[91] H. Irschik, A boundary-integral equation method for bending of orthotropic plates, Int. J. Solid. Struct. 20 (3) (1984) 245–255.
[92] C. Hwu, Anisotropic Elastic Plates, Springer, New York, 2010.
[93] J.N. Reddy, Mechanics of Laminated Composite Plates and Shells: Theory and Analysis, second ed., CRC Press, New York, 2004.
[94] S.G. Lekhniskii, Anisotropic Plates, Gordon and Breach Science Publishers, New York, 1968.

PROBLEMS

3.1 Compute the buckling load of the plate of Fig. P3.1. Data: $h=0.2\,m$, $E=2.1\times10^{7}\mathrm{kN/m^2}$, and $\nu=0.2$. The sides 1-2 and 5-6 are movable in the plane of the plate.

3.2 Analyze the continuous plate of Fig. P3.2. Data: uniform load $f_0=10\mathrm{kN/m^2}$, $h=0.2\,\mathrm{m}$, $E=2.1\times10^{6}\mathrm{kN/m^2}$, and $\nu=0.2$. Large deflections will be considered. Except the clamped edge, all other edges are movable in the plane of the plate.

3.3 On the base of the numerical solution presented in Section 3.7.3 write the computer code in FORTAN or MATLAB language for the analysis of thick plates.

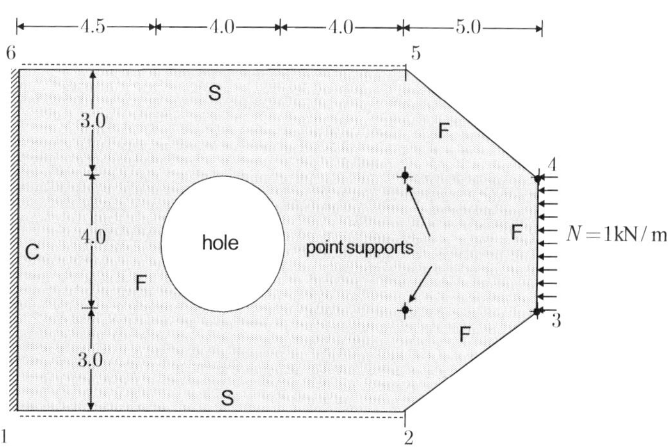

FIGURE P3.1 Plate of composite shape with a hole and internal point supports.

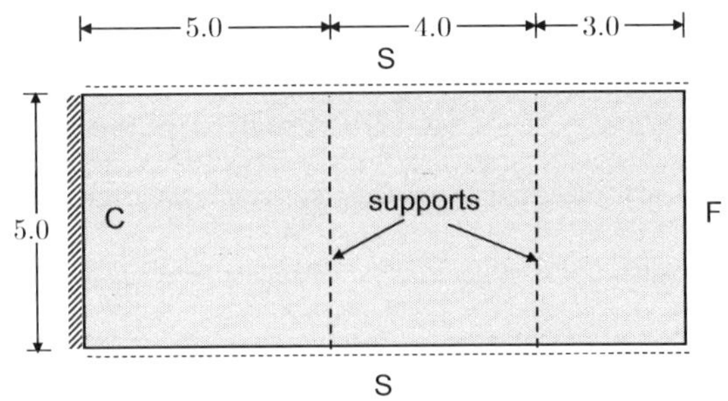

FIGURE P3.2 Continuous plate.

3.4 Prove Eq. (3.175).

3.5 Analyze the mushroom slab of Fig. P3.3a. Data: uniform load $f_0 = 10\text{kN/m}^2$, $h = 0.15\,\text{m}$, $E = 2.1 \times 10^6\text{kN/m}^2$, and $\nu = 0.2$. The plate is simply supported along the boundary. The detail of the column support is shown in Fig. P3.3b.

3.6 Analyze the three-layer cross-ply $(0°/90°/0°)$ anisotropic plate of Fig. P3.4 under the uniform load $f_0 = 5\,\text{kN/m}^2$. The layers are identical with $h = 0.08\,\text{m}$, $E_1 = 19.2\,\text{MPa}$, $E_2 = 1.56\,\text{MPa}$, $G_{12} = 0.82\,\text{MPa}$, $\nu_{12} = 0.24$.

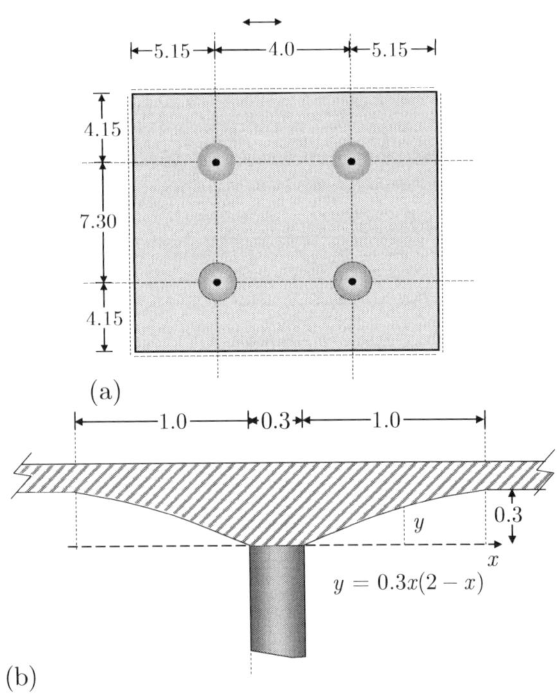

FIGURE P3.3 Mushroom plate. (a) Geometry. (b) Detail of the column support.

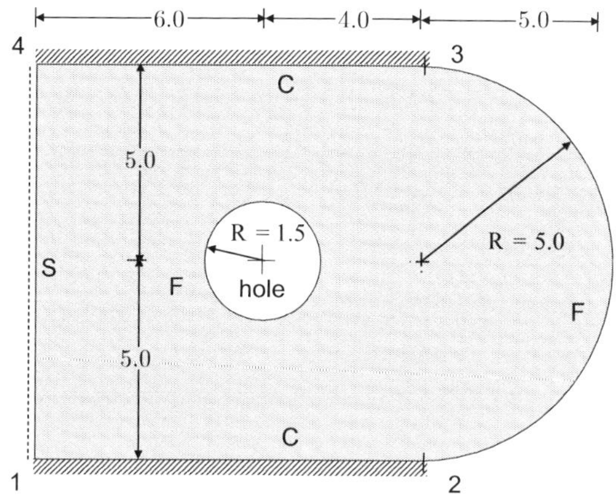

FIGURE P3.4 Anisotropic plate of composite boundary.

3.7 Analyze the three-layer $(45°/0°/-45°)$ clamped anisotropic plate of Fig. P3.5 under the uniform load $f_0 = 10\,\mathrm{kN/m^2}$. The layers are identical with $h = 0.08\,\mathrm{m}$, $E_1 = 19.2\,\mathrm{MPa}$, $E_2 = 1.56\,\mathrm{MPa}$, $G_{12} = 0.82\,\mathrm{MPa}$, $\nu_{12} = 0.24$. The boundaries of the plate are defined by the curves: Outer boundary $\frac{x^4}{a^4} + \frac{y^4}{b^4} = 1\ (a = 5,\ b = 4)$; inner boundary $x^2 + y^2 = r^2$ $(r = 1.5)$.

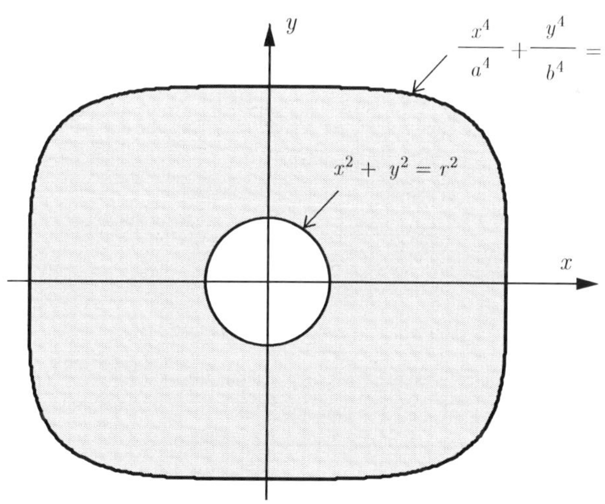

FIGURE P3.5 Anisotropic plate with a hole.

Chapter | four

BEM for Dynamic Analysis of Plates

CHAPTER OUTLINE

The Boundary Element Method for Plate Analysis

4.1 DIRECT BEM FOR THE DYNAMIC PLATE PROBLEM

The equation for the dynamic response of a thin plate may result from the equation of the static problem if the inertia force $-\rho h\ddot{w}$ is included in the transverse load; ρ is the mass density of the material. Hence, the dynamic problem for the plate in terms of the displacement $w = w(\mathbf{x}, t)$, $\mathbf{x}: \{x, y\}$ is described by the equation

$$\nabla^4 w + \frac{\rho h}{D}\ddot{w} = \frac{f(\mathbf{x}, t)}{D}, \quad \mathbf{x} \in \Omega,\ t > 0 \tag{4.1}$$

with boundary conditions (see Section 2.2.6)

$$Vw + k_T w = V_n^* \text{ or } w = w^* \quad \text{on } \Gamma \tag{4.2a}$$

$$Mw - k_R w_{,n} = M_n^* \text{ or } w_{,n} = w_{,n}^* \quad \text{on } \Gamma \tag{4.2b}$$

$$k_c^{(k)} w_k - [\![Tw]\!]_k = R_k^* \text{ or } w_k = w_k^* \quad \text{at corner } k \tag{4.2c}$$

and initial conditions

$$w(\mathbf{x},0) = g_1(\mathbf{x}), \ \dot{w}(\mathbf{x},0) = h_1(\mathbf{x}) \quad \mathbf{x} \in \Omega \tag{4.3a,b}$$

It should be noted that the specified boundary quantities w^*, $w_{,n}^*$, M_n^*, V_n^* depend on time in the case of support excitation.

The formulation of the direct BEM for the initial boundary value problem (4.1)-(4.3) requires the knowledge of the dynamic fundamental solution $\tilde{v}(\mathbf{x}, \boldsymbol{\xi}, t)$ of Eq. (4.1), namely, a singular particular solution of the equation

$$\nabla^4 \tilde{v} + \frac{\rho h}{D}\ddot{\tilde{v}} = \delta(\mathbf{x} - \boldsymbol{\xi})\delta(t), \quad \mathbf{x}: \{x, y\}, \boldsymbol{\xi}: \{\xi, \eta\} \tag{4.4}$$

with initial conditions

$$\tilde{v}(\mathbf{x},0) = 0, \ \dot{\tilde{v}}(\mathbf{x},0) = 0 \tag{4.5a,b}$$

The fundamental solution $\widetilde{v}(\mathbf{x}, \boldsymbol{\xi}, t)$ may be obtained by differentiation with respect to time of the function

$$U(\mathbf{x}, \boldsymbol{\xi}, t) = \frac{t}{4k} F\left|\frac{kr^2}{4t}\right| \tag{4.6}$$

where

$$k = \sqrt{\rho h/D} \text{ and } r = |\mathbf{x} - \boldsymbol{\xi}| \tag{4.7}$$

The function F has the form

$$F(z) = \frac{\pi}{2} - \text{Si}(z) - \sin z + z\text{Ci}(z), \quad z = kr^2/4t \tag{4.8}$$

in which

$$\text{Si}(z) = \int_0^z \frac{\sin s}{s} ds, \quad \text{Ci}(z) = \int_0^z \frac{\cos s}{s} ds \quad z \geq 0 \tag{4.9a,b}$$

are the sine and cosine integrals [1].

Indeed, it can be shown that [2]

$$\nabla^4 U + k^2 \ddot{U} = \delta(\mathbf{x} - \boldsymbol{\xi}) H(t) \tag{4.10}$$

where $H(t)$ is the Heaviside function defined as

$$H(t - t_0) = \begin{cases} 0 & t < t_0 \\ 1 & t_0 > t \end{cases}$$

Differentiating Eq. (4.10) with respect to time and taking into account that $\partial H(t)/\partial t = \delta(t)$ we conclude that the function

$$\widetilde{v}(\mathbf{x}, \boldsymbol{\xi}, t) = \frac{\partial U(\mathbf{x}, \boldsymbol{\xi}, t)}{\partial t} \tag{4.11}$$

satisfies Eq. (4.4) and consequently, it is the sought fundamental solution.

The function $\widetilde{v}(\mathbf{x}, \boldsymbol{\xi}, t)$ represents the transverse displacement at point $\mathbf{x}$ of an infinite plate having flexural stiffness D due to a unit concentrated impulsive load at point $\boldsymbol{\xi}$.

The integral representation of the solution of Eq. (4.1) can be derived using the reciprocal theorem for structural dynamics, known as dynamic Betti-Rayleigh reciprocal theorem [3], which states:

If $p_I(t)$ and $p_{II}(t)$ are two loadings, which act separately on a linear dynamic system and produce the responses $u_I(t)$ and $u_{II}(t)$, then the convolution(*) $C_{I,II} = p_I(t) * u_{II}(t)$ of the loading $p_I(t)$ and the response $u_{II}(t)$ equals the convolution $C_{II,I} = p_{II}(t) * u_I(t)$ of the loading $p_{II}(t)$ and the response $u_I(t)$, namely

$$p_I(t) * u_{II}(t) = p_{II}(t) * u_I(t) \tag{4.12}$$

If we consider two elastodynamic states $\tilde{v}$ and w of a plate, produced by two different load systems (see Eqs. (2.66a,b)), then on the basis of the reciprocal theorem, we obtain

$$\begin{aligned} D\int_{\Omega}\left(\tilde{v} * \nabla^4 w - w * \nabla^4 \tilde{v}\right) d\Omega = -\int_{\Gamma}[\tilde{v} * Vw - w * V\tilde{v} \\ -\tilde{v}_{,n} * Mw + w_{,n} * M\tilde{v}]\, ds + \sum_k \left(\tilde{v} * [\![Tw]\!]_k - w * [\![T\tilde{v}]\!]_k\right) \end{aligned} \tag{4.13}$$

Using Eqs. (4.1) and (4.4) and interchanging the role of points $\boldsymbol{\xi}$ and $\mathbf{x}$, we write the domain integral in Eq. (4.13)

---

*We recall that the convolution of two functions $f(t)$ and $g(t)$ is defined as

$$f(t) * g(t) = \int_0^t f(t-\tau)\, g(\tau)\, d\tau \tag{a}$$

The convolution has the commutative property

$$f(t) * g(t) = g(t) * f(t) \tag{b}$$

Using the Leibnitz rule for the differentiation of an integral with respect to a parameter involved in the integration limits, Eq. (a) gives

$$\frac{d}{dt}[f * g] = \dot{f} * g + f_0 g \tag{c}$$

$$\frac{d^2}{dt^2}[f * g] = \ddot{f} * g + \dot{f}_0 g + f_0 \dot{g} \tag{d}$$

Similarly,

$$\frac{d^2}{dt^2}[g * f] = \ddot{g} * f + \dot{g}_0 f + g_0 \dot{f} \tag{e}$$

Inasmuch as the left-hand sides of Eqs. (d) and (e) are equal we obtain

$$\ddot{f} * g + \dot{f}_0 g + f_0 \dot{g} = \ddot{g} * f + \dot{g}_0 f + g_0 \dot{f} \tag{f}$$

$$D\int_{\Omega}\left(\widetilde{v}*\nabla^{4}w-w*\nabla^{4}\widetilde{v}\right)d\Omega=\int_{\Omega}\left[\widetilde{v}*f-Dw*\delta(\boldsymbol{\xi}-\mathbf{x})\delta(t)-Dk^{2}\left(\ddot{w}*\widetilde{v}-\ddot{\widetilde{v}}*w\right)\right]d\Omega \tag{4.14}$$

Taking into account the identity (f) in the footnote of page 214 with $\widetilde{v}(\mathbf{x},0)=0$, $\dot{\widetilde{v}}(\mathbf{x},0)=0$, and

$$\begin{aligned}\int_{\Omega}w(\boldsymbol{\xi},t)*[\delta(\boldsymbol{\xi}-\mathbf{x})\delta(t)]\,d\Omega &=w(\mathbf{x},t)*\delta(t)\\ &=\int_{0}^{T}w(\mathbf{x},t-\tau)\delta(\tau)*d\tau\\ &=w(\mathbf{x},t)\end{aligned} \tag{4.15}$$

we obtain the integral representation of the solution

$$\begin{aligned}w(\mathbf{x},t)=&\frac{1}{D}\int_{\Omega}\widetilde{v}*f(t)\,d\Omega+k^{2}\int_{\Omega}\left[\widetilde{v}\dot{w}(\boldsymbol{\xi},0)+\dot{\widetilde{v}}w(\boldsymbol{\xi},0)\right]d\Omega\\ &+\frac{1}{D}\int_{\Gamma}[\widetilde{v}*Vw-w*V\widetilde{v}-\widetilde{v}_{,n}*Mw+w_{,n}*M\widetilde{v}]\,ds \quad \mathbf{x}\in\Omega,\ t>0\\ &-\frac{1}{D}\sum_{k}\left(\widetilde{v}*[\![Tw]\!]_{k}-w*[\![T\widetilde{v}]\!]_{k}\right)\end{aligned} \tag{4.16}$$

Letting point $\mathbf{x}$ approach the boundary Γ and applying a limiting process, we obtain the first boundary integral equation

$$\begin{aligned}\frac{1}{2}w(\mathbf{x},t)=&\frac{1}{D}\int_{\Omega}\widetilde{v}*f(t)\,d\Omega+k^{2}\int_{\Omega}\left[\widetilde{v}\dot{w}(\boldsymbol{\xi},0)+\dot{\widetilde{v}}w(\boldsymbol{\xi},0)\right]d\Omega\\ &+\frac{1}{D}\int_{\Gamma}[\widetilde{v}*Vw-w*V\widetilde{v}-\widetilde{v}_{,n}*Mw+w_{,n}*M\widetilde{v}] \quad \mathbf{x}\in\Gamma,\ t>0\\ &-\frac{1}{D}\sum_{k}\left(\widetilde{v}*[\![Tw]\!]_{k}-w*[\![T\widetilde{v}]\!]_{k}\right)\end{aligned} \tag{4.17a}$$

The second boundary integral equation is derived by differentiating Eq. (4.16) with respect to a direction $\boldsymbol{\nu}$ and then letting point $\mathbf{x}$ approach the boundary and direction $\boldsymbol{\nu}$ coincide with the normal to the boundary. Thus, we obtain

$$\begin{aligned}\frac{1}{2}\frac{\partial w(\mathbf{x},t)}{\partial\nu}=&\frac{1}{D}\int_{\Omega}\widetilde{v}_{1}*f(t)\,d\Omega+k^{2}\int_{\Omega}\left[\widetilde{v}_{1}\dot{w}(\boldsymbol{\xi},0)+\dot{\widetilde{v}}_{1}w(\boldsymbol{\xi},0)\right]d\Omega\\ &+\frac{1}{D}\int_{\Gamma}[\widetilde{v}_{1}*Vw-w*V\widetilde{v}_{1}-\widetilde{v}_{1,n}*Mw+w_{,n}*M\widetilde{v}_{1}]\,ds \quad \mathbf{x}\in\Gamma,\ t>0\\ &-\frac{1}{D}\sum_{k}\left(\widetilde{v}_{1}*[\![Tw]\!]_{k}-w*[\![T\widetilde{v}_{1}]\!]_{k}\right)\end{aligned} \tag{4.17b}$$

in which $\widetilde{v}_1 = \widetilde{v}_{,\nu}(\mathbf{x}, \boldsymbol{\xi}, t)$. Note that Eqs. (4.17a,b) are valid for points where the boundary is smooth.

The limiting procedures for the derivation of the boundary integral equations are similar to those presented in Section 2.3.4, but more difficult because of the mathematical complexity of the dynamic fundamental solution and the manipulation of the convolutions. For corner points, namely, for points where the boundary is not smooth, the limiting procedure becomes even more difficult.

The boundary integral equations (4.17a,b) are solved to yield the unspecified boundary quantities. Apparently, this can be done only numerically using both spatial and temporal discretization. The spatial discretization refers to the evaluation of the boundary and domain integrals, while the temporal discretization refers to the total time of motion, which is divided into equal small time intervals Δt. Thus, the use of constant elements to approximate the boundary integrals and linear triangular elements to approximate the domain integrals converts the integral equations into a system of linear equations of motion, which can be solved by a time-step integration method. Evidently, the complexity of the fundamental solution considerably increases the difficulty of the numerical solution. For this reason this BEM formulation for the dynamic plate problem, though very elegant from the mathematical point of view, was very early abandoned as devoid of practical significance for engineering analysis.

For other plate problems such as those discussed in Chapter 3, the dynamic fundamental solution has not been established and the presented formulation cannot be applied for BEM dynamic analysis of plates. Since the BEM is a computational method for realistic engineering problems, the researchers sought other simpler BEM formulations for the solution of dynamic plate problems. The solution becomes simple when the plate is subjected to a harmonic load [4]. In this case the dynamic problem is converted into a static problem with known fundamental solution. Another method that has been widely used is the Laplace transform method. In the transformed domain the problem is similar to that of harmonic load and thus can be solved. The inverse Laplace transform gives the solution in the time domain [5]. But even this method has a limited application, because of the need to use numerical inversion of the Laplace transform, and applies only to classical plate problem for which the fundamental solution in the transformed domain is known. Another approach is to use the static fundamental solution with domain discretization, namely, the D/BEM. This method, which also allows the solution of nonlinear problems, has effectively prevailed. Various formulations have been reported with each new one aiming at improving the prior one and removing its drawbacks [6,7]. Special methods have been developed as well, in which either the Green function is established numerically [8] or a flexibility matrix with respect to a set of domain mass points is computed [9]. The literature on the dynamic plate problem is extensive. Since the review of the published work on the subject is not the aim of this section, we refer to the comprehensive review article by Providakis and Beskos [10]. Here, our presentation for the plate dynamic problem will be limited to the AEM, which, as we will see, solves all plate problems, linear or nonlinear, either as a boundary-only BEM or as D/BEM. The AEM, beside the accuracy of the obtained results, does not need special formulation for each plate problem. The structure of the computer program is the same for all dynamic problems, linear or nonlinear. The AEM has been employed to solve many dynamic plate problems such as plates with variable thickness [11,12], plates reinforced with beams [13–15], stability and post-buckling plate problems and plates under nonconservative loads (linear and nonlinear flutter instability) [16,17], nonlinear vibrations of viscoelastic plates described with integer or fractional derivative models [18], as well as plate thickness optimization problems under constant material volume to maximize the aerodynamic pressure [19].

4.2 AEM FOR THE DYNAMIC PLATE PROBLEM

4.2.1 Equations for the nonlinear dynamic problem

For the economy of the presentation, we will present the AEM for the nonlinear dynamic plate problem. The linear problem will occur by omitting the nonlinear terms. The equations result from the equations of the static problem (3.99) and (3.101) if the membrane $-\rho h\ddot{u}$, $-\rho h\ddot{v}$ and the transverse $-\rho h\ddot{w}$ inertia forces are included in the external load. Thus, if the transverse damping force $-c\dot{w}$ is also included, we have the following equations:

(i) For the transverse displacement

$$\begin{aligned} D\nabla^4 w - \left(N_x w_{,xx} + 2N_{xy} w_{,xy} + N_y w_{,yy}\right) \\ + (p_x - \rho h\ddot{u}) w_{,x} + \left(p_y - \rho h\ddot{v}\right) w_{,y} + c\dot{w} + \rho h\ddot{w} = f(\mathbf{x}, t) \end{aligned} \quad \mathbf{x}\in\Omega, t>0 \tag{4.18}$$

with boundary conditions

$$Vw + N_n w_{,n} + N_t w_{,t} + k_T w = V_n^* \ \text{ or } \ w = w^* \quad \text{on } \Gamma \tag{4.19a}$$

$$Mw - k_R w_{,n} = M_n^* \ \text{ or } \ w_{,n} = w_{,n}^* \quad \text{on } \Gamma \tag{4.19b}$$

$$k_c^{(k)} w^{(k)} - [\![Tw]\!]_k = R_k^* \ \text{ or } \ w^{(k)} = w_k^* \quad \text{at corner } k \tag{4.19c}$$

and initial conditions

$$w(\mathbf{x},0) = g_1(\mathbf{x}), \quad \dot{w}(\mathbf{x},0) = h_1(\mathbf{x}) \quad \mathbf{x}\in\Omega \tag{4.20a,b}$$

(ii) For the membrane displacements

$$\begin{aligned} \nabla^2 u + \frac{1+\nu}{1-\nu}\left(u_{,x} + v_{,y}\right)_{,x} + w_{,x}\left(\frac{2}{1-\nu} w_{,xx} + w_{,yy}\right) \\ + \frac{1+\nu}{1-\nu} w_{,xy} w_{,y} - \frac{\rho\ddot{u}}{G} + \frac{1}{Gh} p_x(\mathbf{x}, t) = 0 \end{aligned} \quad \mathbf{x}\in\Omega, t>0 \tag{4.21a}$$

$$\nabla^2 v + \frac{1+\nu}{1-\nu}\left(u_{,x}+v_{,y}\right)_{,y} + w_{,y}\left(\frac{2}{1-\nu}w_{,yy}+w_{,xx}\right) + \frac{1+\nu}{1-\nu}w_{,xy}w_{,x} - \frac{\rho\ddot{v}}{G} + \frac{1}{Gh}p_y(\mathbf{x},t) = 0 \quad \mathbf{x}\in\Omega, t>0 \tag{4.21b}$$

with boundary conditions

$$N_n = N_n^* \ \text{ or } \ u_n = u_n^* \quad \text{on } \Gamma \tag{4.22a}$$

$$N_t = N_t^* \ \text{ or } \ u_t = u_t^* \quad \text{on } \Gamma \tag{4.22b}$$

and initial conditions

$$u(\mathbf{x},0) = g_2(\mathbf{x}), \qquad \dot{u}(\mathbf{x},0) = h_2(\mathbf{x}) \quad \mathbf{x}\in\Omega \tag{4.23a,b}$$

$$v(\mathbf{x},0) = g_3(\mathbf{x}), \qquad \dot{v}(\mathbf{x},0) = h_3(\mathbf{x}) \quad \mathbf{x}\in\Omega \tag{4.24a,b}$$

where $g_i(\mathbf{x})$, $h_i(\mathbf{x})$ $(i=1, 2, 3)$ are specified functions.

The quantities N_x, N_y, N_{xy} represent the membrane forces given by the relations (3.103).

4.2.2 AEM solution for the nonlinear dynamic problem

The analog equations of Eqs. (4.18) and (4.21) are

$$\nabla^4 w = b(\mathbf{x},t), \quad \nabla^2 u = b^{(1)}(\mathbf{x},t) \quad \nabla^2 v = b^{(2)}(\mathbf{x},t) \quad \mathbf{x}\in\Omega, t>0 \tag{4.25a,b,c}$$

The fictitious loads depend now on time, which is regarded as a parameter, i.e., the equations are quasi-static and are treated as static equations at each instant. This implies that the principle of the analog equation converts a time dependent problem into an infinite sequence of elliptic problems in time.

Application of Eqs. (4.18) and (4.21) at the M domain nodal points give

$$D\mathbf{b} - \left(\mathbf{N}_x\mathbf{w}_{,xx} + 2\mathbf{N}_{xy}\mathbf{w}_{,xy} + \mathbf{N}_y\mathbf{w}_{,yy}\right) + \mathbf{p}_x\mathbf{w}_{,x} + \mathbf{p}_y\mathbf{w}_{,y} + c\dot{\mathbf{w}} + \rho h\left(\ddot{\mathbf{w}} - \ddot{\mathbf{u}}\mathbf{w}_{,x} - \ddot{\mathbf{v}}\mathbf{w}_{,y}\right) = \mathbf{f}(t) \tag{4.26a}$$

$$\mathbf{b}^{(1)} + \frac{1+\nu}{1-\nu}\left(\mathbf{u}_{,x}+\mathbf{v}_{,y}\right)_{,x} + \mathbf{w}_{,x}\left(\frac{2}{1-\nu}\mathbf{w}_{,xx}+\mathbf{w}_{,yy}\right) + \frac{1+\nu}{1-\nu}\mathbf{w}_{,xy}\mathbf{w}_{,y} - \frac{\rho\ddot{\mathbf{u}}}{G} + \frac{1}{Gh}\mathbf{p}_x = 0 \tag{4.26b}$$

$$\mathbf{b}^{(2)} + \frac{1+\nu}{1-\nu}\left(\mathbf{u}_{,x}+\mathbf{v}_{,y}\right)_{,y} + \mathbf{w}_{,y}\left(\frac{2}{1-\nu}\mathbf{w}_{,yy}+\mathbf{w}_{,xx}\right) + \frac{1+\nu}{1-\nu}\mathbf{w}_{,xy}\mathbf{w}_{,x} - \frac{\rho\ddot{\mathbf{v}}}{G} + \frac{1}{Gh}\mathbf{p}_y = 0 \tag{4.26c}$$

while Eqs. (3.103) give

$$\mathbf{N}_x = C\left[\mathbf{u}_{,x} + \frac{1}{2}\mathbf{w}_{,x}\mathbf{w}_{,x} + \nu\left(\mathbf{v}_{,y} + \frac{1}{2}\mathbf{w}_{,y}\mathbf{w}_{,y}\right)\right] \tag{4.27a}$$

$$\mathbf{N}_y = C\left[\mathbf{v}_{,y} + \frac{1}{2}\mathbf{w}_{,y}\mathbf{w}_{,y} + \nu\left(\mathbf{u}_{,x} + \frac{1}{2}\mathbf{w}_{,x}\mathbf{w}_{,x}\right)\right] \tag{4.27b}$$

$$\mathbf{N}_{xy} = C\frac{1-\nu}{2}\left(\mathbf{u}_{,y} + \mathbf{v}_{,x} + \mathbf{w}_{,x}\mathbf{w}_{,y}\right) \tag{4.27c}$$

Equations (4.26) with the help of Eqs. (3.46) and (3.76) give the following three nonlinear equations of motion in terms of the components of the vectors $\mathbf{b}(t)$, $\mathbf{b}^{(1)}(t)$, $\mathbf{b}^{(2)}(t)$:

$$\mathbf{M}\ddot{\mathbf{b}} + \mathbf{C}\dot{\mathbf{b}} + \mathbf{H}\left(\mathbf{b}, \mathbf{b}^{(1)}, \ddot{\mathbf{b}}^{(1)}, \mathbf{b}^{(2)}, \ddot{\mathbf{b}}^{(2)}\right) = \mathbf{f}(t) \tag{4.28a}$$

$$\mathbf{M}^{(1)}\ddot{\mathbf{b}}^{(1)} + \mathbf{A}_1\mathbf{b}^{(1)} + \mathbf{B}_1\mathbf{b}^{(2)} + \mathbf{H}_1(\mathbf{b}) = \mathbf{p}_1(t) \tag{4.28b}$$

$$\mathbf{M}^{(2)}\ddot{\mathbf{b}}^{(2)} + \mathbf{A}_2\mathbf{b}^{(1)} + \mathbf{B}_2\mathbf{b}^{(2)} + \mathbf{H}_2(\mathbf{b}) = \mathbf{p}_2(t) \tag{4.28c}$$

where

$\mathbf{M}, \mathbf{M}^{(1)}, \mathbf{M}^{(2)}, \mathbf{C}$ are $M \times M$ known matrices representing generalized masses and damping.

$\mathbf{H}\left(\mathbf{b}, \mathbf{b}^{(1)}, \ddot{\mathbf{b}}^{(1)}, \mathbf{b}^{(2)}, \ddot{\mathbf{b}}^{(2)}\right)$ is a generalized stiffness vector nonlinearly depending on the components of $\mathbf{b}, \mathbf{b}^{(1)}, \ddot{\mathbf{b}}^{(1)}, \mathbf{b}^{(2)}, \ddot{\mathbf{b}}^{(2)}$.

$\mathbf{H}_1(\mathbf{b})$, $\mathbf{H}_2(\mathbf{b})$ are generalized stiffness matrices nonlinearly depending on the components of $\mathbf{b}$.

$\mathbf{f}$, $\mathbf{p}_1$, $\mathbf{p}_2$ are known load vectors.

The initial conditions $\mathbf{b}(0)$, $\mathbf{b}^{(i)}(0)$ and $\dot{\mathbf{b}}(0)$, $\dot{\mathbf{b}}^{(i)}(0)$ are obtained from Eqs. (4.20), (4.23) and (4.24) with the help of Eqs. (3.46) and (3.76), after solving them for $\mathbf{b}$, $\mathbf{b}^{(i)}$.

For $f(t) = p_x(t) = p_y(t) = 0$ and time-independent boundary conditions, the motion is due to the nonzero initial conditions and the plate performs free vibrations.

The solution of Eqs. (4.28a,b,c) gives the fictitious loads at time t, which are then inserted into Eqs. (3.46) and (3.76) to give the displacements and their derivatives. Subsequently, the stress resultants are obtained from Eqs. (2.19), (2.21), and (4.27).

4.2.3 AEM solution for the linear dynamic problem

The linear problem results from Eqs. (4.26) by omitting the nonlinear terms, which after neglecting the membrane inertia forces read

$$D\mathbf{b}-\left(\mathbf{N}_x\mathbf{w}_{,xx}+2\mathbf{N}_{xy}\mathbf{w}_{,xy}+\mathbf{N}_y\mathbf{w}_{,yy}\right)+\mathbf{p}_x\mathbf{w}_{,x}+\mathbf{p}_y\mathbf{w}_{,y}+\mathbf{c}\dot{\mathbf{w}}+\rho h\ddot{\mathbf{w}}=\mathbf{f}(t) \tag{4.29a}$$

and

$$\mathbf{b}^{(1)}+\frac{1+\nu}{1-\nu}\left(\mathbf{u}_{,x}+\mathbf{v}_{,y}\right)_{,x}+\frac{1}{Gh}\mathbf{p}_x=0 \tag{4.29b}$$

$$\mathbf{b}^{(2)}+\frac{1+\nu}{1-\nu}\left(\mathbf{u}_{,x}+\mathbf{v}_{,y}\right)_{,y}+\frac{1}{Gh}\mathbf{p}_y=0 \tag{4.29c}$$

Eqs. (4.29b,c) are solved independently of Eq. (4.29a) using the procedure described in Section 3.3.2.3.

The stress resultants $\mathbf{N}_x$, $\mathbf{N}_y$, $\mathbf{N}_{xy}$ are computed using the linearized Eqs. (4.27), i.e.,

$$\mathbf{N}_x=C\left(\mathbf{u}_{,x}+\nu\mathbf{v}_{,y}\right) \tag{4.30a}$$

$$\mathbf{N}_y=C\left(\mathbf{v}_{,y}+\nu\mathbf{u}_{,x}\right) \tag{4.30b}$$

$$\mathbf{N}_{xy}=C\frac{1-\nu}{2}\left(\mathbf{u}_{,y}+\mathbf{v}_{,x}\right) \tag{4.30c}$$

and are used in Eq. (4.29a), which with the help of Eq. (3.46) takes the form

$$\mathbf{M}\ddot{\mathbf{b}}+\mathbf{C}\dot{\mathbf{b}}+\mathbf{K}\mathbf{b}=\mathbf{f}(t) \tag{4.31}$$

When $\mathbf{f}(t)=0$ and $\mathbf{C}=0$, the plate performs undamped free vibrations. Setting $\mathbf{b}(t)=\boldsymbol{\beta}e^{-i\omega t}$ in Eq. (4.31), where $\boldsymbol{\beta}$ is a constant vector, we obtain the linear algebraic eigenvalue problem

$$(\mathbf{K}-\omega^2\mathbf{M})\boldsymbol{\beta}=0 \tag{4.32}$$

which gives the eigenfrequencies ω_i and the eigenvectors $\boldsymbol{\beta}_i$, $i=1,2,\cdots,M$. Subsequently, setting $\mathbf{b}=\boldsymbol{\beta}_i$ and $\mathbf{c}=0$ in Eq. (3.46), we obtain the mode shapes.

4.2.4 Numerical solution of the semi-discretized equations of motion

The solution of large systems of semi-discretized equations of motion is an essential problem in structural dynamics, especially for the equations resulting from dynamic BEM analysis, where the mass, damping, and stiffness matrices are not symmetric. Eqs. (4.28) and (4.31) can be solved by using any of the known time-step integration methods, such as the central difference method, the Newmark's mean acceleration method, the analog equation method, etc. In any case the user of these methods should be aware of the applicability of the employed method as well as of its advantages and drawbacks. These methods are not presented here as they are described in detail in books on structural dynamics, such as [3,20,21]. The method employed here for obtaining numerical results in the example problems is a new time-step integration method developed recently [22]. This method is simple to implement. It is self-starting, second-order accurate for linear systems, unconditionally stable, and it does not exhibit amplitude decay or period elongation. It performs well when large deformations and long time durations are considered in the transient response analysis of structures. It can be used as a practical method for integration of the equations of motion in cases where widely used time integration procedures, e.g., Newmark's, do not conserve energy and momentum and become unstable. Another important advantage of this method is that it can be readily extended to solve evolution equations of order higher than two, such as those arising from the static or dynamic response of viscoelastic structures described with multi-term differential viscoelastic models [23,24]. The basic ingredients of the method are presented in Appendix C.

4.2.5 Flutter instability of plates

The flutter of structures and in particular of plates arises when a plate is subjected to nonconservative loads or, as they are usually called, to follower loads, namely, to loads whose direction changes during the motion following the deformed configuration. The loads may be membrane forces, which remain tangential to the deformed middle surface of the plate or transverse loads, which remain normal to it such as aerodynamic loads due to the air flow. With increasing the nonconservative load, it reaches a critical value, for which the plate becomes unstable as it performs large amplitude vibrations. However, in the presence of damping the plate reaches a periodic motion, referred to as limit cycle. The consequence of this response is the failure of the structure. This destabilizing phenomenon is referred to as flutter [25,26]. Because of the great importance, especially in the aerospace structures, the flutter problem has been the subject of extensive investigation. The related literature is rich but mostly for beams [27–30]. However, for plates, the related literature is rather limited [31–33]. Apparently, because it is too difficult to solve the governing differential equations. The BEM has been employed recently and has successfully solved linear [16] and nonlinear [17] flutter problems for plates of arbitrary geometry and boundary conditions. The equations pertaining to the plate flutter problem are presented in the following subsections.

4.2.5.1 NONLINEAR FLUTTER PROBLEM

We consider a plate subjected to the nonconservative membrane load densities $p_x(x, y)$, $p_y(x, y)$ and to the dynamic pressure Δp due to air flow. Moreover, the nonconservative loads P_n^*, P_t^* may act along the free edges (Fig. (4.1). The aerodynamic pressure is given by the relation [33]

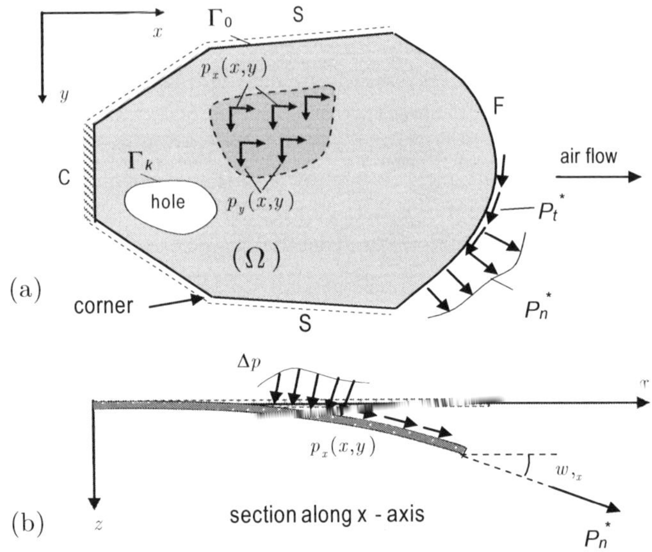

FIGURE 4.1 (a) Geometry, nonconservative loading, and support conditions of the plate; (b) section along the x axis.

$$\Delta p = -\left(q_x w_{,x} + q_y w_{,y} + c\dot{w}\right) \tag{4.33}$$

which represents a good approximation for large supersonic Mach numbers. The quantities q_x, q_y are parameters depending on air density, air velocity the velocity and direction of the air flow, while c represents the external damping. Internal damping η may be present.

The governing equations and the associated boundary conditions are derived using the Hamilton principle, which by omitting the membrane inertia forces are written in terms of the displacements w(**x**,t), u(**x**,t), v(**x**,t) as [17]

(i) For the transverse displacement

$$\begin{aligned} D\nabla^4 w - \left(N_x w_{,xx} + 2N_{xy} w_{,xy} + N_y w_{,yy}\right) + q_x w_{,x} \\ + q_y w_{,y} + \rho h\ddot{w} + c\dot{w} + \eta D\nabla^4 \dot{w} = 0 \end{aligned} \quad \text{in } \Omega \tag{4.34}$$

with boundary conditions

$$Vw + k_T w = 0 \ \text{ or } \ w = 0 \quad \text{on } \Gamma \tag{4.35a}$$

$$Mw - k_R w_{,n} = 0 \ \text{ or } \ w_{,n} = 0 \quad \text{on } \Gamma \tag{4.35b}$$

$$k_c^{(k)} w^{(k)} - [\![Tw]\!]_k = 0 \ \text{ or } \ w^{(k)} = 0 \quad \text{at corner } k \tag{4.35c}$$

and initial conditions

$$w(\mathbf{x},0)=g_1(\mathbf{x}),\ \dot{w}(\mathbf{x},0)=h_1(\mathbf{x}) \quad \text{in } \Omega \tag{4.36a,b}$$

(ii) For the membrane displacements

$$\begin{aligned}\nabla^2 u+\frac{1+\nu}{1-\nu}\left(u_{,x}+v_{,y}\right)_{,x}+w_{,x}\left(\frac{2}{1-\nu}w_{,xx}+w_{,yy}\right)&\\ +\frac{1+\nu}{1-\nu}w_{,xy}w_{,y}+\frac{p_x}{Gh}=0&\end{aligned} \quad \text{in } \Omega \tag{4.37a}$$

$$\begin{aligned}\nabla^2 v+\frac{1+\nu}{1-\nu}\left(u_{,x}+v_{,y}\right)_{,y}+w_{,y}\left(\frac{2}{1-\nu}w_{,yy}+w_{,xx}\right)&\\ +\frac{1+\nu}{1-\nu}w_{,xy}w_{,x}+\frac{p_y}{Gh}=0&\end{aligned} \quad \text{in } \Omega \tag{4.37b}$$

with boundary conditions

$$N_n=P_n^* \quad \text{or} \quad u_n=u_n^* \quad \text{on } \Gamma \tag{4.38a}$$

$$N_t=P_t^* \quad \text{or} \quad u_t=u_t^* \quad \text{on } \Gamma \tag{4.38b}$$

The membrane forces are given by Eqs. (3.103). Note that the components of the external loading p_x, p_y and the forces N_n, N_t do not appear in Eqs. (4.34) and (4.35a), respectively.

The application of the AEM yields the following system of algebraic equations for the fictitious forces:

$$\mathbf{M}\ddot{\mathbf{b}}+\mathbf{C}\dot{\mathbf{b}}+\mathbf{H}\left(\mathbf{b},\mathbf{b}^{(1)},\mathbf{b}^{(2)}\right)=0 \tag{4.39a}$$

$$\mathbf{A}_1\mathbf{b}^{(1)}+\mathbf{B}_1\mathbf{b}^{(2)}+\mathbf{H}_1(\mathbf{b})=\mathbf{p}_1(t) \tag{4.39b}$$

$$\mathbf{A}_2\mathbf{b}^{(1)}+\mathbf{B}_2\mathbf{b}^{(2)}+\mathbf{H}_2(\mathbf{b})=\mathbf{p}_2(t) \tag{4.39c}$$

We observe that Eqs. (4.39b,c) are quasi-static and linear with respect to $\mathbf{b}^{(1)}$, $\mathbf{b}^{(2)}$ and can be solved for these vectors, which are then substituted in Eq. (4.39a) to give

$$\mathbf{M}\ddot{\mathbf{b}}+\mathbf{C}\dot{\mathbf{b}}+\mathbf{S}(\mathbf{b})=0 \tag{4.40}$$

Equation (4.40) is solved with initial conditions $\mathbf{b}(0)$, $\dot{\mathbf{b}}(0)$.

A detailed description of the AEM for the flutter problem is given in [16,17].

4.2.5.2 LINEAR FLUTTER PROBLEM

The problem is linearized if the nonlinear terms in Eqs. (4.37a,b) are neglected. Then the resulting equations are solved independently of Eq. (4.34) to give the membrane displacements and subsequently the membrane forces with the help of Eqs. (4.30). In this case Eq. (4.39a) becomes

$$\mathbf{M}\ddot{\mathbf{b}} + \mathbf{C}\dot{\mathbf{b}} + (\mathbf{K} + \mathbf{F})\mathbf{b} = 0 \tag{4.41}$$

where **K** and **F** are the matrices of the generalized stiffnesses. Note that **F** expresses the geometrical stiffness.

The flutter problem is solved by considering harmonic vibrations. Thus we set

$$\mathbf{b}(t) = \boldsymbol{\beta} e^{i\lambda t} \tag{4.42}$$

where $\boldsymbol{\beta}$ is a constant vector and λ the eigenfrequency of the vibration. Substituting for $\mathbf{b}$ the expression (4.42) in Eq. (4.41), we obtain the quadratic eigenvalue problem

$$\left(-\lambda^2\mathbf{M} + \lambda i\mathbf{C} + \mathbf{K} + \mathbf{F}\right)\boldsymbol{\beta} = 0 \tag{4.43}$$

The eigenvalues of the problem (4.43) are in general complex, $\lambda = a + ib$. With increasing the load two types of instability may occur: (a) Divergence type, which appears when the eigenvalue is real, i.e., $\mathrm{Im}(\lambda) = 0$, and becomes zero (Example 4.3) and (b) Flutter type, which appears when the eigenvalue is complex and its imaginary part becomes negative, i.e., $\mathrm{Im}(\lambda) = b < 0$ (Example 4.4). In the divergence type the plate becomes unstable and reaches a new stable position, while in the flutter type the plate becomes unstable with a vibratory motion of increasing amplitude. The value of the load that produces the divergence instability is the buckling load, while the load that produces the flutter instability is the flutter load. In both cases, as the displacements of the plate become large, the membrane forces take a considerable value and the amplitude of the motion remains bounded. Moreover, in flutter phenomenon the presence of damping plays a key role, leading to a nonlinear vibration of constant amplitude, which is called the limit cycle [17].

4.2.6 Examples

The AEM for dynamic plate analysis as developed in previous sections has been employed for the solution of many dynamic plate problems. In the following some representative examples are given, which demonstrate the efficiency of the method to treat difficult dynamic plate problems arising in engineering.

EXAMPLE 4.1

In this example, the dynamic response of the plate of Fig. 4.2 is studied. The support conditions are shown in the figure. The plate is loaded by the suddenly applied uniform load $f = 0.1H(t)\,\mathrm{kN/m^2}$. The plate has a thickness $h = 0.2\,\mathrm{m}$,

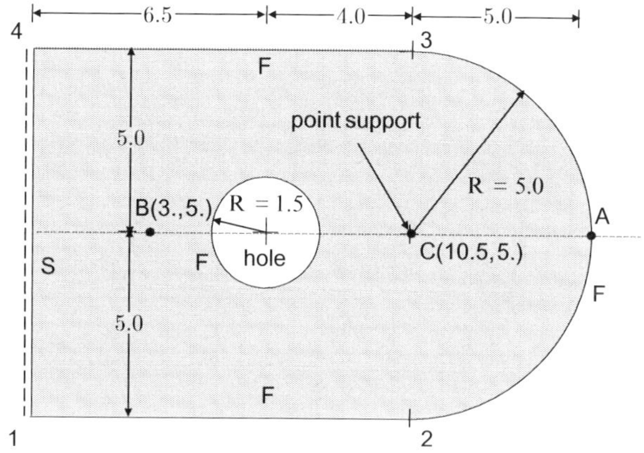

FIGURE 4.2 Plate with complicate boundary in Example 4.1.

density $\rho = 10\,\text{kNm}^{-4}\text{s}^2$, and elastic constants $E = 2.1 \times 10^6\text{kN/m}^2$, $\nu = 0.2$. The results were obtained with $N = 900$ boundary elements and $M = 336$ domain nodal points resulting from 512 linear triangular elements. The boundary elements are distributed as follows: N1 = 150 on each of the edges 1-2 and 3-4, N2 = 300 on the circular edge 2-3, N3 = 100 on the edge 4-1, and N4 = 200 on the circumference of the circular hole. The discretization is shown in Fig. 4.3. Fig. 4.4 shows the time history of the displacement w at point A. Moreover, Fig. 4.5 shows the time history of the bending moment M_x at point B, while Fig. 4.6 presents the time history of the reaction P at the point support C. Finally, Figs. 4.7 through 4.10 show the first four vibration mode shapes with the respective eigenfrequencies. The results are compared with a FEM solution using 6432 quadrilateral elements.

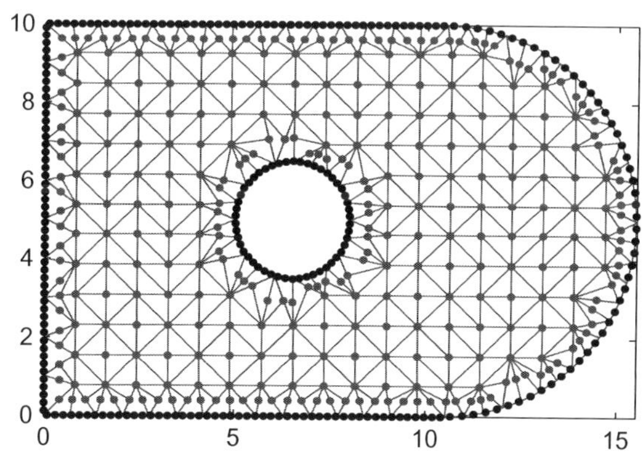

FIGURE 4.3 Boundary and domain discretization in Example 4.1; ($N = 900$, $M = 336$).

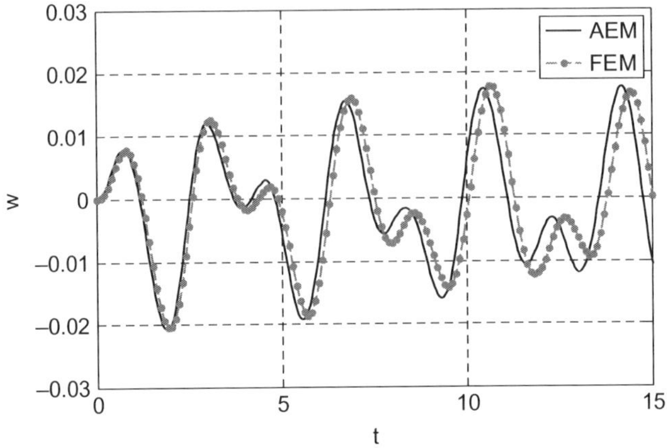

FIGURE 4.4 Time history of the displacement w at point A in Example 4.1.

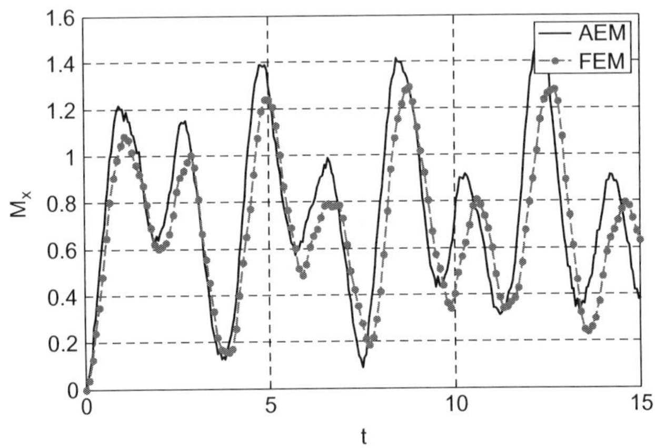

FIGURE 4.5 Time history of the bending moment M_x at point B in Example 4.1.

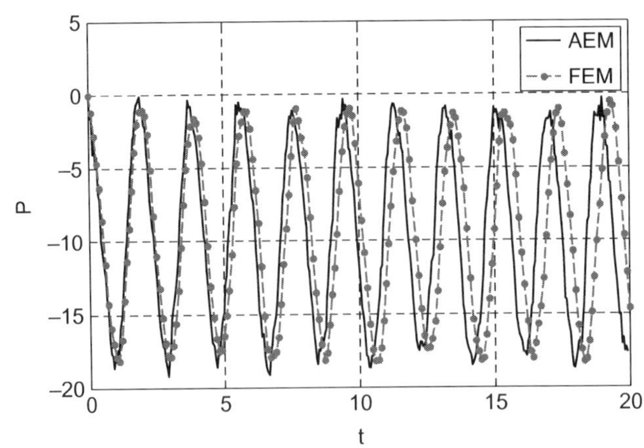

FIGURE 4.6 Time history of the reaction P at the point support C in Example 4.1.

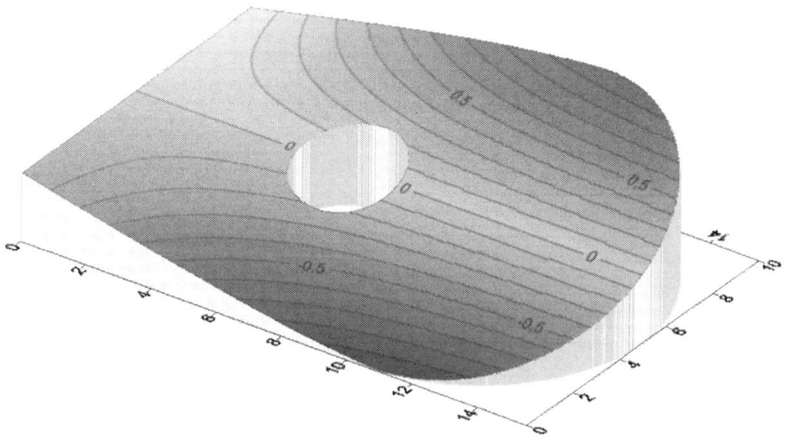

FIGURE 4.7 First vibration mode in Example 4.1; $\lambda_1 = 1.363\,(\text{FEM}: 1.347)$.

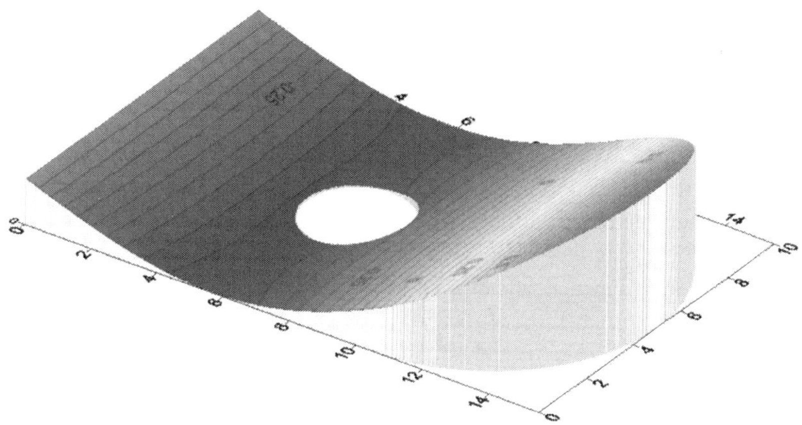

FIGURE 4.8 Second vibration mode in Example 4.1; $\lambda_2 = 1.784\,(\text{FEM}: 1.777)$.

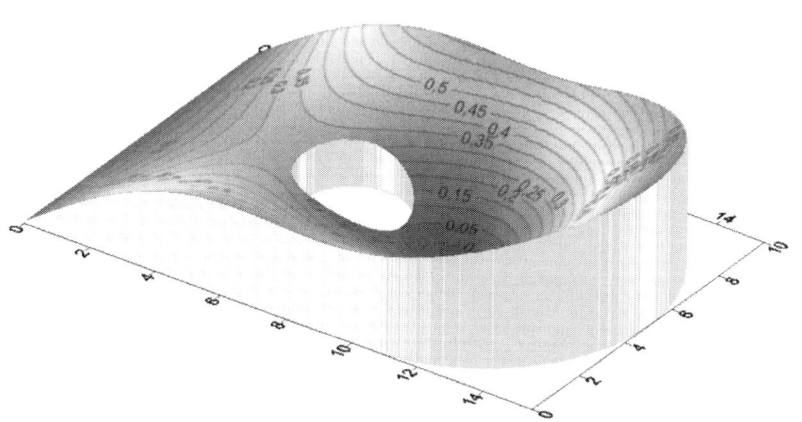

FIGURE 4.9 Third vibration mode in Example 4.1; $\lambda_3 = 3.328\,(\text{FEM}: 3.307)$.

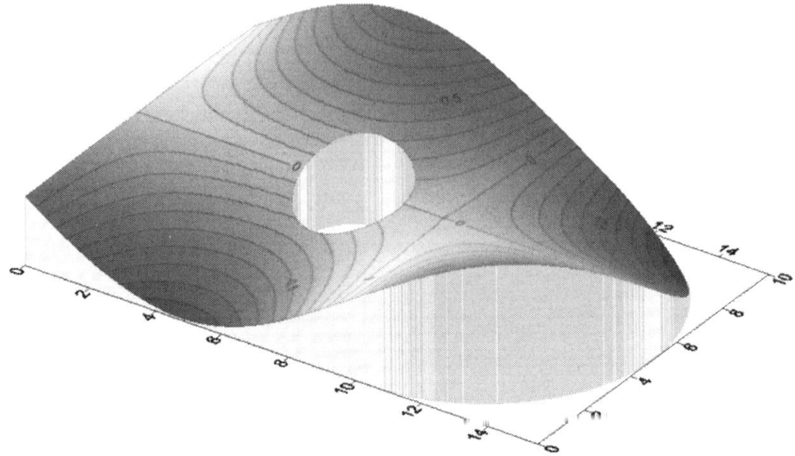

FIGURE 4.10 Fourth vibration mode in Example 4.1; $\lambda_4 = 4.761\,(\mathrm{FEM}:\ 4.713)$.

EXAMPLE 4.2

In this example the nonlinear vibrations of a square plate with side length $a = 4\mathrm{m}$, $0 \le x,\ y \le a$ and thickness $h = 0.1\mathrm{m}$ are studied. All edges are simply supported ($w = Mw = 0$) and immovable in the plane of the plate ($u = v = 0$). The employed material constants are $E = 21 \times 10^6\ \mathrm{kN/m^2}$, $\rho = 10\,\mathrm{kNm^{-4}s^2}$, $\nu = 0.3$. The plate is subjected to the two uniform loads (i) $f = 100H(t)$ and (ii) $f = 100\sin(100t)$. The results were obtained with $N = 204$ boundary elements and $M = 137$ domain nodal points resulting from 216 linear triangular elements. The discretization is shown in Fig 4.11. Fig. 4.12 and Fig. 4.13 show the time histories of the transverse displacement w at the center of the plate and the membrane displacement u at point (1, 2) for load cases (i) and (ii), respectively. The system of the nonlinear

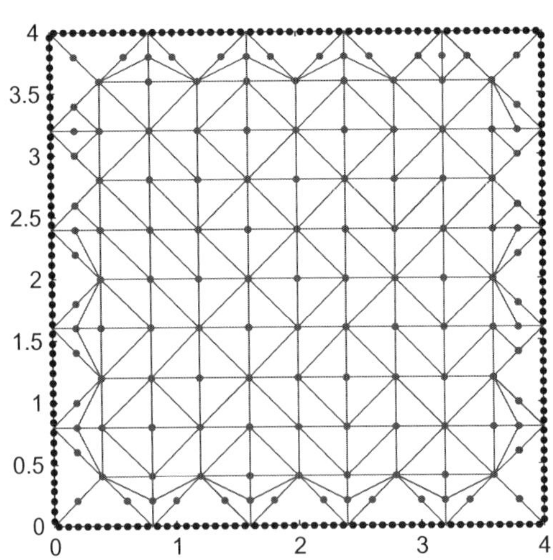

FIGURE 4.11 Boundary and domain discretization in Example 4.2; ($N = 204, M = 137$).

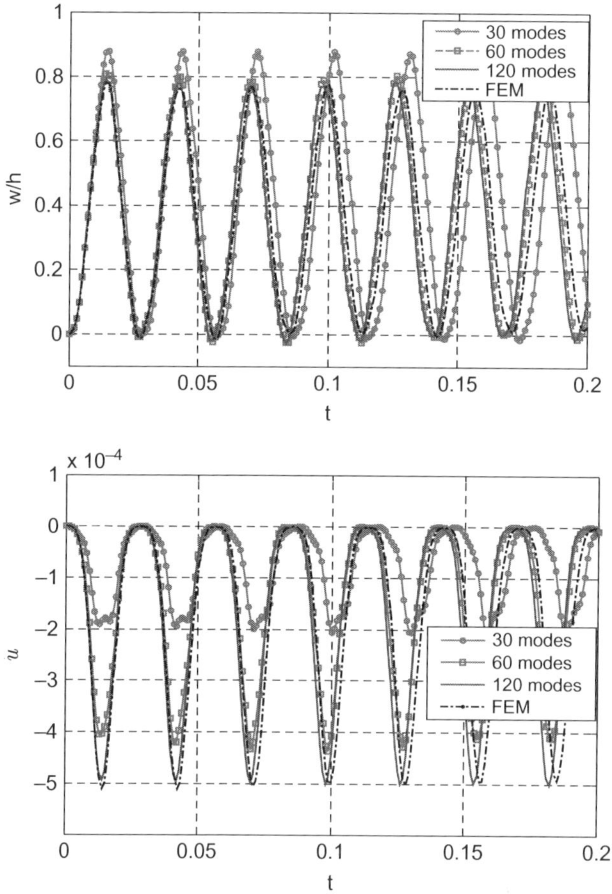

FIGURE 4.12 Time history of the transverse central displacements w and membrane displacement u at point (1, 2) in Example 4.2; $f = 100H(t)$.

equations of motion was solved after reduction of the degrees of freedom by the Ritz method using the linear vibration modes as Ritz vectors [3]. The results obtained with 120 modes (out of 3×137) are in agreement with those obtained by FEM with 1600 rectangular elements. Apparently, acceptable results are obtained using 60 modes.

EXAMPLE 4.3

The stability of the square plate of Fig. 4.14 is investigated in this example. The edges $y = 0, 4$ are simply supported, while the other two edges ($x = 0, 4$) are free. At the free edges a uniformly distributed membrane load P is applied and produces buckling. Two types of membrane boundary conditions at the supported edges are examined: (i) $N_y = 0$, $N_{yx} = 0$, i.e., movable edges, and (ii) $u = 0, v = 0$, i.e., immovable edges. In the first case the produced membrane forces are constant, i.e., $N_x = N$, $N_y = N_{xy} = 0$, while in the second case they are variable and

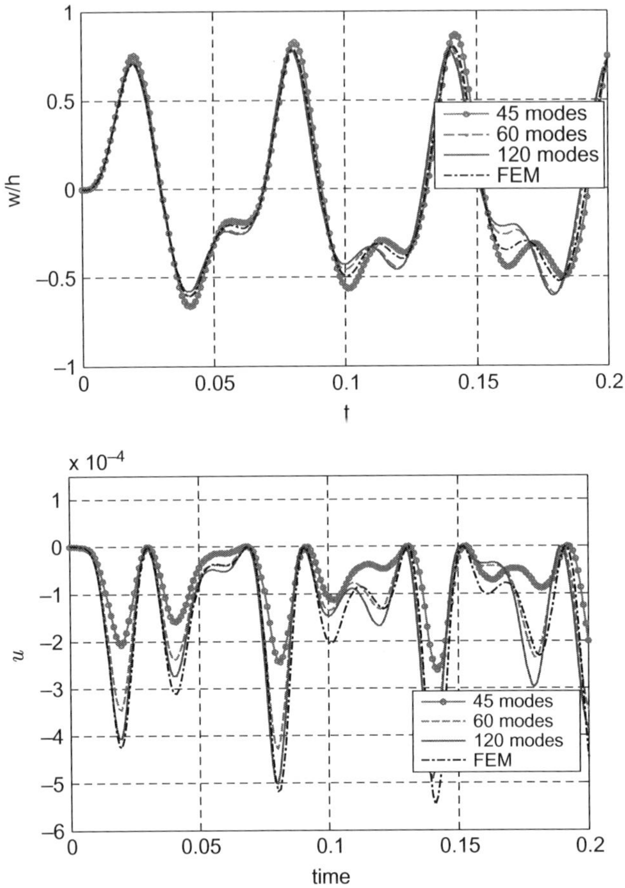

FIGURE 4.13 Time history of the transverse central displacements w and membrane displacement u at point (1, 2) in Example 4.2; $f = 100\sin(100t)$.

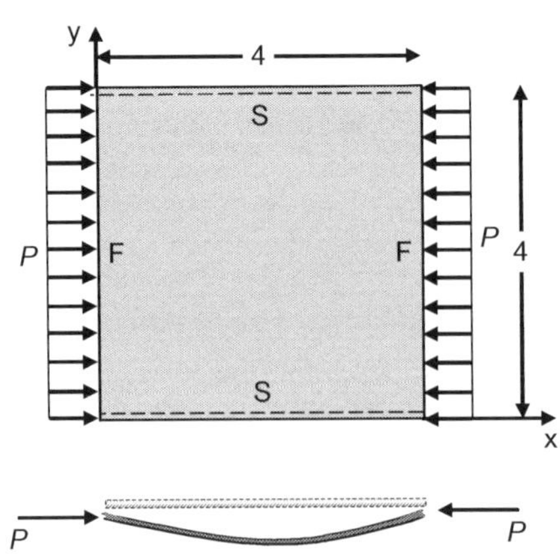

FIGURE 4.14 Square plate under uniform compressive edge load in Example 4.3.

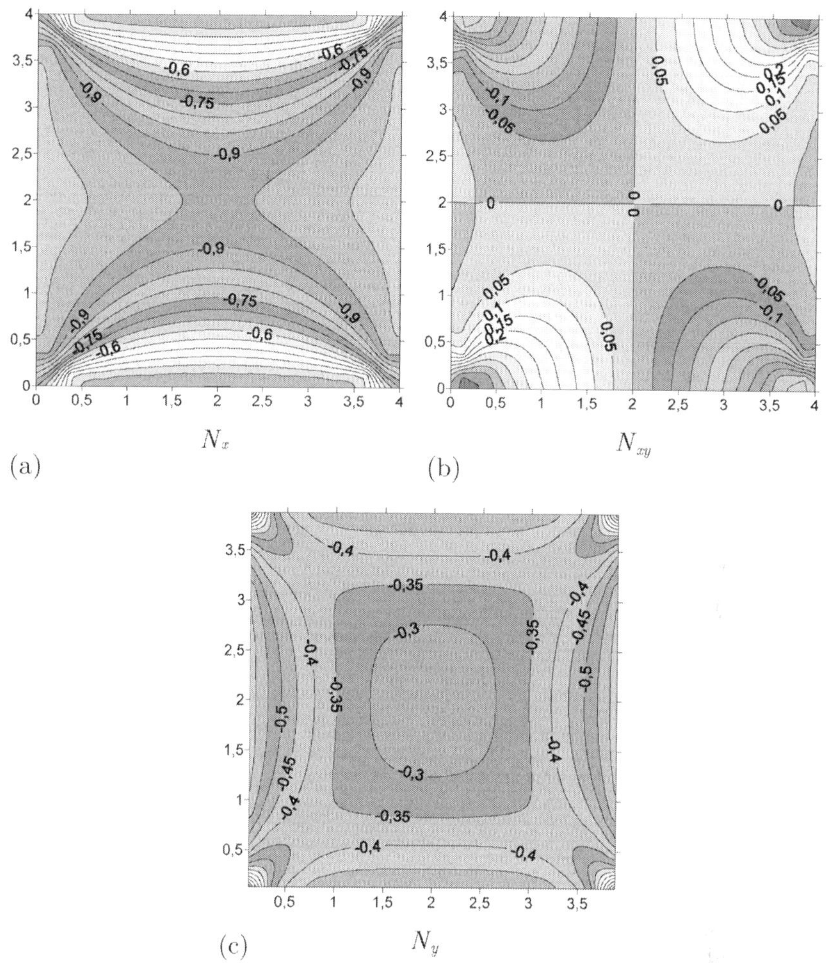

FIGURE 4.15 Contour curves of the membrane forces N_x, N_{xy}, N_y in case (ii) in Example 4.3.

result from the solution of the plane stress problem, which is solved independently. The contours of the computed membrane forces for case (ii) are shown in Fig. 4.15. The employed data are $E=30\times10^6\,\mathrm{kN/m^2}$, $h=0.1\mathrm{m}$, $\nu=0.5$, $\rho=10\,\mathrm{kNm^{-4}s^2}$. The first buckling load P_{cr} as of divergent type is computed using the dynamic criterion, namely, it results as the value of P for which the first eigenfrequency vanishes. The higher-order buckling loads are obtained as the values of P for which the corresponding eigenfrequencies vanish. In Fig. 4.16 the load-frequency curves of the first two modes for the two support cases are shown. Table 4.1 presents the first buckling load for various values of the employed boundary and domain nodal points. The computed second buckling load for case (i) is $(P_{cr})_2=4206$, while for case (ii) it is $(P_{cr})_2=2962$. The corresponding values obtained with the FEM are 4095 and 2921.

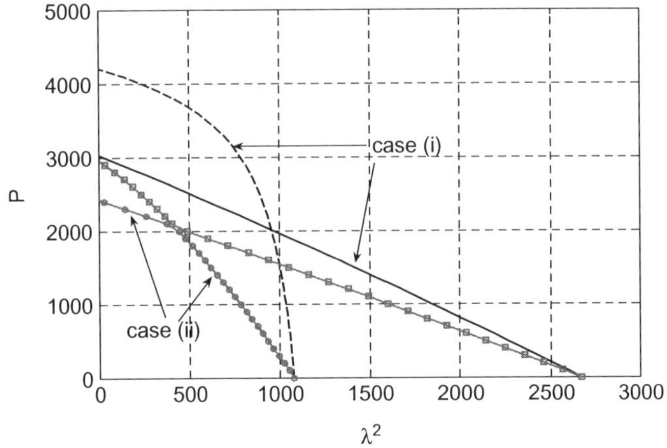

FIGURE 4.16 Load–frequency curves in Example 4.3.

TABLE 4.1 First buckling load in Example 4.3 for various values of *N* and *M*.

Boundary points N	Domain points M	Case (i) $(P_{cr})_1$	Case (ii) $(P_{cr})_1$
200	40	3034	2522
	190	3033	2426
	319	3035	2420
400	40	3038	2523
	190	3037	2427
	320	3038	2423
600	40	3039	2523
	190	3038	2428
	320	3040	2423
Exact [34]		3081	–
FEM (1600 elements)		2965	2364

EXAMPLE 4.4

In this example the linear and nonlinear flutter instability of the cantilever plate of Fig. 4.17 is investigated. The plate is subjected to the steady supersonic airflow with parameters $q_x=0$, $q_y=0.031\mathrm{v}^2$, where v is the velocity of the air. The employed data are $E=210\times10^6\,\mathrm{kN/m^2}$, $h=0.01\,\mathrm{m}$, $\nu=0.3$, $\rho=7.55\,\mathrm{kNm^{-4}s^2}$. The results were obtained with $N=385$ boundary elements and $M=125$ domain nodal points resulting from 195 linear triangular elements. The discretization is shown in Fig. 4.18. The system of the nonlinear equations of motion was solved after reduction of the degrees of freedom from 125 to 20 using the Ritz method. The first 20 mode shapes of the linear flutter problem

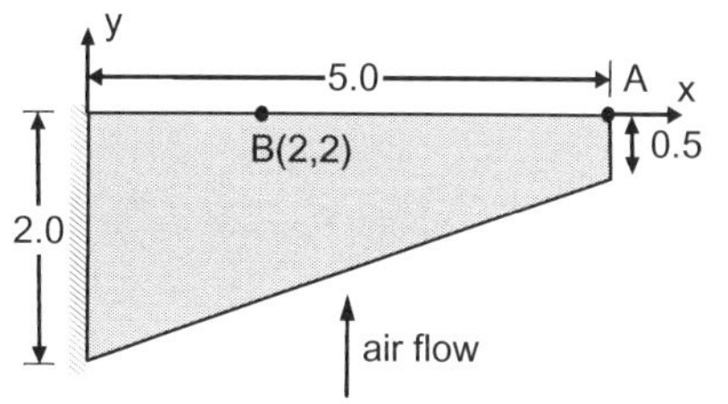

FIGURE 4.17 Cantilever plate subjected to supersonic air velocity in Example 4.4.

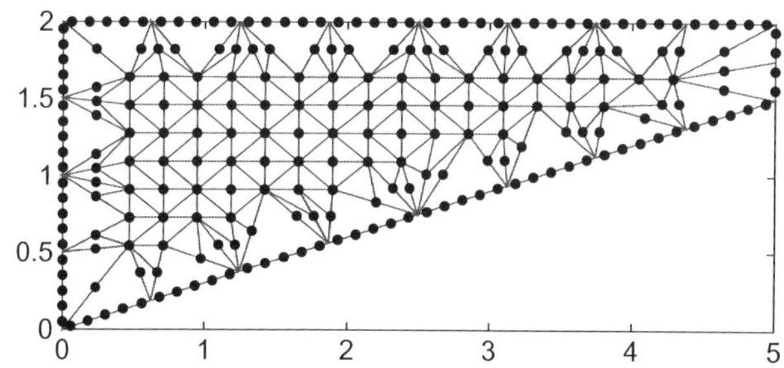

FIGURE 4.18 Discretization in Example 4.4.

were employed as Ritz vectors [17]. Figs. 4.19 through 4.21 present the variation of the eigenfrequencies of the linear problem versus the velocity of the air without damping, with external (viscous) damping and internal (structural) damping, respectively. The computed critical velocity is $\mathrm{v}_{cr}=38.9\ \mathrm{m/s}$ in the absence of damping, $\mathrm{v}_{cr}=40.5\ \mathrm{m/s}$ with external damping $c=0.2$, $\mathrm{v}_{cr}=42.3\ \mathrm{m/s}$ with external damping $c=0.4$, and $\mathrm{v}_{cr}=26.7\ \mathrm{m/s}$ with internal damping $\eta=0.01$. Fig. 4.22 shows the maximum displacement at point A (5.0,2.0) in the absence of damping and the amplitude of the limit cycle in the presence of internal or external damping for various values of air velocity. Moreover, Fig. 4.23 presents the phase plane at point A for $\mathrm{v}=43\ \mathrm{m/s}$ and $c=0.2$. The limit cycle at this point is a Lamé curve with $n=2$ (ellipse). Finally, Fig. 4.24 shows the phase plane and the limit cycle of the displacement w at point $B(2,2)$ in the presence of external damping $c=0.2$.

EXAMPLE 4.5

An airplane wing is modeled with a triangular cantilever plate. The plate is subjected to aerodynamic pressure due to supersonic air flow in the x direction (Fig. 4.25). The initial plate thickness is $h_0=0.1\,\mathrm{m}$ and the material constants $E=21\times10^6\mathrm{kN/m^2}$, $\nu=0.3$, and $\rho=7.55\ \mathrm{kNm^{-4}s^2}$. We seek the optimum thickness distribution $h(x,y)$, for which the critical pressure is maximized, while the volume of the plate is kept constant. The computed critical pressure parameter for constant thickness is $q_x^0=123267\,\mathrm{kN/m^2}$. The optimization was performed with thickness bounds $0.05<h(x,y)<0.2$ and constraint parameter $\varepsilon=0.0001$ (see Section 3.6). In the absence of damping the obtained maximum

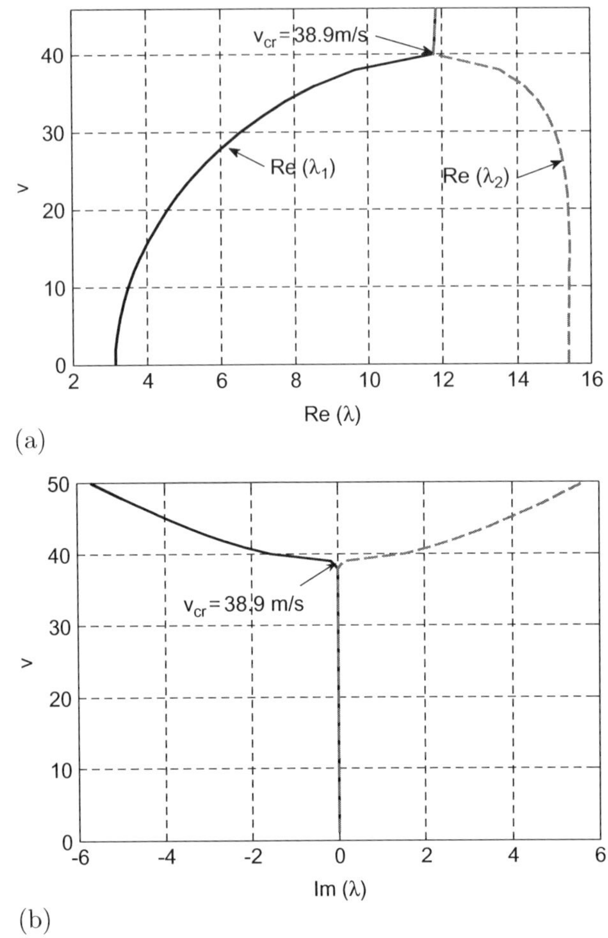

FIGURE 4.19 Variation of (a) $\mathrm{Re}(\lambda)$ and (b) $\mathrm{Im}(\lambda)$ versus the air velocity v in Example 4.4. Flutter without damping, $\mathrm{Re}(\lambda_1) = \mathrm{Re}(\lambda_2)$.

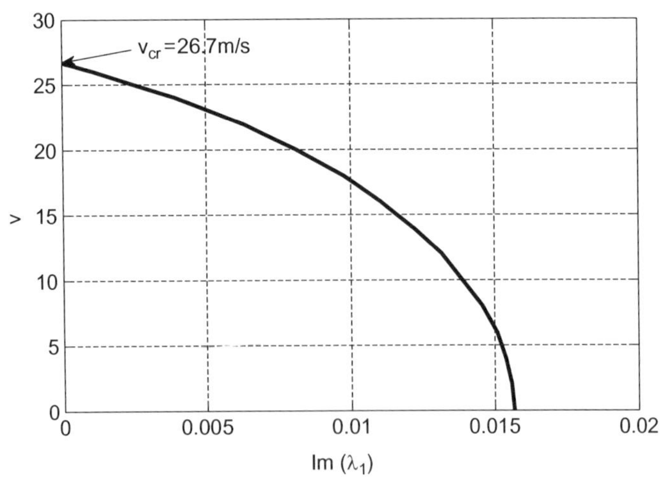

FIGURE 4.20 Variation of $\mathrm{Im}(\lambda_1)$ versus the air velocity v in Example 4.4. Flutter with structural damping ($\eta = 0.01, \mathrm{v}_{cr} = 26.7\mathrm{m/s}$), $\mathrm{Im}(\lambda_1) = 0$.

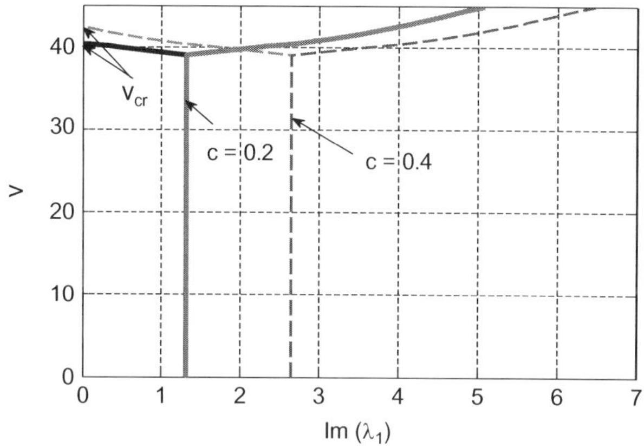

FIGURE 4.21 Variation of $\mathrm{Im}(\lambda_1)$ versus the air velocity v in Example 4.4. Flutter with viscous damping ($c = 0.2$, $v_{cr} = 40.5$ m/s, $c = 0.4$, $v_{cr} = 42.3$ m/s), $\mathrm{Im}(\lambda_1) = 0$.

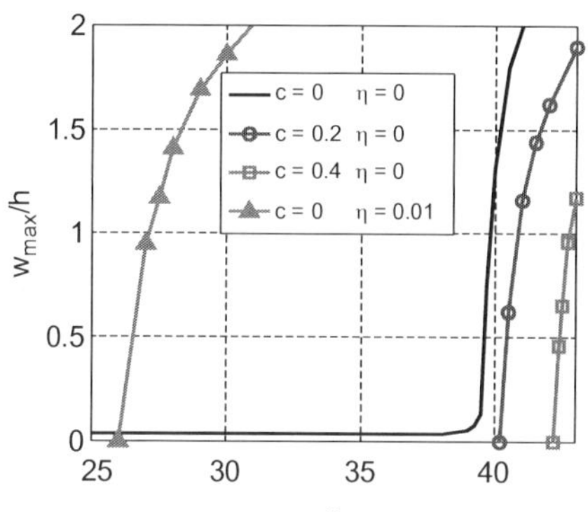

FIGURE 4.22 Maximum deflection at corner point A in the absence of damping (linear flutter) and amplitude of the limit cycles in the presence of damping (nonlinear flutter) versus air velocity v in Example 4.4.

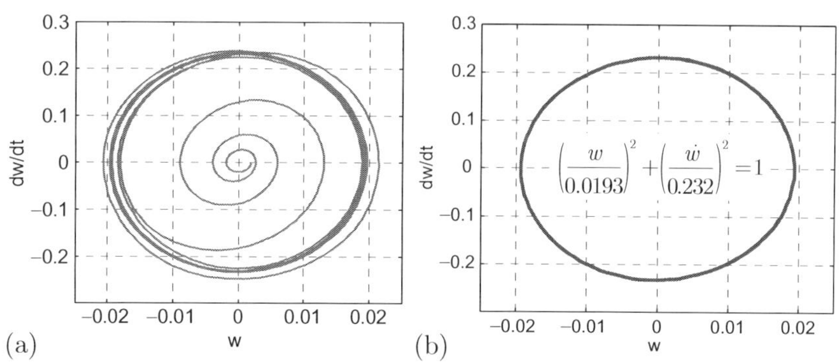

FIGURE 4.23 Phase-plane plot at point A with external damping: (a) $0 < t < 8$ sec, (b) limit cycle for $t > 7$ sec ($v = 43$ m/s, 20 Ritz vectors) in Example 4.4.

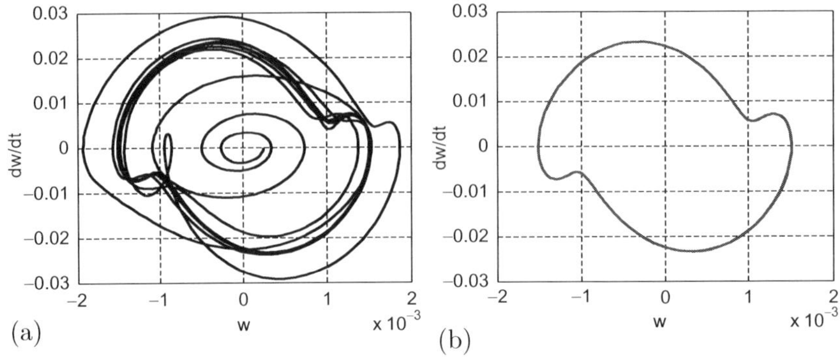

FIGURE 4.24 Phase-plane plot at point B(2, 2) with external damping $c = 0.2$ (a) $0 < t < 8$ sec, (b) limit cycle for $t > 7$ sec ($v = 43$ m/s, 20 Ritz modes) in Example 4.4.

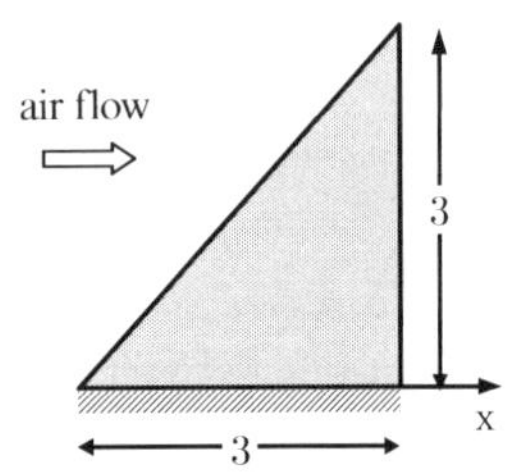

FIGURE 4.25 Triangular cantilever plate in Example 4.5.

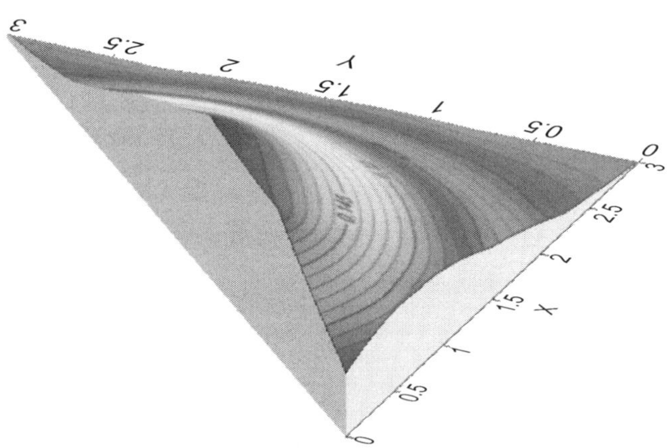

FIGURE 4.26 Optimum thickness distribution in Example 4.5.

aerodynamic pressure parameter is $q_x^{\max} = 507928\,\text{kN/m}^2$, which is about four times greater than q_x^0. Fig. 4.26 shows the optimum thickness distribution, while Fig. 4.27 shows the curves of the aerodynamic parameter q_x versus the eigenfrequencies, and the critical values.

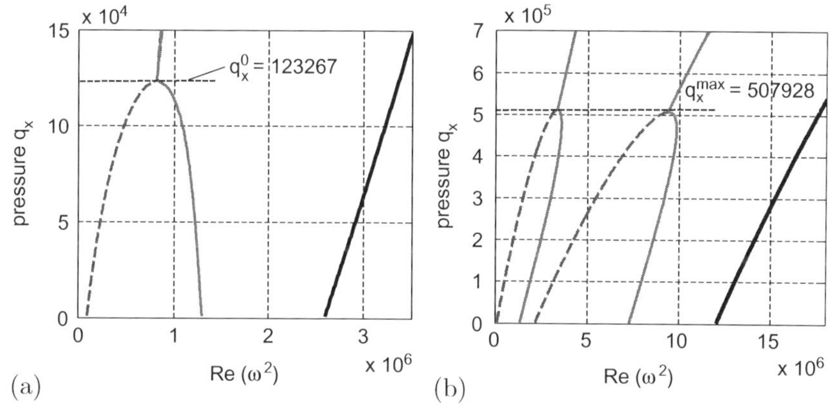

FIGURE 4.27 Aerodynamic pressure parameter q_x versus eigenfrequency and critical values for (a) constant thickness and (b) for optimum thickness distribution in Example 4.5.

4.3 VIBRATIONS OF THIN ANISOTROPIC PLATES

4.3.1 Equations of motion and their solutions

The equations result from the equations of the static problem (3.184) if the inertia force $-\rho h\ddot{w}$ and damping force $-c\dot{w}$ are included in the external load. Thus, the dynamic response of the anisotropic plate in terms of the transverse deflection $w(\mathbf{x}, t)$ is described by the following equation:

$$\begin{aligned} D_{11}\frac{\partial^4 w}{\partial x^4} &+ 4D_{16}\frac{\partial^4 w}{\partial x^3 \partial y} + 2(D_{12}+2D_{66})\frac{\partial^4 w}{\partial x^2 \partial y^2} \\ &+ 4D_{26}\frac{\partial^4 w}{\partial x \partial y^3} + D_{22}\frac{\partial^4 w}{\partial y^4} + c\dot{w} + \rho h\ddot{w} = f(\mathbf{x}; t) \end{aligned} \qquad \mathbf{x}\in\Omega,\ t>0 \tag{4.44}$$

under the boundary conditions (3.182) and the initial conditions (4.20).

The initial boundary value problem is solved by the AEM. In this case the analog equation is

$$\nabla^4 w = b(\mathbf{x}, t) \quad \mathbf{x}\in\Omega, t>0 \tag{4.45}$$

Applying the procedure described in Section 3.8.3, we obtain the semi-discretized equations of motion for the fictitious load

$$\mathbf{M}\ddot{\mathbf{b}} + \mathbf{C}\dot{\mathbf{b}} + \mathbf{K}\mathbf{b} = \mathbf{f}(t) \tag{4.46}$$

The solution of Eqs. (4.46) gives the fictitious loads at time t, which are then inserted into Eqs. (3.46) to give the displacements and their derivatives, which are then used in Eqs. (3.171a) to obtain the bending and twisting moments.

4.3.2 Examples

EXAMPLE 4.6

The dynamic response of a rectangular simply supported orthotropic plate with dimensions $a\times b$ is analyzed under the uniformly distributed load $f = q_0 \cos\Omega t$. The plate admits an exact solution [35]:

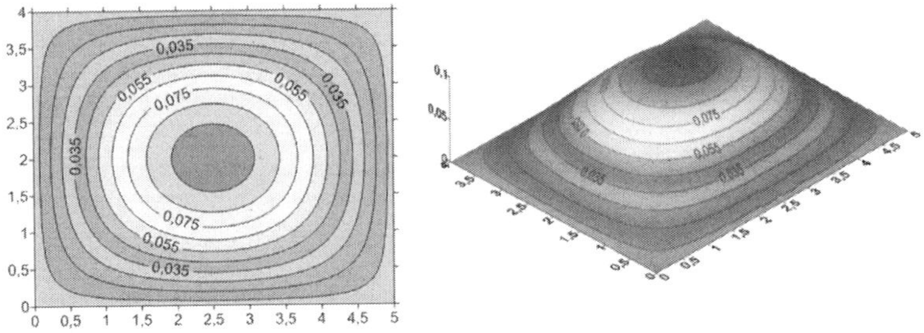

1st vibration mode $\omega_1 = 145.30$

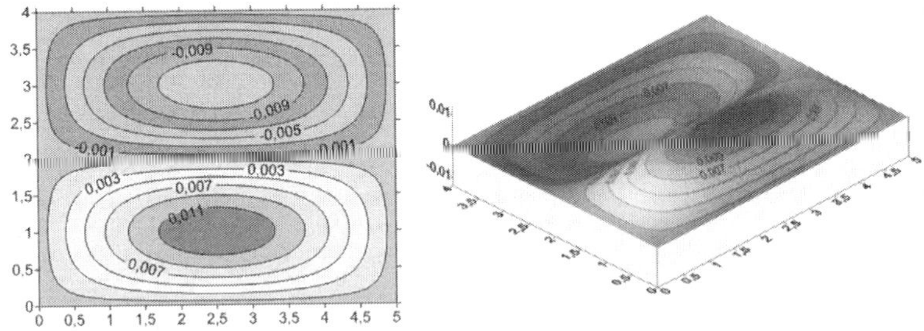

2nd vibration mode $\omega_2 = 290.58$

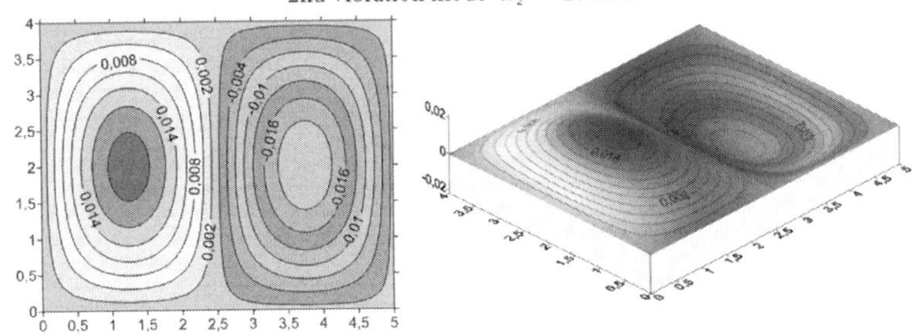

3nd vibration mode $\omega_3 = 489.79$

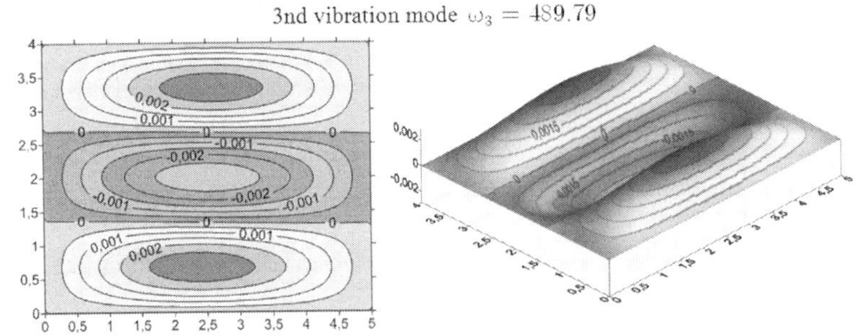

4th vibration mode $\omega_4 = 570.95$

FIGURE 4.28 The first four mode shapes of the orthotropic plate in Example 4.6.

$$w = \frac{1}{\rho h}\sum_{m=1}^{\infty}\sum_{m=1}^{\infty}\left[\frac{c_{mn}}{\omega_{mn}^2 - \Omega^2}(\cos\Omega \mathrm{t} - \cos\omega_{mn}t)\sin\frac{m\pi x}{a}\sin\frac{n\pi y}{b}\right] \tag{4.47}$$

where

$$c_{mn} = \frac{4q_0}{ab}(\cos m\pi - 1)(\cos n\pi - 1) \tag{4.48}$$

TABLE 4.2 Eigenfrequencies of the rectangular simply supported orthotropic plate in Example 4.6.

	AEM				
ω_i	M = 209	M = 279	M = 347	M = 423	Exact [35]
1	145.68	145.52	145.43	145.30	144.68
2	288.52	289.88	290.81	290.58	287.71
3	492.26	491.48	490.43	489.79	486.26
4	554.60	563.18	571.51	570.95	564.10
5	590.70	591.53	590.39	589.30	578.72
6	796.44	816.32	819.38	819.21	794.60
7	938.16	953.48	977.12	977.00	962.18

The numerical results have been obtained with the data: $h = 0.1m$, $a = 5.0m$, $b = 4.0m$ $\rho = 2.0\,\mathrm{kNm^{-4}s^2}$, $E_1 = 206.8GPa$, $E_2 = E_1/10$, $G_{12} = E_2/3$, $\nu = 0.30$. The computed eigenfrequencies obtained with $N = 280$ boundary elements and various domain discretizations are shown in Table 4.2 as compared with the exact ones:

$$\omega_{mn} = \frac{\pi^2}{b^2}\sqrt{\frac{1}{\rho h}}\sqrt{D_{11}\left(\frac{m}{c}\right)^4 + 2(D_{12} + 2D_{66})n^2\left(\frac{m}{c}\right)^2 + D_{22}n^4}, \; c = a/b \tag{4.49}$$

Figure 4.28 presents the first four mode shapes together with the corresponding eigenfrequencies. The time history of the displacement w, velocity $\dot{w}$, and bending moments M_x and M_y at the center of the plate are shown in Fig. 4.29. The results have been obtained with $N = 280$ boundary elements and $M = 423$ domain nodal points resulting from 744 linear triangular elements. The discretized equations of motion were solved after reducing the degrees of freedom from 423 to 100 by the Ritz method using the vibration modes as Ritz vectors.

EXAMPLE 4.7

A 16-layer clamped circular anisotropic plate having thickness $h = 0.08m$ and radius $a = 2.0m$ is analyzed. The layers, which are orthotropic with thickness $h/16$ and material constants $\rho = 2.0\,\mathrm{kNm^{-4}s^2}$, $E_1 = 206.8\,GPa$, $E_2 = E_1/15.4$, $G_{12} = 0.79E_2$, $\nu_{12} = 0.30$, $\nu_{21} = \nu_{12}E_2/E_1$, are placed symmetrically $[\theta/-\theta/\theta/-\theta/\theta/-\theta/\theta/-\theta]_{symmetric}$; $\theta = \measuredangle x, \bar{x}$ with $\bar{x}, \bar{y}$ being the principal axes of the orthotropic layer. Table 4.3 shows the computed six first frequency parameters $\lambda = \omega a^2\sqrt{12\rho(1-\nu_{12}\nu_{21})/E_1h^2}$ for various angles θ. Moreover, Fig. 4.30 presents the four first mode shapes. The results have been obtained using $N = 300$ boundary elements and $M = 224$ domain nodal points, resulting from 386 linear triangular elements.

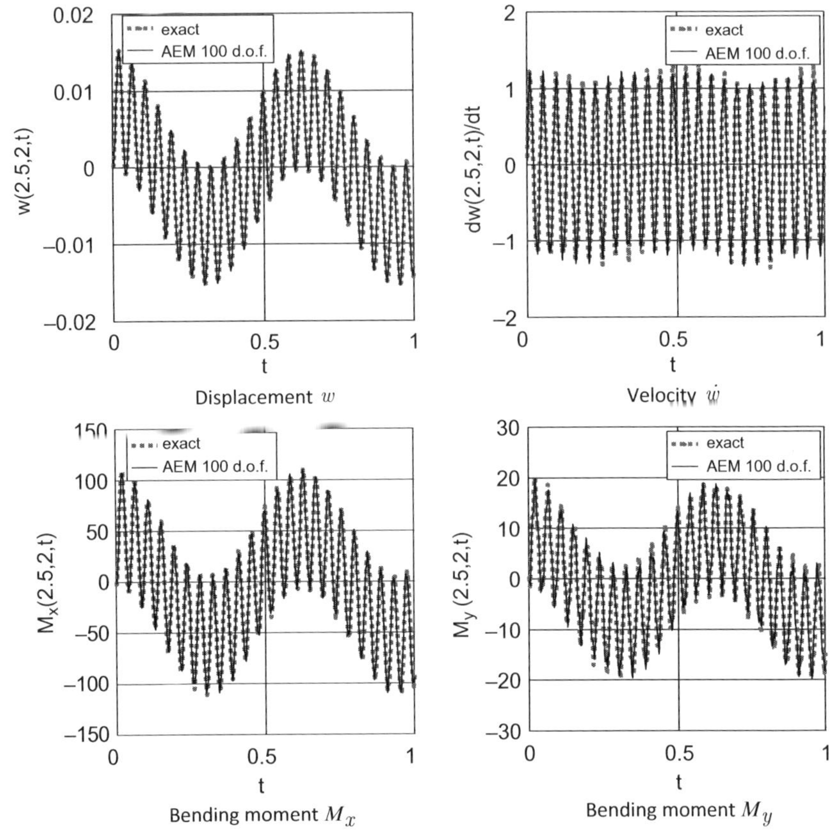

FIGURE 4.29 Time history of displacement w, velocity $\dot{w}$, and bending moments M_x, M_y at the center of the orthotropic plate in Example 4.6.

TABLE 4.3 Frequency parameter λ of symmetrically 16-layer clamped circular anisotropic plates in Example 4.7. Upper row: AEM; Lower row: Ref [36].

Plate and notation	Mode	θ 0°	15°	30°	45°
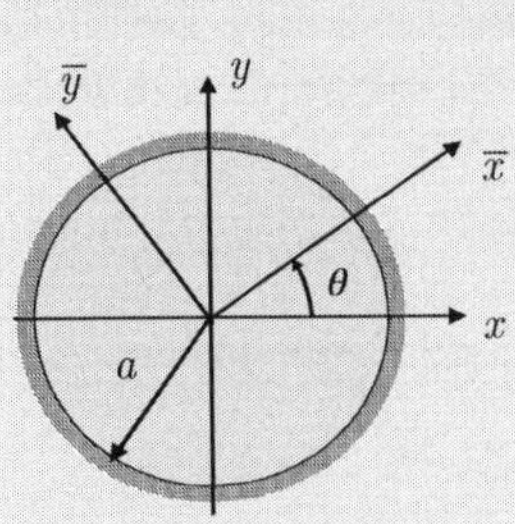	1	26.598 26.619	26.674 26.713	26.789 26.821	26.852 26.862
	2	37.695 36.922	40.245 39.902	47.038 46.935	53.184 52.966
	3	53.087 52.097	58.553 57.978	63.524 63.385	58.811 58.569
	4	68.811 68.665	67.702 67.522	73.268 72.695	86.869 86.034
	5	70.107 72.323	80.855 81.167	93.635 93.078	93.573 92.831
	6	85.001 82.957	88.262 86.987	105.965 104.820	108.939 107.930

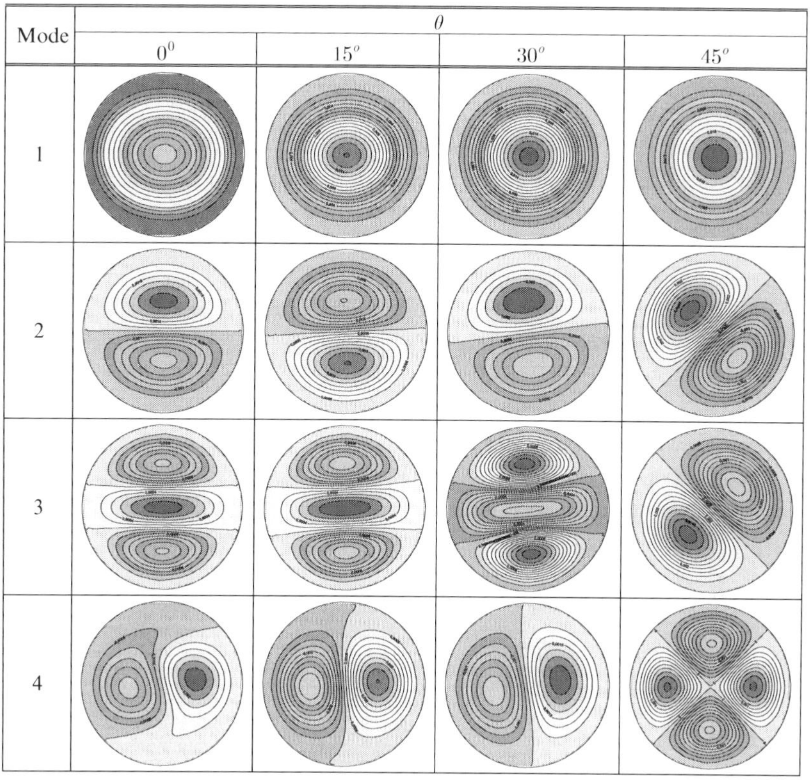

FIGURE 4.30 Mode shapes for the 16-layer clamped circular anisotropic plates in Example 4.7.

4.4 VISCOELASTIC PLATES

4.4.1 Viscoelastic plates

Thin plates of viscoelastic material are extensively used as structural members in many modern engineering applications such as aircrafts, ships, and other industrial structures. Viscoelastic materials, such as polymers, exhibit both elastic and viscous character and have the advantage of light weight and high strength. There have been proposed various linear and nonlinear models for the mechanical behavior of these materials, e.g., Maxwell, Voigt, Zener, and multi-parameter models. The constitutive equations can be presented in differential or in hereditary integral form. Recently, many researchers have shown that differential viscoelastic models with fractional derivatives are in better agreement with the experimental results than the integer derivative models [37–39]. The parameters of the differential models of either integer or fractional type should satisfy certain restrictions so that non-negative internal work and rate of energy dissipation is predicted. Such restrictions have been derived by Bagley and Torvic [40].

Due to the complexity of the viscoelastic models, which include time as an independent variable, exact analytical solutions have been obtained for only a few simplified problems [41,42]. Rigorous predictions of the viscoelastic behavior usually rely on numerical approaches such as the finite difference method, the finite element method (FEM), and the boundary element method (BEM). Their potential as a computational tool in viscoelasticity has been demonstrated in the context of certain linear models for both quasi-static and dynamic problems. The BEM has been employed for the solution of various viscoelastic problems using either the correspondence principle, which allows us to obtain the elastic solution in the Laplace transform space and then in the time domain by numerical inversion of Laplace transform [43–45] or by direct integration of the equations of motion in the time domain [46,47].

Various BEM formulations of viscoelastic problems are presented in the review article [48]. The BEM for linear and nonlinear viscoelastic plate problems has been employed by few researchers using the correspondence principle [49] or by direct integration of the equations of motion using the AEM [18,50]. The multi-term differential models lead to semi-discretized evolution type equations, which may be of order greater than 2, and consequently special care is required to establish the redundant initial conditions. In the following, we present the BEM for viscoelastic plates described with multi-term differential models of both integer and fractional order. The solution of the resulting integer-order semi-discretized evolution equations is obtained using the time-step integration method presented in Appendix C, while of the fractional order differential equations using the numerical method presented in [22].

4.4.2 Problem statement for multi-term differential models of integer order

The general viscoelastic constitutive equations described with integer-order multi-term differential model for the two-dimensional isotropic body under plane stress are given as [41]

$$\mathrm{P}\boldsymbol{\sigma} = \mathbf{C}\mathrm{Q}\boldsymbol{\varepsilon} \tag{4.50}$$

where

$$\boldsymbol{\sigma} = \begin{Bmatrix} \sigma_x \\ \sigma_y \\ \tau_{xy} \end{Bmatrix}, \boldsymbol{\varepsilon} = \begin{Bmatrix} \varepsilon_x \\ \varepsilon_y \\ \gamma_{xy} \end{Bmatrix}, \mathbf{C} = \frac{E}{1-\nu^2}\begin{bmatrix} 1 & \nu & 0 \\ \nu & 1 & 0 \\ 0 & 0 & \frac{1}{2}(1-\nu) \end{bmatrix} \tag{4.51}$$

and P, Q are differential operators defined as

$$\mathrm{P} = \sum_{k=0}^{l} p_k \frac{d^k}{dt^k}, \quad \mathrm{Q} = \sum_{k=0}^{m} q_k \frac{d^k}{dt^k} \tag{4.52a,b}$$

where p_k, q_k are the viscoelastic parameters, which should satisfy certain conditions resulting from second law of thermodynamics. Apparently, for the

Kelvin-Voigt model it is $l=0$, $m=1$, $p_0=1$, $q_0=1$, $q_1=\eta$ and the constitutive equation (4.50) reduces to

$$\boldsymbol{\sigma}=\mathbf{C}(\varepsilon+\eta\dot{\varepsilon}) \tag{4.53}$$

while for the three-parameter (standard solid) model it is $l=m=1$, $p_0=1$, $p_1=\dfrac{\eta b}{1+b}$, $q_0=\dfrac{b}{1+b}$, $q_1=\dfrac{\eta b}{1+b}$, with η and b being parameters of the material, and the constitutive equation (4.50) becomes [17]

$$\boldsymbol{\sigma}+p_1\dot{\boldsymbol{\sigma}}=\mathbf{C}(q_0\varepsilon+q_1\dot{\varepsilon}) \tag{4.54}$$

The constitutive equations for the plate result from Eq. (4.50) by virtue of Eqs. (2.9), which in the absence of membrane stretching reads

$$\mathrm{P}\begin{Bmatrix}\sigma_x\\ \sigma_y\\ \tau_{xy}\end{Bmatrix}=\frac{E}{1-\nu^2}\begin{bmatrix}1 & \nu & 0\\ \nu & 1 & 0\\ 0 & 0 & \dfrac{1}{2}(1-\nu)\end{bmatrix}\mathrm{Q}\begin{Bmatrix}-zw_{,xx}\\ -zw_{,yy}\\ -zw_{,xy}\end{Bmatrix} \tag{4.55}$$

Introducing Eqs. (4.55) in Eqs. (2.15) and integrating through the thickness yields

$$\mathrm{P}M_x=-D\mathrm{Q}\left(w_{,xx}+\nu w_{,yy}\right) \tag{4.56a}$$

$$\mathrm{P}M_y=-D\mathrm{Q}\left(w_{,yy}+\nu w_{,xx}\right) \tag{4.56b}$$

$$\mathrm{P}M_{xy}=D(1-\nu)\mathrm{Q}w_{,xy} \tag{4.56c}$$

The plate equilibrium equation (2.39) with inclusion of transverse inertia force $-\rho h\ddot{w}$ becomes

$$M_{x,xx}-2M_{xy,xy}+M_{y,yy}=-f+\rho h\ddot{w} \quad \text{in } \Omega \tag{4.57}$$

Applying now the operator P to Eq. (4.57), we have

$$\mathrm{P}M_{x,xx}-2\mathrm{P}M_{xy,xy}+\mathrm{P}M_{y,yy}=-\mathrm{P}f+\rho h\mathrm{P}\ddot{w} \quad \text{in } \Omega \tag{4.58}$$

which with the help of Eqs. (4.56) yields

$$D\mathrm{Q}\nabla^4 w+\rho h\mathrm{P}\ddot{w}=\mathrm{P}f \quad \text{in } \Omega \tag{4.59}$$

Equation (4.59) is the equation of motion of the thin viscoelastic plate for the integer-order multi-term differential viscoelastic model with initial conditions given by Eqs. (4.3a,b).

The boundary conditions may result if the operator P is applied to the boundary stress resultants Mw, Vw:

$$\mathrm{P}Mw = M_n^* \tag{4.60a}$$

$$\mathrm{P}Vw = V_n^* \tag{4.60b}$$

Evidently, the force boundary conditions lead to differential equations with respect to time, which must be solved together with the plate equation, Eq. (4.59). For example, if the plate is simply supported the boundary conditions should be

$$w = 0, \quad \mathrm{P}Mw = 0 \tag{4.61}$$

It has been shown that, if the prescribed boundary quantities M_n^*, V_n^* do not depend on time, then Eqs. (4.60) are also time independent. Consequently, the boundary conditions of Eqs. (4.2) are also valid in this case. The same boundary conditions can be employed, even when M_n^*, V_n^* depend on time, since the resulting error from this simplification is small. This assumption, though not fully justified, is adopted for the numerical examples.

4.4.3 AEM solution

Application of the AEM as described in Section 3.3.2 leads to the following semi-discretized equation of motion for the fictitious load vector $\mathbf{b}(t)$, which results from the application of Eq. (4.59) at the M domain nodal points:

$$D\sum_{k=0}^{m} q_k \frac{d^k\mathbf{b}}{dt^k} + \rho h \mathbf{S} \sum_{k=0}^{l} p_k \frac{d^{k+2}\mathbf{b}}{dt^{k+2}} = \sum_{k=0}^{l} p_k \frac{d^k\mathbf{f}}{dt^k} \tag{4.62}$$

The initial conditions $\mathbf{b}(0)$, $\dot{\mathbf{b}}(0)$ are obtained from Eqs. (4.20) with the help of Eq. (3.46) after solving it for $\mathbf{b}$:

$$\mathbf{b}(0) = \mathbf{S}^{-1}(\mathbf{g}_1 - \mathbf{c}), \quad \dot{\mathbf{b}}(0) = \mathbf{S}^{-1}\mathbf{h}_1 \tag{4.63a,b}$$

For the Kelvin-Voigt model, Eq. (4.62) reduces to

$$D\mathbf{b} + D\eta\dot{\mathbf{b}} + \rho h \mathbf{S}\ddot{\mathbf{b}} = \mathbf{f} \tag{4.64}$$

while for the three-parameter model it reduces to

$$Dq_0\mathbf{b} + Dq_1\dot{\mathbf{b}} + \rho h p_0 \mathbf{S}\ddot{\mathbf{b}} + \rho h p_1 \mathbf{S}\dddot{\mathbf{b}} = p_0\mathbf{f} + p_1\dot{\mathbf{f}} \tag{4.65}$$

Apparently, the order of Eq. (4.62) is equal to $\max\{m, l+2\} \geq 2$. If the order is equal to two, as in Eq. (4.64), then the physical initial conditions are sufficient to solve the equation. If it is greater than two, $\max\{m, l+2\}-2$ additional initial conditions are needed in order to solve the initial value problem. These initial conditions cannot be specified arbitrarily, but they must be expressed in terms of the physical ones. Certain techniques for the establishment of the redundant initial conditions have been reported, but they depend on the specific viscoelastic model [41]. For Eq. (4.65), a third initial condition is needed. It can be obtained from Eq. (4.57) for $t=0$ with the assumption that the structure (plate) is initially unstrained. This yields

$$\ddot{w}(\mathbf{x},0)=\frac{f(\mathbf{x},0)}{\rho h} \tag{4.66}$$

which can be used to determine $\ddot{\mathbf{b}}(0)$ and solve Eq. (4.65).

A general method for establishing the redundant initial conditions has been proposed by Nerantzaki and Babouskos [23].

Equation of motion (4.62) is solved using the method described in Appendix C after adjusting it for equations of order higher than two.

4.4.4 Problem statement for multi-term differential models of fractional order

The viscoelastic constitutive equations described with the multi-term differential model of fractional-order result from Eq. (4.50) if the integer-order derivatives in the operators P, Q are replaced with fractional-order derivatives. Thus, we have

$$\mathrm{P}=\sum_{k=0}^{l} p_k D^{\alpha_k}, \quad \mathrm{Q}=\sum_{k=0}^{m} q_k D^{\alpha_k} \tag{4.67a,b}$$

in which D^{a_k} is the fractional derivative of order α_k, with $\alpha_0=0$, and p_k, q_k are the viscoelastic parameters, which should satisfy certain conditions resulting from the second law of thermodynamics [51,52]. Different types of fractional derivatives have been introduced [53]. The Riemann-Liouville and the Caputo fractional derivatives are usually employed in viscoelasticity. An advantage of the Caputo derivative is that it allows the application of initial conditions having direct physical meaning. Therefore, the use of this derivative is more convenient to treat viscoelastic problems and obtain numerical results. The Caputo derivative is defined as

$$D_c^{\alpha} u(t)=\begin{cases} \dfrac{1}{\Gamma(m-\alpha)}\displaystyle\int_0^t \frac{u^{(m)}(\tau)}{(t-\tau)^{\alpha+1-m}}\,d\tau, & m-1<\alpha<m \\ \dfrac{d^m}{dt^m}u(t), & m=\alpha \end{cases} \tag{4.68}$$

Apparently, the fractional Kelvin-Voigt model becomes

$$\boldsymbol{\sigma} = \mathbf{C}\left(\varepsilon + \eta D_c^{\alpha} \varepsilon\right), \quad 0 < \alpha \leq 1 \tag{4.69}$$

and the three-parameter model

$$\boldsymbol{\sigma} + p_1 D_c^{\alpha} \dot{\boldsymbol{\sigma}} = \mathbf{C}\left(q_0 \varepsilon + q_1 D_c^{\alpha} \varepsilon\right), \quad 0 < \alpha \leq 1 \tag{4.70}$$

The plate equation (4.59) also applies here with the operators P, Q given now by Eqs. (4.67a,b)

Application of the AEM leads to the following semi-discretized equation of motion for the fictitious load vector $\mathbf{b}(t)$:

$$D \sum_{k=0}^{m} q_k D_c^{\alpha_k} \mathbf{b} + \rho h \mathbf{S} \sum_{k=0}^{l} p_k D_c^{\alpha_k + 2} \mathbf{b} = \sum_{k=0}^{l} p_k D_c^{\alpha_k} \mathbf{f} \tag{4.71}$$

with initial conditions (4.63).

For the fractional Kelvin-Voigt model, Eq. (4.71) reduces to

$$D\mathbf{b} + D\eta D_c^{\alpha} \mathbf{b} + \rho h \mathbf{S} \ddot{\mathbf{b}} = \mathbf{f}, \quad 0 < \alpha \leq 1 \tag{4.72}$$

while for the three-parameter model it reduces to

$$D q_0 \mathbf{b} + D q_1 D_c^{\alpha} \mathbf{b} + \rho h p_0 \mathbf{S} D_c^2 \mathbf{b} + \rho h p_1 \mathbf{S} D_c^{2+\alpha} \mathbf{b} = p_0 \mathbf{f} + p_1 D_c^{\alpha} \mathbf{f}, \quad 0 < \alpha \leq 1 \tag{4.73}$$

Apparently, the two initial conditions (4.63a,b) are sufficient to solve Eq. (4.72). However, the order of Eq. (4.73) is $2 < 2 + \alpha \leq 3$ and thus one additional initial condition is required. It can be obtained from Eq. (4.57) for $t = 0$ with the assumption that the structure (plate) is initially unstrained. Thus, we have

$$\ddot{w}(\mathbf{x},0) = \frac{f(\mathbf{x},0)}{\rho h} \tag{4.74}$$

4.4.5 Examples

EXAMPLE 4.8

In this example, the dynamic response of the viscoelastic plate of Fig. 4.31 under the suddenly applied load $f = H(t)$ is investigated. The support conditions are shown in the figure. The plate has thickness $h = 0.1\,\mathrm{m}$, density $\rho = 2.5\,\mathrm{kNm^{-4}s^2}$, and elastic constants $E = 21\,GPa$, $\nu = 0.25$. The results were obtained with $N =$ 300 boundary elements and $M = 200$ domain nodal points resulting from 333 linear triangular elements. The discretization is shown in Fig. 4.32.

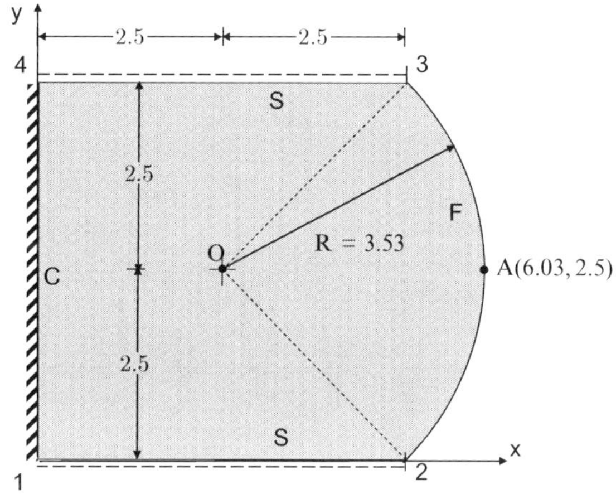

FIGURE 4.31 Geometry and boundary conditions of the plate in Example 4.8.

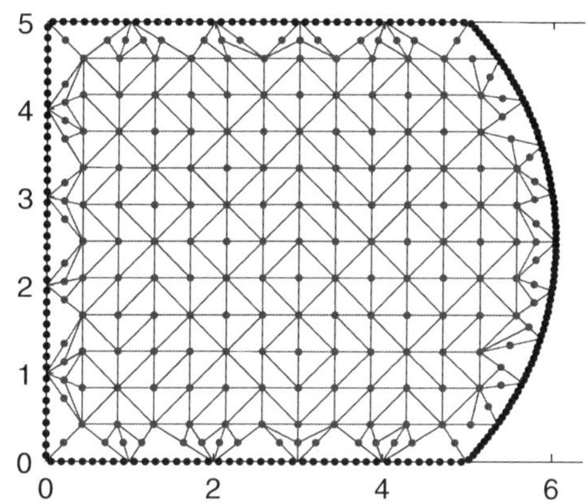

FIGURE 4.32 Boundary and domain discretization of the plate in Example 4.8.

First, the response of the elastic plate is studied. The elastic analysis gives the linear modes that may be used as Ritz vectors to reduce the degrees of freedom of discretized equations of motion. This is shown in Fig. 4.33, which gives the time history of the deflection at point A. Apparently, a few modes are enough to give very good results. Table 4.4 shows the first six eigenfrequencies for two different discretizations as compared with those obtained using the FEM with 2956 quadrilateral elements.

Next, the response of the viscoelastic plate is studied. Fig. 4.34 presents the time history of the displacement at point A for the fractional Kelvin-Voigt model with viscoelastic parameter $\eta = 0.1$. The analysis has been performed for various values of the order of the fractional derivative. Since there is no available solution in the literature for the problem at hand, a comparison of the results was not possible. The accuracy of the solution procedure has been attested by solving the problem for a very small value of the order α, say $\alpha = 0.01$, of the fractional derivative and comparing it with the elastic solution with $E^* = (1 + \eta)\, E$, since

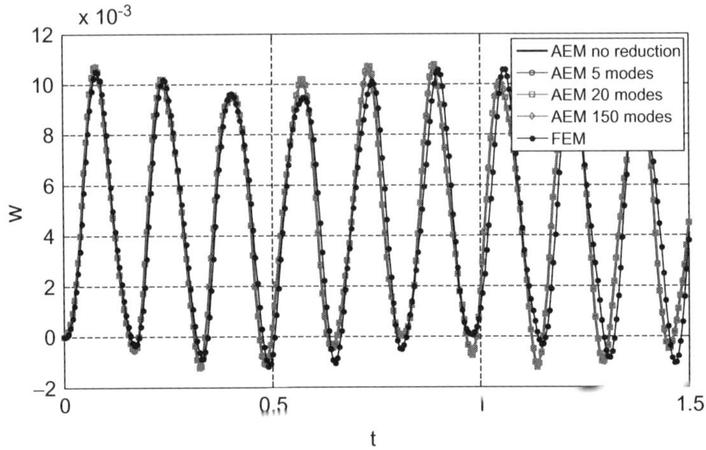

FIGURE 4.33 Time history of displacement at point *A* of the elastic plate in Example 4.8.

TABLE 4.4 Eigenfrequencies ω_i of the plate in Example 4.8.

ω_i	AEM N=300 M=200	AEM N=430 M=324	FEM
ω_1	38.89	38.58	38.76
ω_2	86.81	86.42	85.39
ω_3	138.52	137.33	136.60
ω_4	184.62	183.50	180.28
ω_5	194.39	192.94	189.57
ω_6	301.06	298.19	289.84

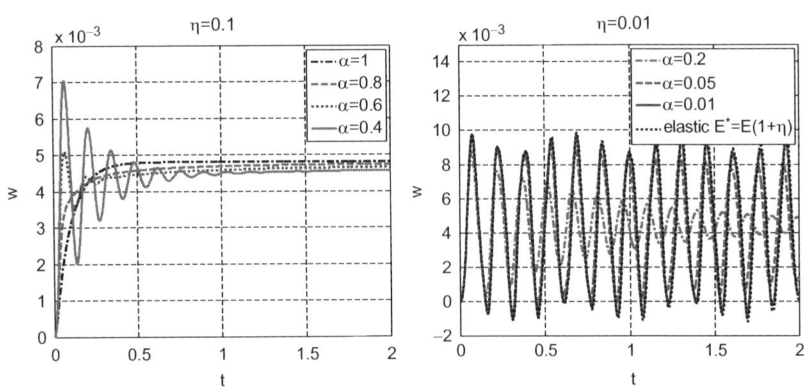

FIGURE 4.34 Time history of displacement at point *A* of the plate with Kelvin-Voigt model for different orders of the fractional derivative in Example 4.8.

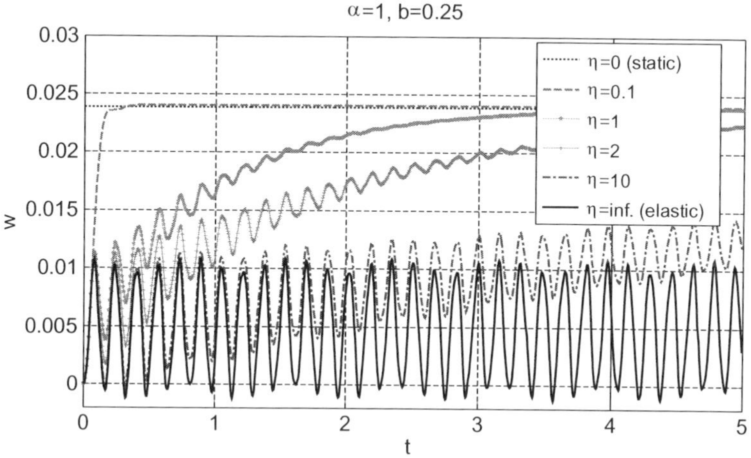

FIGURE 4.35 Time history of displacement at point A of the plate with standard solid model for different values of the viscoelastic parameter η in Example 4.8.

$\lim_{\alpha\to 0} D_c^\alpha \varepsilon = \varepsilon$ and Eq. (4.53) becomes $\boldsymbol{\sigma} = (1+\eta)\,\mathbf{C}\varepsilon$. Fig. 4.35 shows the time history of displacement at point A for the three-parameter (standard solid) model with $\alpha = 1$, $b = 0.25$ and different values of the viscoelastic parameter η. Note that the dynamic response approaches the elastic static with $E' = Eb/(1+b)$ for large values of t. Moreover, Fig. 4.36 presents the time history of the deflection at the same point for the three-parameter model with $\eta = 1$, $b = 0.25$ and different values of the order a of the fractional derivative. We observe that the viscoelastic response approaches the elastic one with $E^* = Eb(1+\eta)/[1+(1+\eta)]$ as α becomes small. This results from Eq. (4.70), which gives $\boldsymbol{\sigma} = b(1+\eta)/[1+b(1+\eta)]\mathbf{C}\varepsilon$ since $\lim_{\alpha\to 0} D_c^\alpha \varepsilon = \varepsilon$. Finally, Fig. 4.37 and Fig. 4.38 present the time history of bending moment M_y at point O of the plate for the

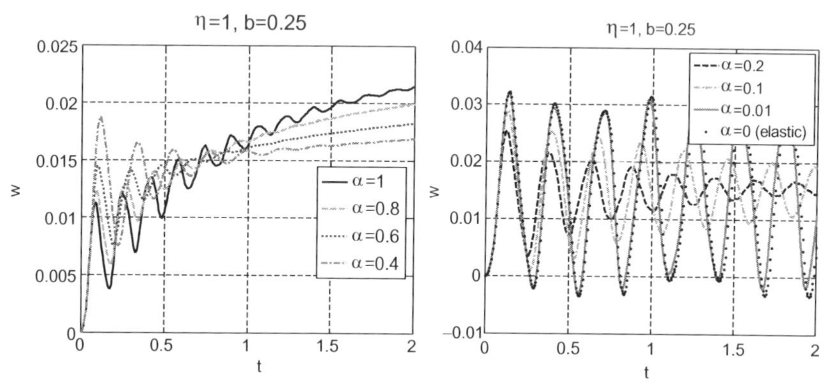

FIGURE 4.36 Time history of displacement at point A of the plate with standard solid model for different orders of the fractional derivative in Example 4.8.

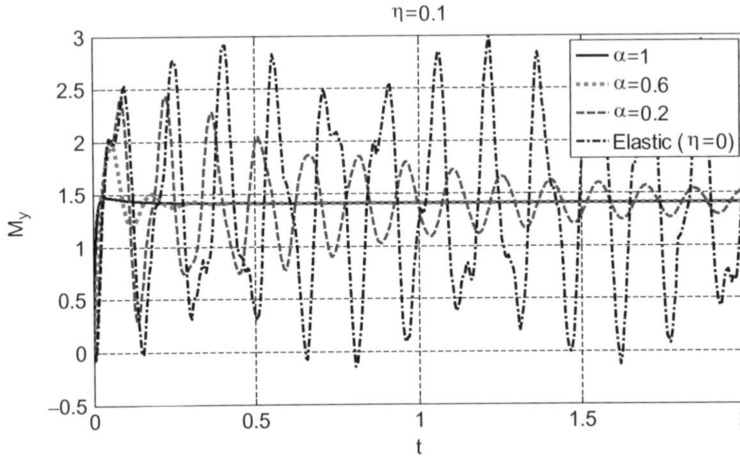

FIGURE 4.37 Time history of bending moment M_y at point O of the plate for the Kelvin-Voigt model in Example 4.8.

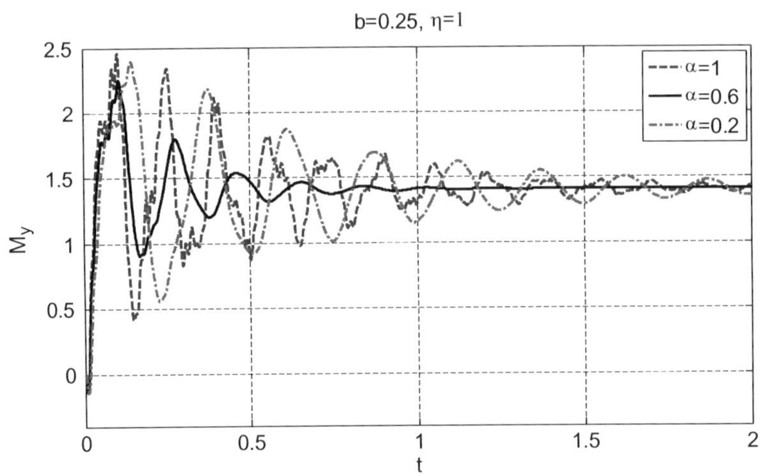

FIGURE 4.38 Time history of bending moment M_y at point O of the plate for the standard solid model in Example 4.8.

Kelvin-Voigt and standard solid models, respectively. Note that the bending moment is computed from Eq. (4.56b), which for the considered models becomes: Kelvin-Voigt

$$M_y = -D(\nu w_{,xx} + w_{,yy}) - \eta D(\nu D_c^\alpha w_{,xx} + D_c^\alpha w_{,yy}) \tag{4.75a}$$

Standard solid model

$$M_y + p_1 D_c^\alpha M_y = -q_0 D(\nu w_{,xx} + w_{,yy}) - q_1 D(\nu D_c^\alpha w_{,xx} + D_c^\alpha w_{,yy}) \tag{4.75b}$$

Apparently, the computation of the bending moment for the standard solid model requires the solution of the differential equation (4.75b). As anticipated, for both models the bending moment approaches the elastic static value $M_y^{static} = 1.41\,\text{kNm}$.

4.5 REFERENCES

[1] M. Abramowitz, I. Stegun (Eds.), Handbook of Mathematical Functions, 10th ed., Dover, New York, 1972.

[2] G. Bezine, D. Gamby, Étude des Mouvements Transitoires de Flexion d' une Plaque par la Méthode des Équations Intégrales de Frontière, Journal de Mécanique Appliquée 1 (3) (1982) 451–466.

[3] J.T. Katsikadelis, Dynamics of Structures, Symmetria Publications, Athens, 2012.

[4] D.E. Beskos, Dynamic analysis of plates, in: D.E. Beskos (Ed.), Boundary Element Analysis of Plates and Shells, Springer-Verlag, Berlin, 1991, pp. 35–92.

[5] C.P. Providakis, D.E. Beskos, Free and forced vibrations of plates by the boundary elements, Comput. Meth. Appl. Mech. Eng. 4 (1989) 231–250.

[6] J.T. Katsikadelis, E.J. Sapountzakis, E.G. Zorba, A BEM approach to static and dynamic analysis of plates with internal supports, Comput. Mech. 7 (1) (1990) 31–40.

[7] J.T. Katsikadelis, E.J. Sapountzakis, A BEM solution to dynamic analysis of plates with variable thickness, in: C.A. Brebbia, J.J. Connor (Eds.), Advances in Boundary Elements, vol. 3, Springer Verlag, Berlin, 1989, pp. 285–302.

[8] J.T. Katsikadelis, E.J. Sapountzakis, Numerical evaluation of the Green's function for the biharmonic equation using BEM with applications to static and dynamic analysis of plates, in: C.A. Brebbia, W.K. Wendland, G. Kuhn (Eds.), Boundary Elements IX, vol. 2, Springer, Berlin, 1987, pp. 51–67.

[9] J.T. Katsikadelis, C.B. Kandilas, A flexibility matrix solution of the vibration problem of plates based on the boundary element method, Acta Mech. 83 (1–2) (1990) 51–60.

[10] C.P. Providakis, D.E. Beskos, Dynamic analysis of plates by boundary elements, Appl. Mech. Rev. 52 (7) (1999) 213–236.

[11] M.S. Nerantzaki, J.T. Katsikadelis, An analog equation solution to dynamic analysis of plates with variable thickness, Eng. Anal. Bound. Elem. 17 (2 Special Issue) (1996) 145–152.

[12] M.S. Nerantzaki, J.T. Katsikadelis, Nonlinear dynamic analysis of circular plates with varying thickness, Arch. Appl. Mech. 77 (6) (2007) 381–391.

[13] E.J. Sapountzakis, J.T. Katsikadelis, Dynamic analysis of elastic plates reinforced with beams of doubly-symmetrical cross section, Comput. Mech. 3 (5) (1999) 430–439.

[14] E.J. Sapountzakis, J.T. Katsikadelis, Creep and shrinkage effect on the dynamic analysis of reinforced concrete slab-and-beam structures, J. Sound Vib. 260 (3) (2003) 403–416.

[15] E.J. Sapountzakis, J.T. Katsikadelis, Influence of the inplane boundary conditions on the vibration frequencies and buckling load of ribbed plates, Int. J. Struct. Stabil. Dynam. 2 (2002) 25–43.

[16] N. Babouskos, J.T. Katsikadelis, Flutter instability of damped plates under combined conservative and nonconservative loads, Arch. Appl. Mech. 79 (2009) 541–556.

[17] J.T. Katsikadelis, N. Babouskos, Nonlinear flutter instability of thin damped plates. An AEM solution, J. Mech. Mater. Struct. 4 (7–8) (2009) 1394–1414.

[18] N. Babouskos, J.T. Katsikadelis, Nonlinear vibrations of viscoelastic plates of fractional derivative type. An AEM solution, Open Mech. J. 4 (2010) 8–20.

[19] N. Babouskos, J.T. Katsikadelis, Plate buckling optimization under conservative and nonconservative loads with BEM, in: Proc. 9th HSTAM International Congress on Mechanics, Limassol, Cyprus, 12–14 July, 2010, 2010.

[20] T.J.R. Hughes, The Finite Element Method, Prentice Hall Inc, Englewood Cliffs, NJ, USA, 1987.

[21] K.J. Bathe, Finite Element Procedures, Prentice Hall, Englewood Cliffs, NJ, 1996.

[22] J.T. Katsikadelis, A new direct time integration method for the equations of motion in structural dynamics. Z. angew. Math. Mech. (2013) doi: http://dx.doi.org/10.1002/zamm.201200245.
[23] M.S. Nerantzaki, N.G. Babouskos, Analysis of inhomogeneous anisotropic viscoelastic bodies described with fractional differential constitutive models, Comput. Math. Appl. 62 (3) (2011) 945–960.
[24] M.S. Nerantzaki, N.G. Babouskos, Vibrations of inhomogeneous anisotropic viscoelastic bodies described with fractional derivative models, Eng. Anal. Bound. Elem. 36 (2012) 1894–1907.
[25] G. Herrmann, Stability of equilibrium of elastic systems subjected to nonconservative forces, Appl. Mech. Rev. 20 (1967) 103–108.
[26] H. Leipholz, Stability of Elastic Systems, Noordhoff, The Netherlands, 1980.
[27] A.N. Kounadis, J.T. Katsikadelis, On the discontinuity of the flutter load for various types of cantilevers, Int. J. Solid. Struct. 16 (4) (1980) 375–383.
[28] A.P. Seyranian, A.A. Mailybaev, Multiparameter Stability Theory with Mechanical Applications, World Scientific Publishing Company, 2004.
[29] J.T. Katsikadelis, G.C. Tsiatas, Non-linear dynamic stability of damped Beck's column with variable cross-section, Int. J. Nonlin. Mech. 42 (1) (2007) 164–171.
[30] J.T. Katsikadelis, C.G. Tsiatas, Optimum design of structures subjected to follower forces, Int. J. Mech. Sci. 49 (11) (2007) 1204–1212.
[31] S. Adali, Stability of a rectangular plate under nonconservative and conservative forces, Int. J. Solid. Struct. 18 (1982) 1043–1052.
[32] E.H. Dowell, Nonlinear oscillations of a fluttering plate, AIAA J. 4 (7) (1966) 1267–1275.
[33] X. Guo, C. Mei, Using aeroelastic modes for nonlinear flutter at arbitrary supersonic yawed angle, AIAA J. 41 (2) (2003) 272–279.
[34] Q.H. Zuo, H.L. Schreyer, Flutter and divergence instability of nonconservative beams and plates, Int. J. Solid. Struct. 33 (9) (1996) 1355–1367.
[35] S.G. Lekhniskii, Anisotropic Plates, Gordon and Breach Science Publishers, New York, 1968.
[36] G. Davi, A. Milazzo, A meshfree method for transverse vibrations of anisotropic plates, Int. J. Solid. Struct. 40 (2003) 5229–5249.
[37] G. Haneczok, M. Weller, A fractional model of viscoelastic relaxation, Mater. Sci. Eng. 370 (2004) 209–212.
[38] F.C. Meral, T.G. Royston, R. Magin, Fractional calculus in viscoelasticity: an experimental study, Comm. Nonlinear Sci. Numer. Simulat. 15 (2010) 939–945.
[39] K. Adolfsson, M. Enelund, P. Olsson, On the fractional order model of viscoelasticity, Mech. Time-Depend. Mat. 9 (2005) 15–34.
[40] R.L. Bagley, P.J. Torvik, On the fractional calculus model of viscoelastic behavior, J. Rheol. 30 (1) (1986) 133–155.
[41] W. Flügge, Viscoelasticity, second revised ed., Springer-Verlag, Berlin/Heidelberg, 1975.
[42] S. Srinivas, A.K. Rao, An exact analysis of free vibrations of simply-supported viscoelastic plates, J. Sound Vib. 19 (3) (1971) 251–259.
[43] F.J. Rizzo, D.J. Shippy, An application of the correspondence principle of linear viscoelasticity theory, SIAM J. Appl. Math. 21 (1971) 321–330.
[44] D. Polyzos, A.A. Stamos, D.E. Beskos, BEM computation of DSIF in cracked viscoelastic plates, Comm. Numer. Meth. Eng. 10 (1994) 81–87.
[45] L. Gaul, M. Schanz, A comparative study of three boundary element approaches to calculate the transient response of viscoelastic solids with unbounded domains, Comput. Meth. Appl. Mech. Eng. 179 (1999) 111–123.
[46] M. Schanz, A boundary element formulation in time domain for viscoelastic solids, Comm. Numer. Meth. Eng. 15 (1999) 799–809.

[47] A.D. Mesquita, H.B. Coda, Boundary integral equation method for general viscoelastic analysis, Int. J. Solid. Struct. 39 (2002) 2643–2664.
[48] S. Syngellakis, Boundary element methods for polymer analysis, Eng. Anal. Bound. Elem. 27 (2003) 125–135.
[49] R. Ding, Z.Y. Zhu, C.J. Cheng, Boundary element method for solving dynamical response of viscoelastic thin plate, Appl. Math. Mech. (Engl. Ed.) 18 (1997) 229–235.
[50] J.T. Katsikadelis, N.G. Babouskos, Postbuckling analysis of viscoelastic plates with fractional derivative models, Eng. Anal. Bound. Elem. 34 (2010) 1038–1048.
[51] T.M. Atanackovic, A modified zener model of a viscoelastic body, Contin. Mech. Thermodyn. 14 (2002) 137–148.
[52] R.M. Christensen, Theory of Viscoelasticity, Academic Press, New York, 1971.
[53] I. Podlubny, Fractional Differential Equations, Academic Press, New York, 1999.

PROBLEMS

4.1 Study the stability of the plate of Fig. P4.1 using the dynamic criterion. The edge 1-4 is simply supported and immovable in its plane, while all other edges are free ($M_n^* = V_n^* = 0$). The plate has also a point support at its center. The plate is subjected to the uniform membrane load N_n^*. Data: $h = 0.08\text{m}$, $E = 3 \times 10^7 \text{kN/m}^2$, $\nu = 0.2$, $\rho = 10\,\text{kNm}^{-4}\text{s}^2$.

4.2 Compute the buckling load of the plate of Fig. P4.2. Data: $h = 0.20\text{m}$, $E = 3 \times 10^7 \text{kN/m}^2$, $\nu = 0.2$, $\rho = 10\,\text{kNm}^{-4}\text{s}^2$. The sides 1-2 and 5-6 are movable in the plane of the plate.

4.3 Analyze the dynamic response of a clamped elliptic plate with semi-axes $a = 3.0m$, $b = 2.0m$ under the load $f(t) = D(\alpha + \beta t)$. Data: $h = 0.1\text{m}$, $\alpha = 1$, $\beta = 2$, $E = 3 \times 10^7 \text{kN/m}^2$, $\nu = 0.2$, $\rho = 10\,\text{kNm}^{-4}\text{s}^2$. Show that the

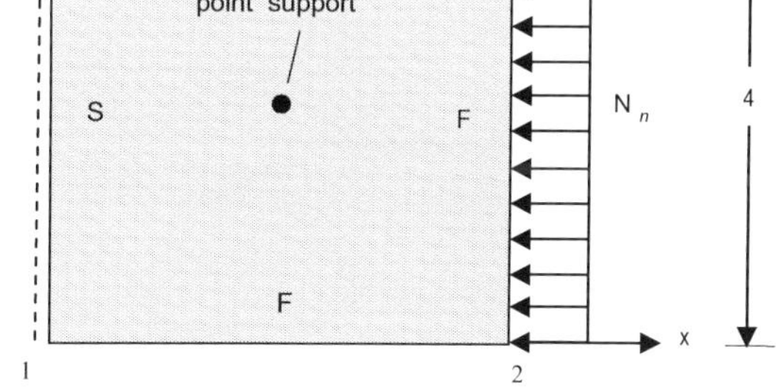

FIGURE P4.1 Square plate with boundary conditions.

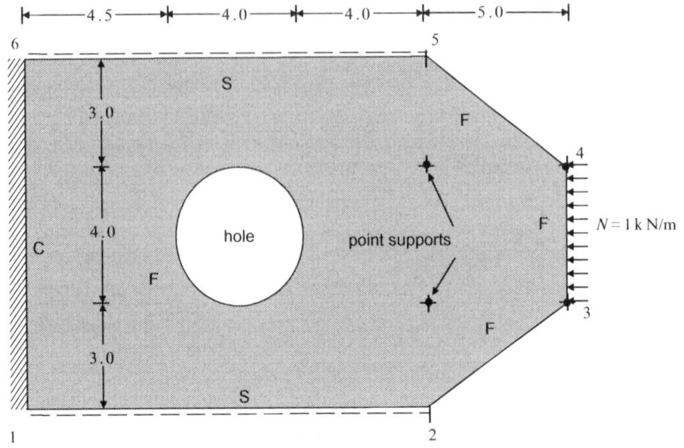

FIGURE P4.2 Plate of composite shape with a hole and internal point supports.

problem admits an exact solution $w = \dfrac{c^4}{8(3c^4+2c^2+3)}\left(1-\dfrac{x^2}{a^2}-\dfrac{y^2}{b^2}\right)^2(\alpha+\beta t)$, $c = a/b$ and compare the results.

4.4 Analyze the dynamic response of the viscoelastic plate of Fig. P4.3, when the viscoelastic material is described by the differential (i) Kelvin-Voigt model with $\eta = 0.01$, $\alpha = 1$ and (ii) The plate data are: thickness $h = 0.1$m, density $\rho = 2.4\,\mathrm{kNm^{-4}s^2}$, elastic constants $E = 30$ GPa, $v = 0.25$, load $f = 20H(t)\,\mathrm{kN/m^2}$; semi-axes $a = 4$, $b = 3$.

4.5 Study the dynamic response of the three-layer $(45^0/0^0/-45^0)$ clamped anisotropic plate of Fig. P4.4 under the uniform load $f_0 = 10H(t)$ kN/m^2. The layers are identical with $h = 0.08$m $E_1 = 19.2$ MPa, $E_2 = 1.56$ MPa, $G_{12} = 0.82$ MPa, $\nu_{12} = 0.24$, and $\rho = 10\,\mathrm{kNm^{-4}s^2}$. The boundaries of the plate are defined by the curves: Outer boundary: $\frac{x^4}{a^4}+\frac{y^4}{b^4} = 1$ $(a = 5,\ b = 4)$; inner boundary: $x^2+y^2 = r^2$ $(r = 1.5)$.

4.6 Derive the equation of the viscoelastic anisotropic plate described with the multi-term differential model.

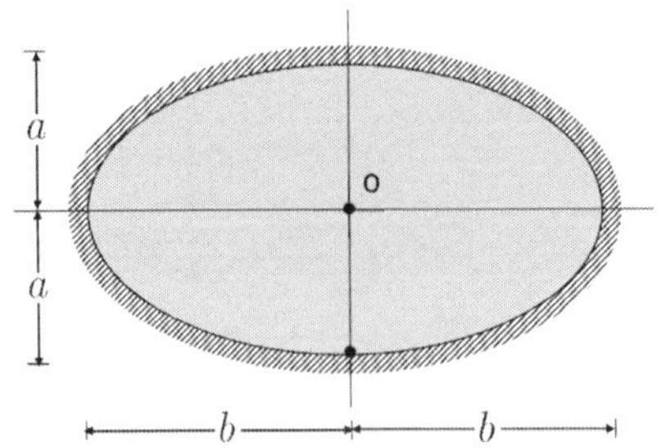

FIGURE P4.3 Elliptic clamped plate.

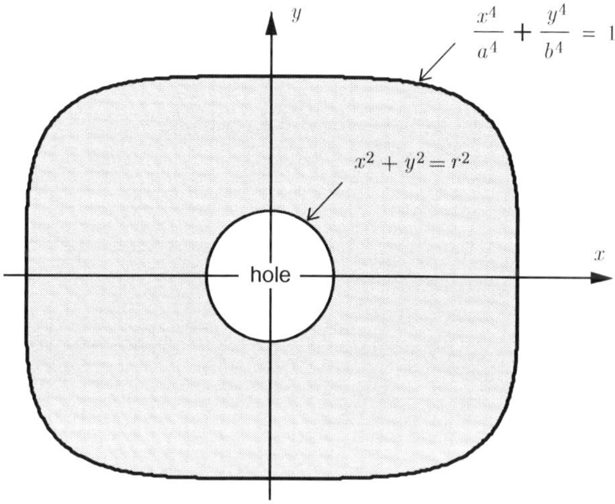

FIGURE P4.4 Anisotropic plate of composite boundary.

Chapter | five

BEM for Large Deflection Analysis of Membranes

CHAPTER OUTLINE

5.1 INTRODUCTION

Membranes are surface structures of very small thickness h, whose middle surface is a flat surface, *flat membranes*, or a nonflat surface, *space membranes*.

The Boundary Element Method for Plate Analysis

In this chapter we will present the BEM for the static and dynamic analysis of boundary-supported flat membranes. The material may be elastic or viscoelastic. Since the thickness of the membrane is very small, its bending stiffness is negligible. In the analysis of flat membranes we distinguish two stages: (i) the *prestress stage* and (ii) the *in-service stage*.

The prestress is induced either by boundary displacements or boundary forces. The external transverse load $f(x, y)$ is applied in the in-service stage and produces transverse displacement and membrane stretching at the middle surface beside that due to initial prestress. The transverse displacement renders the flat membrane a space surface $w(x, y)$. In this deformed state the membrane forces give transverse components, which equilibrate the load $f(x, y)$. This functioning of the membrane creates its bearing capacity. Linear theory assumes that the additional stretching of the middle surface due to the in-service load is negligibly small, while the membrane forces are a priori known and do not change during the out-of-plane deformation of the membrane (see [1], Section 6.3). Nevertheless, if the transverse load increases beyond a certain value, the additional stretching of the membrane cannot be neglected. The consequence of this is that the resulting differential equations governing the equilibrium of the membrane are coupled and nonlinear.

In the following analysis large transverse deflections are studied, resulting from nonlinear kinematical relations, where only the squares of the rotations of the middle surface and their products, i.e., $w_{,x}^2, w_{,y}^2, w_{,x}\ w_{,y}$, are retained, while the strain components continue to remain small compared to the unity ($\varepsilon_x, \varepsilon_y, \gamma_{xy} << 1$). Thus, the kinematical relations (2.6) apply to this case, too. This means that we treat moderately large deflections of membranes.

The solution of the nonlinear partial differential equations describing the nonlinear response of the membrane is a very difficult mathematical problem. Hence, the existing analytical solutions are very few and are limited to circular membranes under axisymmetric loading [2], where the problem becomes one-dimensional. The approximate and numerical solutions until 1980 include only axisymmetric circular membranes [3,4] and rectangular membranes [5–9]. In these solutions, the prestress, which is a basic ingredient of the structural membranes, was not encountered. Föppl [5] formulated the equations for the membrane in terms of the stress function, which seemingly simplifies the solution, but it has the inherent drawback of the inability to treat displacement boundary conditions. Besides, this formulation cannot treat membranes with holes. The actual solution to the problem was achieved numerically by the FEM [10]. The BEM was applied first with domain discretization (D/BEM) in 1996 [11] and as boundary-only in 2001 [12] after the development of the BEM for nonlinear problems [13]. Then several publications appeared which studied the static and dynamic response of isotropic, anisotropic, and nonhomogeneous elastic membranes [14–16].

The ponding of a liquid on a boundary-supported elastic membrane and the ponding on a floating elastic membrane, both highly nonlinear fluid-structure interaction problems, were also investigated by the BEM [17,18]. The minimal surface problem, known also as Plateau's or soap bubble problem, is also included in the membrane problems [19]. The minimal surface is taken as the initial shape for space membranes. Another related problem is the determination of a surface with assigned distribution of the Gaussian curvature, a problem arising in the shell theory [13]. Regarding the space membranes, the derivation of the field equations and their solution are extremely difficult mathematical problems. Here, the initial surface is not a plane, but an unknown surface passing through a given space curve or supported on flexible elastic cables. The BEM solution for the space membrane problem was developed recently [20–22].

5.2 STATIC ANALYSIS OF ELASTIC MEMBRANES

5.2.1 Differential equations for the nonlinear problem of the elastic membrane

We consider a flat elastic membrane of constant thickness h occupying the multiply connected domain Ω of the xy plane with boundary $\Gamma = \cup_{i=0}^{i=K}\Gamma_i$, where $\Gamma_i\,(i=1,\,2,\,\ldots,\,K)$ are K nonintersecting curves surrounded by Γ_0 (Fig. 5.1). The membrane is subjected to the transverse load $f(x,\,y)$ and it is elastically supported on the boundary. The membrane forces N_n^*, N_t^* and the transverse force V_n^* may be applied along the boundary. The deformation of the membrane is described by the membrane (inplane) displacements $u(x,\,y)$, $v(x,\,y)$ and the transverse displacement $w(x,\,y)$.

The equations are derived using the energy method. The total potential energy Π of the membrane results as the sum of the membrane strain energy U_m, the elastic energy U_s of the yielding support, and the potential of the external forces V

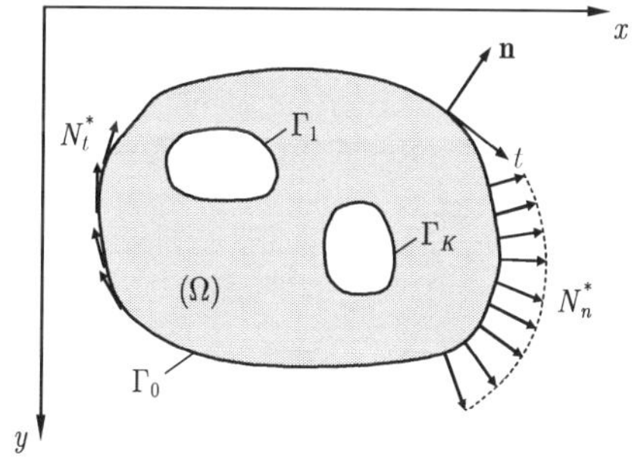

FIGURE 5.1 Geometry and support conditions of the elastic membrane.

$$\Pi = U_s + U_m + V \tag{5.1}$$

The elastic energy of the yielding support with stiffness $k_T(s)$ is written as

$$U_s = \frac{1}{2}\int_\Gamma k_T w^2\, ds \tag{5.2}$$

and the potential of the external forces is expressed as

$$V = -\int_\Omega f w d\Omega - \int_\Gamma \left(N_n^* u_n + N_t^* u_t + V_n^* w\right) ds \tag{5.3}$$

Regarding the expression of U_m we may distinguish two cases:

(a) *The membrane forces* N_n^*, N_t^* *and the transverse load* f *are imposed simultaneously.*

Taking into account that the stress and strain components are constant through the thickness of the membrane, we have

$$U_m = \frac{h}{2}\int_\Omega \left(\sigma_x\varepsilon_x + \sigma_y\varepsilon_y + \tau_{xy}\gamma_{xy}\right) d\Omega \tag{5.4}$$

where

$$\varepsilon_x = u_{,x} + \frac{1}{2} w_{,x}^2 \tag{5.5a}$$

$$\varepsilon_y = v_{,y} + \frac{1}{2} w_{,y}^2 \tag{5.5b}$$

$$\gamma_{xy} = u_{,y} + v_{,x} + w_{,x} w_{,y} \tag{5.5c}$$

$$\sigma_x = \frac{C}{h}\left(\varepsilon_{,x} + \nu\varepsilon_{,y}\right) \tag{5.6a}$$

$$\sigma_y = \frac{C}{h}\left(\varepsilon_{,y} + \nu\varepsilon_{,x}\right) \tag{5.6b}$$

$$\tau_{xy} = \frac{C}{h}\frac{1-\nu}{2}\gamma_{xy} \tag{5.6c}$$

where $C = Eh/(1-\nu^2)$ is the *stiffness* of the membrane.

Introducing Eqs. (5.5) and (5.6) in Eq. (5.4) gives

$$\begin{aligned} U_m = \frac{C}{2}\int_\Omega \Bigg[& \left(u_{,x} + \frac{1}{2} w_{,x}^2\right)^2 + \left(v_{,y} + \frac{1}{2} w_{,y}^2\right)^2 \\ & + 2\nu\left(u_{,x} + \frac{1}{2} w_{,x}^2\right)\left(v_{,y} + \frac{1}{2} w_{,y}^2\right) + \frac{1-\nu}{2}\left(u_{,y} + v_{,x} + w_{,x}\, w_{,y}\right)^2 \Bigg] d\Omega \end{aligned} \tag{5.7}$$

The equilibrium of the membrane requires the vanishing of the first variation of the total potential, $\delta\Pi = 0$. Thus, working as in Section 2.2.5 we obtain the differential equations:

(i) For the transverse displacement

$$N_x w_{,xx} + 2N_{xy} w_{,xy} + N_y w_{,yy} = -f \quad \text{in } \Omega \tag{5.8}$$

with associated boundary conditions

$$N_n w_{,n} + N_t w_{,t} + k_T w = V_n^* \quad \text{or} \quad w = w^* \quad \text{on } \Gamma \tag{5.9}$$

(ii) For the inplane displacements

$$\begin{aligned}\nabla^2 u + \frac{1+\nu}{1-\nu}\left(u_{,x} + v_{,y}\right)_{,x} + w_{,x}\left(\frac{2}{1-\nu} w_{,xx} + w_{,yy}\right) \\ + \frac{1+\nu}{1-\nu} w_{,xy} w_{,y} = 0\end{aligned} \quad \text{in } \Omega \tag{5.10a}$$

$$\begin{aligned}\nabla^2 v + \frac{1+\nu}{1-\nu}\left(u_{,x} + v_{,y}\right)_{,y} + w_{,y}\left(\frac{2}{1-\nu} w_{,yy} + w_{,xx}\right) \\ + \frac{1+\nu}{1-\nu} w_{,xy} w_{,x} = 0\end{aligned} \quad \text{in } \Omega \tag{5.10b}$$

with associated boundary conditions

$$N_n = N_n^* \quad \text{or} \quad u_n = u_n^* \quad \text{on } \Gamma \tag{5.11a}$$

$$N_t = N_t^* \quad \text{or} \quad u_t = u_t^* \quad \text{on } \Gamma \tag{5.11b}$$

The quantities N_x, N_y, N_{xy} in Eq. (5.8) represent the membrane forces produced in the interior of the membrane and are given by the relations [cf. Eqs. (3.103)]

$$N_x = C\left[u_{,x} + \frac{1}{2}w_{,x}^2 + \nu\left(v_{,y} + \frac{1}{2}w_{,y}^2\right)\right] \tag{5.12a}$$

$$N_y = C\left[v_{,y} + \frac{1}{2}w_{,y}^2 + \nu\left(u_{,x} + \frac{1}{2}w_{,x}^2\right)\right] \tag{5.12b}$$

$$N_{xy} = C\frac{1-\nu}{2}\left(u_{,y} + v_{,x} + w_{,x}w_{,y}\right) \tag{5.12c}$$

It should be noted that Eqs. (5.8) and (5.10a,b) are coupled and give nontrivial displacements u, v even when no external membrane forces are applied.

(b) *The membrane forces N_n^*, N_t^* are applied as prestress before the transverse load f.*
In this case the initial membrane forces N_x^0, N_y^0, N_{xy}^0 are known and result from the solution of the linear plane stress problem under the imposed boundary forces N_n^*, N_t^* or the boundary displacements u_n^*, u_t^*. The membrane strain energy U_m^0 due to these forces is given by the relation

$$U_m^0 = \int_\Omega \left(N_x^0 \varepsilon_x + N_y^0 \varepsilon_y + N_{xy}^0 \gamma_{xy} \right) d\Omega \tag{5.13}$$

There is another part of U_m, which is due to the additional displacements u, v produced by the transverse deflection. This part is given by Eq. (5.7). The potential of the external loads is given by Eq. (5.3), where now $N_n^* = N_t^* = 0$. The total potential energy reads

$$\Pi = U_s + U_m^0 + U_m + V \tag{5.14}$$

The condition $\delta\Pi = 0$ yields the equations:
(i) For the transverse displacement

$$\left(N_x^0 + N_x\right) w_{,xx} + 2\left(N_{xy}^0 + N_{xy}\right) w_{,xy} + \left(N_y^0 + N_y\right) w_{,yy} = -f \quad \text{in } \Omega \tag{5.15}$$

with boundary conditions given by Eqs. (5.9).
(ii) For the inplane displacements

$$\nabla^2 u + \frac{1+\nu}{1-\nu}\left(u_{,x} + v_{,y}\right)_{,x} + w_{,x}\left(\frac{2}{1-\nu} w_{,xx} + w_{,yy}\right) + \frac{1+\nu}{1-\nu} w_{,xy} w_{,y} = 0 \quad \text{in } \Omega \tag{5.16a}$$

$$\nabla^2 v + \frac{1+\nu}{1-\nu}\left(u_{,x} + v_{,y}\right)_{,y} + w_{,y}\left(\frac{2}{1-\nu} w_{,yy} + w_{,xx}\right) + \frac{1+\nu}{1-\nu} w_{,xy} w_{,x} = 0 \quad \text{in } \Omega \tag{5.16b}$$

with the homogeneous boundary conditions

$$u_n = 0 \quad \text{on } \Gamma \tag{5.17a}$$

$$u_t = 0 \quad \text{on } \Gamma \tag{5.17b}$$

The membrane forces N_x, N_y, N_{xy} in Eq. (5.15) are given by Eqs. (5.12). The displacements u, v resulting from the solution of the boundary value problem (5.15) and (5.16) express the additional membrane displacements produced by the out-of-plane deformation of the membrane. Eqs. (5.12) can be used to evaluate the additional membrane forces. The total membrane forces result by adding N_x^0, N_y^0, N_{xy}^0.

Attention should be paid when applying the prestress, which must produce extension of the membrane to avoid wrinkling. This is ensured if the produced principal membrane forces N_1^0, N_2^0 are tensile forces. This condition is satisfied if

$$N_{1,2}^0 = \frac{N_x^0 + N_y^0}{2} \pm \sqrt{\left(\frac{N_x^0 - N_y^0}{2}\right)^2 + \left(N_{xy}^0\right)^2} > 0 \tag{5.18}$$

Apparently, the equations of the membrane may result directly as a special case of the plate equations for large deflections, i.e., Eqs. (3.99) and (3.101) or (3.107) and (3.108), by setting $D=0$ and keeping only the boundary condition (3.100a) with $Vw=0$. The membrane problem could also be treated as a plate problem for a plate with very small thickness by considering the plate-bending equation as a fourth-order degenerate differential equation (see [23], Chapter 5).

5.2.2 AEM solution

The large deflections of membranes are governed by three coupled nonlinear second-order partial differential equations in terms of the displacements, namely, Eqs. (5.8), (5.10a,b) or (5.15), (5.16a,b) depending on the case. They can be solved efficiently using the AEM either as a boundary-only method or with domain discretization. We distinguish the following two cases:

(a) *The loads N_n^*, N_t^* and f are imposed simultaneously.*

The problem is described by Eqs. (5.8) and (5.10a,b), which are of the second order. Therefore, according to the principle of the analog equation, they can be replaced by three Poisson's equations:

$$\nabla^2 w = b^{(1)}(\mathbf{x}) \tag{5.19a}$$

$$\nabla^2 u = b^{(2)}(\mathbf{x}) \tag{5.19b}$$

$$\nabla^2 v = b^{(3)}(\mathbf{x}) \tag{5.19c}$$

where $\mathbf{b}^{(i)}(\mathbf{x})$, $\mathbf{x}:\{x, y\} \in \Omega$ $(i=1, 2, 3)$ designate unknown fictitious loads to be determined.

Using the procedure described in Section 3.3.2.3, we obtain

$$\mathbf{Hw} = \mathbf{Gw}_n + \mathbf{Db}^{(1)} \tag{5.20a}$$

$$\mathbf{H}\mathbf{u} = \mathbf{G}\mathbf{u}_n + \mathbf{D}\mathbf{b}^{(2)} \tag{5.20b}$$

$$\mathbf{H}\mathbf{v} = \mathbf{G}\mathbf{v}_n + \mathbf{D}\mathbf{b}^{(3)} \tag{5.20c}$$

in which

$$\mathbf{H} = -\left[v_n^*\right]^{ij} + \frac{1}{2}\delta_{ij} \tag{5.21a}$$

$$\mathbf{G} = -\left[v^*\right]^{ij} \tag{5.21b}$$

$$\mathbf{D} = \left[D^{ik}\right] \tag{5.21c}$$

The boundary conditions (5.9) and (5.11) may be combined as

$$\alpha_1 w + \alpha_2 (N_n w_{,n} + N_t w_{,t} + k_T w) = \alpha_3 \tag{5.22a}$$

$$\beta_1 u_n + \beta_2 N_n = \beta_3 \tag{5.22b}$$

$$\gamma_1 u_t + \gamma_2 N_t = \gamma_3 \tag{5.22c}$$

where α_i, β_i, γ_i $(i=1, 2, 3)$ are constants. Eqs. (5.22) can express all admissible support conditions for appropriate values of α_i, β_i, γ_i, e.g.

(i) Fixed support:
$\alpha_1 = \beta_1 = \gamma_1 = 1$, $\alpha_2 = \beta_2 = \gamma_2 = 0$, $\alpha_3 = \beta_3 = \gamma_3 = 0$.

(ii) Immovable support in the plane of the membrane and elastically supported with external transverse load:
$\alpha_1 = 0$, $\beta_1 = \gamma_1 = 1$, $\alpha_2 = 1$, $\beta_2 = \gamma_2 = 0$, $\alpha_3 = V_n^*$, $\beta_3 = \gamma_3 = 0$.

(iii) Prestressed by imposed boundary displacements u_n^*, u_t^* and elastically supported with external transverse load:
$\alpha_1 = 0$, $\beta_1 = \gamma_1 = 1$, $\alpha_2 = 1$, $\beta_2 = \gamma_2 = 0$, $\alpha_3 = V_n^*$, $\beta_3 = u_n^*$, $\gamma_3 = u_t^*$.

Further, applying the boundary conditions (5.22) at the N boundary nodal points, and expressing the tangential derivative $w_{,t}$ in terms of w using finite differences (see Section 3.3.2.1), we obtain

$$\boldsymbol{\alpha}_1 \mathbf{w} + \boldsymbol{\alpha}_2 (\mathbf{N}_n \mathbf{w}_n + \mathbf{N}_t \mathbf{d}\mathbf{w} + \mathbf{k}_T \mathbf{w}) = \boldsymbol{\alpha}_3 \tag{5.23a}$$

$$\boldsymbol{\beta}_1 \mathbf{u}_n + \boldsymbol{\beta}_2 \mathbf{N}_n = \boldsymbol{\beta}_3 \tag{5.23b}$$

$$\boldsymbol{\gamma}_1 \mathbf{u}_t + \boldsymbol{\gamma}_2 \mathbf{N}_t = \boldsymbol{\gamma}_3 \tag{5.23c}$$

Equations (5.20) and (5.23) are combined and solved to express the boundary quantities $\mathbf{w}$, $\mathbf{w}_n$, $\mathbf{u}$, $\mathbf{u}_n$, $\mathbf{v}$, $\mathbf{v}_n$ in terms of the fictitious loads $\mathbf{b}^{(i)}$. We observe that Eqs. (5.23) are nonlinear if any of a_2, β_2, γ_2 equals to one. In this case it is convenient to solve the equations using the procedure described in Section 3.5.4. This dictates to write the boundary forces $\mathbf{N}_n$ and $\mathbf{N}_t$ in the form

$$\mathbf{N}_n = \mathbf{N}_n^l + \mathbf{N}_n^{nl} \tag{5.24a}$$

$$\mathbf{N}_n = \mathbf{N}_n^l + \mathbf{N}_n^{nl} \tag{5.24b}$$

where $\mathbf{N}_n^{nl}$ and $\mathbf{N}_t^{nl}$ are the nonlinear parts of $\mathbf{N}_n$ and $\mathbf{N}_t$, respectively. These quantities are treated as new unknowns. The expressions that relate $\mathbf{N}_n^{nl}$ and $\mathbf{N}_t^{nl}$ to $\mathbf{w}$, $\mathbf{w}_n$ provide the additional required equations (see Section 3.5.4).

In the following we illustrate the case where $a_1 = \beta_1 = \gamma_1 = 1$, $a_2 = \beta_2 = \gamma_2 = 0$, $a_3 = w^*$, $\beta_3 = u_n^*$, $\gamma_3 = u_t^*$, namely, when the displacements are specified on the boundary. Apparently, Eqs. (5.20) and (5.23) give directly $\mathbf{w}_n$, $\mathbf{u}_n$, $\mathbf{v}_n$ in terms of $\mathbf{b}^{(1)}$, $\mathbf{b}^{(2)}$, $\mathbf{b}^{(3)}$, respectively. Then introducing the boundary quantities in the expressions for the displacements and their derivatives at the domain nodal points [cf. Eqs. (3.75)]

$$\mathbf{w}_{,pq} = \mathbf{D}_{,pq}\mathbf{b}^{(1)} - [v^*]_{,pq}\mathbf{w}_n + [v_n^*]_{,pq}\mathbf{w} \tag{5.25a}$$

$$\mathbf{u}_{,pq} = \mathbf{D}_{,pq}\mathbf{b}^{(2)} - [v^*]_{,pq}\mathbf{u}_n + [v_n^*]_{,pq}\mathbf{u} \tag{5.25a}$$

$$\mathbf{v}_{,pq} = \mathbf{D}_{,pq}\mathbf{b}^{(3)} - [v^*]_{,pq}\mathbf{v}_n + [v_n^*]_{,pq}\mathbf{v} \tag{5.25b}$$

we obtain

$$\mathbf{w}_{,pq} = \mathbf{U}_{,pq}^{(1)}\mathbf{b}^{(1)} + \mathbf{e}_{,pq}^{(1)} \tag{5.26a}$$

$$\mathbf{u}_{,pq} = \mathbf{U}_{,pq}^{(2)}\mathbf{b}^{(2)} + \mathbf{e}_{,pq}^{(2)} \tag{5.26b}$$

$$\mathbf{v}_{,pq} = \mathbf{U}_{,pq}^{(3)}\mathbf{b}^{(3)} + \mathbf{e}_{,pq}^{(3)} \tag{5.26c}$$

where $\mathbf{U}_{,pq}^{(i)}$ (i=1,2,3) are $M \times M$ known matrices and $\mathbf{e}_{,pq}^{(i)}$ (i=1,2,3) known vectors ($p,\ q = 0,\ x,\ y$).

Finally, application of Eqs. (5.8) and (5.10) at the M domain nodal points gives

$$\mathbf{N}_x\mathbf{w}_{,xx} + 2\mathbf{N}_{xy}\mathbf{w}_{,xy} + \mathbf{N}_y\mathbf{w}_{,yy} = -\mathbf{f} \tag{5.27a}$$

$$\mathbf{b}^{(2)} + \frac{1+\nu}{1-\nu}\left(\mathbf{u}_{,x} + \mathbf{v}_{,y}\right)_{,x} + \mathbf{w}_{,x}\left(\frac{2}{1-\nu}\mathbf{w}_{,xx} + \mathbf{w}_{,yy}\right) + \frac{1+\nu}{1-\nu}\mathbf{w}_{,xy}\mathbf{w}_{,y} = 0 \tag{5.27b}$$

$$\mathbf{b}^{(3)} + \frac{1+\nu}{1-\nu}\left(\mathbf{u}_{,x} + \mathbf{v}_{,y}\right)_{,y} + \mathbf{w}_{,y}\left(\frac{2}{1-\nu}\mathbf{w}_{,yy} + \mathbf{w}_{,xx}\right) + \frac{1+\nu}{1-\nu}\mathbf{w}_{,xy}\mathbf{w}_{,x} = 0 \tag{5.27c}$$

where

$$\mathbf{N}_x = C\left[\mathbf{u}_{,x} + \frac{1}{2}\mathbf{w}_{,x}\mathbf{w}_{,x} + \nu\left(\mathbf{v}_{,y} + \frac{1}{2}\mathbf{w}_{,y}\mathbf{w}_{,y}\right)\right] \tag{5.28a}$$

$$\mathbf{N}_y = C\left[\mathbf{v}_{,y} + \frac{1}{2}\mathbf{w}_{,y}\mathbf{w}_{,y} + \nu\left(\mathbf{u}_{,x} + \frac{1}{2}\mathbf{w}_{,x}\mathbf{w}_{,x}\right)\right] \tag{5.28b}$$

$$\mathbf{N}_{xy} = C\frac{1-\nu}{2}\left(\mathbf{u}_{,y} + \mathbf{v}_{,x} + \mathbf{w}_{,x}\mathbf{w}_{,y}\right) \tag{5.28c}$$

Introducing the derivatives from Eqs. (5.26) in Eqs. (5.27), we obtain the system of $3M$ nonlinear algebraic equations

$$\mathbf{F}^{(1)}\left(\mathbf{b}^{(1)}, \mathbf{b}^{(2)}, \mathbf{b}^{(3)}\right) = -\mathbf{f} \tag{5.29a}$$

$$\mathbf{A}_{11}\mathbf{b}^{(2)} + \mathbf{A}_{12}\mathbf{b}^{(3)} + \mathbf{F}^{(2)}\left(\mathbf{b}^{(1)}\right) = 0 \tag{5.29b}$$

$$\mathbf{A}_{21}\mathbf{b}^{(2)} + \mathbf{A}_{22}\mathbf{b}^{(3)} + \mathbf{F}^{(3)}\left(\mathbf{b}^{(1)}\right) = 0 \tag{5.29c}$$

in which $\mathbf{A}_{ij}$ $(i, j = 1, 2)$ are $M \times M$ known matrices and $\mathbf{F}^{(i)}$ $(i = 1, 2, 3)$ are vectors, whose elements are nonlinear functions of the components of the vectors $\mathbf{b}^{(i)}$. The solution of Eqs. (5.29) yields the vectors $\mathbf{b}^{(1)}$, $\mathbf{b}^{(2)}$, $\mathbf{b}^{(2)}$. Subsequently, the solution is obtained from the expressions (5.26).

(b) *The membrane forces N_n^*, N_t^* are applied as prestress before the transverse load.*

Equation (5.15) now becomes

$$\left(\mathbf{N}_x^0 + \mathbf{N}_x\right)\mathbf{w}_{,xx} + 2\left(\mathbf{N}_{xy}^0 + \mathbf{N}_{xy}\right)\mathbf{w}_{,xy} + \left(\mathbf{N}_y^0 + \mathbf{N}_y\right)\mathbf{w}_{,yy} = -\mathbf{f} \tag{5.30}$$

where $\mathbf{N}_x^0$, $\mathbf{N}_{xy}^0$, $\mathbf{N}_y^0$ are $M \times M$ known diagonal matrices, while $\mathbf{N}_x$, $\mathbf{N}_{xy}$, $\mathbf{N}_y$ are $M \times M$ unknown diagonal matrices whose elements are given by the expressions (5.28). The subsequent procedure yields a system of the form (5.29), which is solved to yield the fictitious loads.

5.2.3 Examples

EXAMPLE 5.1

In this example a circular membrane of radius a with fixed boundary subjected to the uniform load f is analyzed. The employed data are $a = 5.0\text{m}$, $h = 0.004\text{m}$, $E = 1.1 \times 10^5 \text{kN/m}^2$, $\nu = 0.3$. The obtained numerical results are given in Table 5.1 as compared with those obtained from other solutions. The results given in [7] are considered accurate. We observe that the AEM gives accurate results.

EXAMPLE 5.2

In this example a prestressed square membrane with side a, $\Omega: \{-a/2 \leq x, y \leq a/2\}$, under the uniform load f is analyzed. The prestress is imposed by boundary displacements whose distribution is shown in Fig. 5.2. The employed data are $a = 5.0\text{m}$, $h = 0.002\text{m}$, $E = 1.1 \times 10^5\text{kN/m}^2$, $\nu = 0.3$, $\widetilde{u} = \widetilde{v} = 0.05\text{m}$, and $\widetilde{w} = 0$. The results were obtained for various values of the

TABLE 5.1 Displacements $\overline{w}=w/a$ and membrane forces $\overline{N}_r=N_r(1-\nu^2)/Eh$, $\overline{N}_\phi=N_\phi(1-\nu^2)/Eh$ along the radius of the unprestressed circular membrane in Example 5.1 ($\nu=0.3$, $\overline{f}=f/[a(1-\nu^2)/Eh]=0.015$, $N=100$, $M=50$)

$\frac{r}{a}$	$\overline{w}$			$\overline{N}_r$			$\overline{N}_\phi$		
	[7]	FEM	AEM	[7]	FEM	AEM	[7]	FEM	AEM
0	0.166	0.169	0.166	0.0254	0.0260	0.0254	0.0254	0.0260	0.0254
0.2	0.160	0.164	0.160	0.0252	0.0256	0.0252	0.0243	0.0254	0.0248
0.4	0.142	0.146	0.142	0.0246	0.0244	0.0246	0.0230	0.0238	0.0230
0.6	0.111	0.114	0.111	0.0235	0.0238	0.0235	0.0196	0.0204	0.0196
0.8	0.065	0.065	0.065	0.0219	0.0226	0.0219	0.0143	0.0143	0.0143
1.0	0.000	0.000	0.000	0.0196	0.0211	0.0206	0.0059	0.0063	0.0062

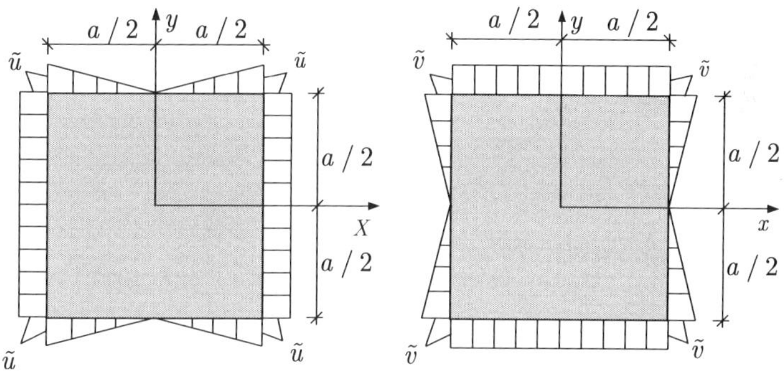

FIGURE 5.2 Distribution of the boundary displacements producing the prestress in Example 5.2.

transverse load using $N=100$ constant boundary elements and $M=153$ domain nodal points resulting from 232 triangular elements. The numerical results using nondimensional quantities $\overline{w}=w/a$, $\overline{f}=f/[a(1-\nu^2)/Eh]$, $\overline{N}=N(1-\nu^2)/Eh$ $\overline{V}_n=V_n(1-\nu^2)/Eh$ are shown in graphical form in Fig. 5.3 through 5.9. From Fig. 5.5, we observe that the relation between $\overline{N}_x$ and $\overline{f}$ becomes linear after a certain value of the load $\overline{f}$. It was also observed that the convergence of the solution of the nonlinear equations (5.29) was improved with increasing prestress. Finally, Fig. 5.10 shows the variation of the strain component ε_x at $x=0$, $y=0$ and $x=a/2$, $y=0$ versus the central displacement in the square membrane of Example 5.2 together with the corresponding quantities at $r=0$ and $r=a$ of the circular membrane of Example 5.1. These results can be used to estimate the limits of validity between the linear theory (small deflections – small strains), the nonlinear intermediate theory (large deflections – small strains) and finite deformations (large deflections – large strains). From this figure, we conclude that the intermediate theory holds for large transverse displacements.

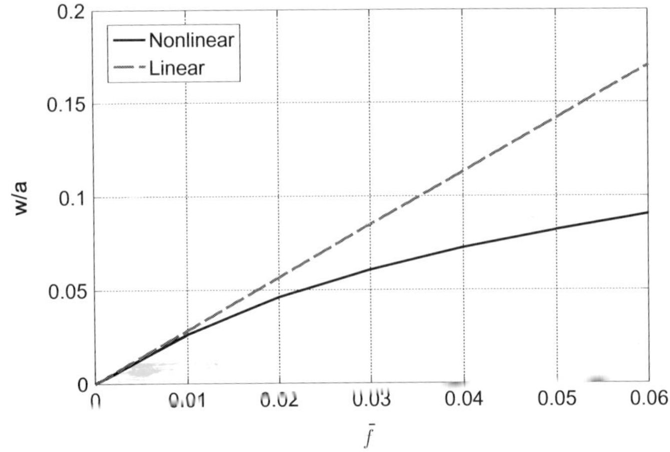

FIGURE 5.3 Central deflection versus the transverse load $\overline{f}$ in Example 5.2.

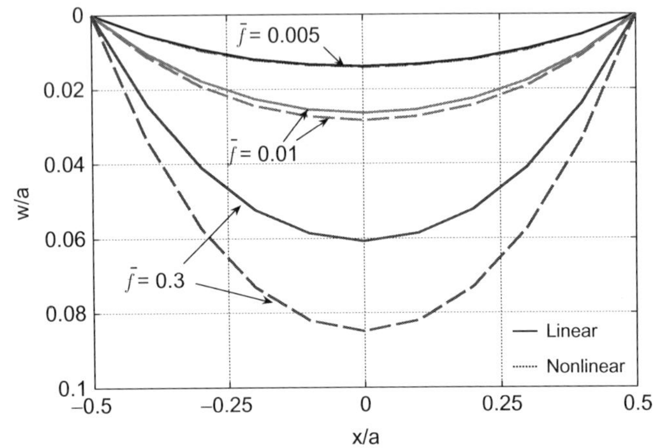

FIGURE 5.4 Displacements along the central line $y=0$ of the membrane for various values of the load $\overline{f}$ in Example 5.2.

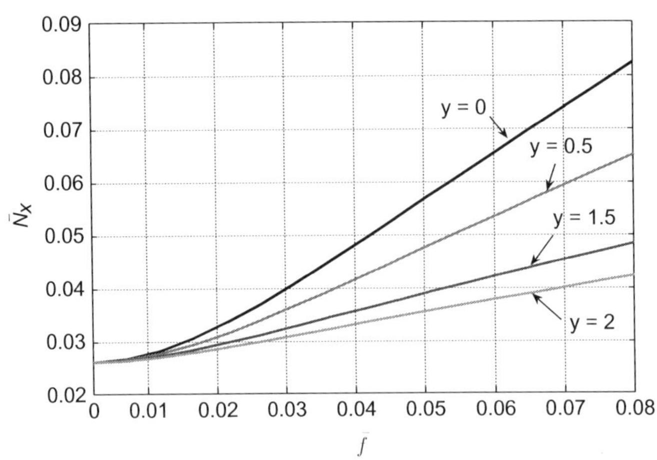

FIGURE 5.5 Membrane force $\overline{N}_x$ versus load $\overline{f}$ at different points along the line $x/a=0.3$ in Example 5.2.

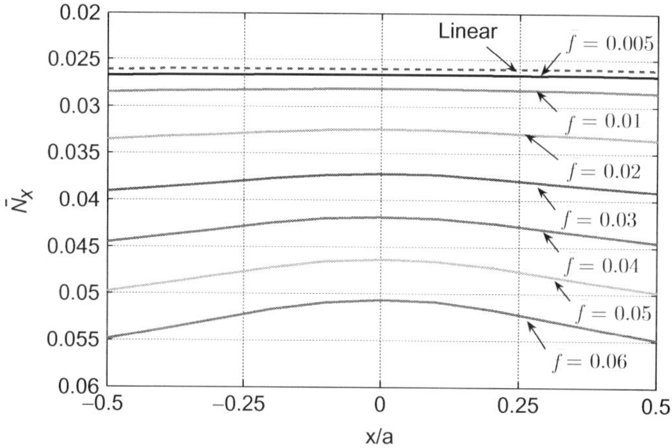

FIGURE 5.6 Membrane force $\overline{N}_x$ along the central line $y = 0$ for various values of the load $\overline{f}$ in Example 5.2.

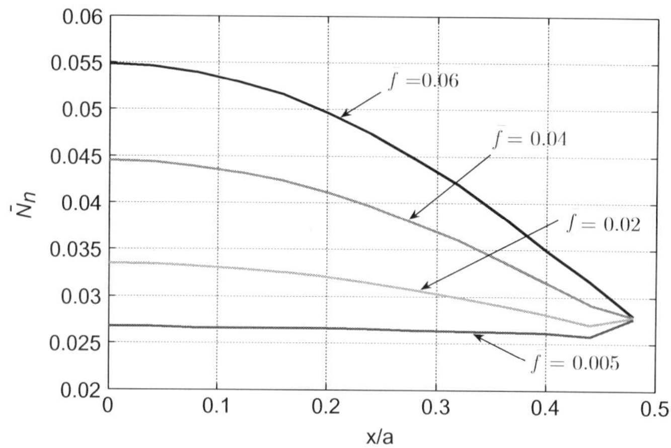

FIGURE 5.7 Normal reaction $\overline{N}_n$ along the edge $y = a/2$ for various values of the load $\overline{f}$ in Example 5.2.

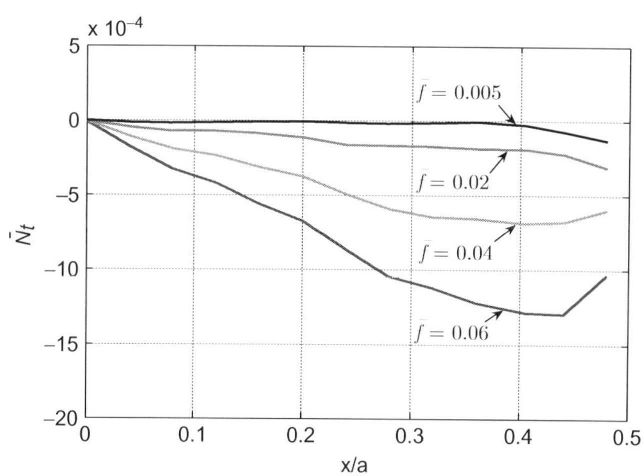

FIGURE 5.8 Tangential reaction $\overline{N}_t$ along the edge $y = a/2$ for various values of the load $\overline{f}$ in Example 5.2.

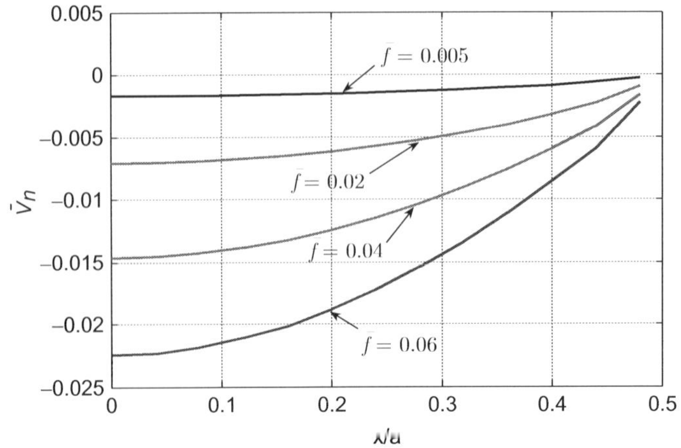

FIGURE 5.9 Transverse reaction $\overline{V}_n$ along the edge $y = a/2$ for various values of the load $\overline{f}$ in Example 5.2.

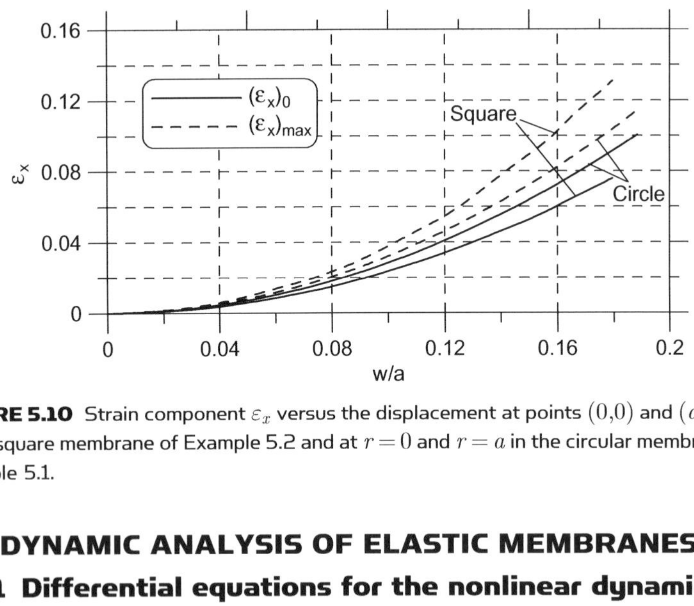

FIGURE 5.10 Strain component ε_x versus the displacement at points $(0,0)$ and $(a/2, 0)$ in the square membrane of Example 5.2 and at $r = 0$ and $r = a$ in the circular membrane of Example 5.1.

5.3 DYNAMIC ANALYSIS OF ELASTIC MEMBRANES

5.3.1 Differential equations for the nonlinear dynamic problem of the elastic membrane

The load and displacements now depend also on time, namely, $f(\mathbf{x}, t)$, $w(\mathbf{x}, t)$, $u(\mathbf{x}, t)$, $v(\mathbf{x}, t)$. The equations governing the response of the nonlinear vibrations of membranes can be obtained from Eqs. (5.8) and (5.10a,b) or (5.15) and (5.16a,b), depending on the case, if the inertia forces are included in the external loads. Thus, the respective initial boundary value problems are stated as follows:

(a) *The membrane forces N_n^*, N_t^* and the transverse load f are imposed simultaneously.*

$$N_x w_{,xx} + 2N_{xy} w_{,xy} + N_y w_{,yy} - \rho h \ddot{w} + \rho h \ddot{u} w_{,x} + \rho h \ddot{v} w_{,y} = -f \quad \text{in } \Omega \tag{5.31a}$$

$$\nabla^2 u + \frac{1+\nu}{1-\nu}\left(u_{,x} + v_{,y}\right)_{,x} + w_{,x}\left(\frac{2}{1-\nu}w_{,xx} + w_{,yy}\right) + \frac{1+\nu}{1-\nu}w_{,xy}w_{,y} - \frac{\rho\ddot{u}}{G} = 0 \quad \text{in } \Omega \tag{5.31b}$$

$$\nabla^2 v + \frac{1+\nu}{1-\nu}\left(u_{,x} + v_{,y}\right)_{,y} + w_{,y}\left(\frac{2}{1-\nu}w_{,yy} + w_{,xx}\right) + \frac{1+\nu}{1-\nu}w_{,xy}w_{,x} - \frac{\rho\ddot{v}}{G} = 0 \quad \text{in } \Omega \tag{5.31c}$$

with boundary conditions

$$N_n w_{,n} + N_t w_{,t} + k_T w = V_n^* \ \text{ or } \ w = w^* \quad \text{on } \Gamma \tag{5.32a}$$

$$N_n = N_n^* \ \text{ or } \ u_n = u_n^* \quad \text{on } \Gamma \tag{5.32b}$$

$$N_t = N_t^* \ \text{ or } \ u_t = u_t^* \quad \text{on } \Gamma \tag{5.32c}$$

and initial conditions

$$w(\mathbf{x},0) = g_1(\mathbf{x}), \quad \dot{w}(\mathbf{x},0) = h_1(\mathbf{x}) \quad \mathbf{x} \in \Omega \tag{5.33a}$$

$$u(\mathbf{x},0) = g_2(\mathbf{x}), \quad \dot{u}(\mathbf{x},0) = h_2(\mathbf{x}) \quad \mathbf{x} \in \Omega \tag{5.33b}$$

$$v(\mathbf{x},0) = g_3(\mathbf{x}), \quad \dot{v}(\mathbf{x},0) = h_3(\mathbf{x}) \quad \mathbf{x} \in \Omega \tag{5.33c}$$

where $g_i(\mathbf{x})$, $h_i(\mathbf{x})$ $(i = 1, 2, 3)$ are specified functions.

(b) *The membrane forces N_n^*, N_t^* are applied as prestress before the transverse load f.*

$$\left(N_x^0 + N_x\right)w_{,xx} + 2\left(N_{xy}^0 + N_{xy}\right)w_{,xy} + \left(N_y^0 + N_y\right)w_{,yy} - \rho h\ddot{w} + \rho h\ddot{u}w_{,x} + \rho h\ddot{v}w_{,y} = -f \quad \text{in } \Omega \tag{5.34a}$$

$$\nabla^2 u + \frac{1+\nu}{1-\nu}\left(u_{,x} + v_{,y}\right)_{,x} + w_{,x}\left(\frac{2}{1-\nu}w_{,xx} + w_{,yy}\right) + \frac{1+\nu}{1-\nu}w_{,xy}w_{,y} - \frac{\rho\ddot{u}}{G} = 0 \quad \text{in } \Omega \tag{5.34b}$$

$$\nabla^2 v + \frac{1+\nu}{1-\nu}\left(u_{,x} + v_{,y}\right)_{,y} + w_{,y}\left(\frac{2}{1-\nu}w_{,yy} + w_{,xx}\right) + \frac{1+\nu}{1-\nu}w_{,xy}w_{,x} - \frac{\rho\ddot{v}}{G} = 0 \quad \text{in } \Omega \tag{5.34c}$$

with boundary conditions

$$N_n w_{,n} + N_t w_{,t} + k_T w = V_n^* \quad \text{or} \quad w = w^* \quad \text{on } \Gamma \tag{5.35a}$$

$$u_n = 0 \quad \text{on } \Gamma \tag{5.35b}$$

$$u_t = 0 \quad \text{on } \Gamma \tag{5.35c}$$

and initial conditions given by Eqs. (5.33). The displacements u and v resulting from the solution of the coupled nonlinear Eqs. (5.34a,b,c), represent the additional displacements at the points of the middle beyond prestress.

5.3.2 AEM solution

The analog equations in this case are

$$\nabla^2 w = b^{(1)}(\mathbf{x}, t) \tag{5.36a}$$

$$\nabla^2 u = b^{(2)}(\mathbf{x}, t) \tag{5.36b}$$

$$\nabla^2 v = b^{(3)}(\mathbf{x}, t) \tag{5.36c}$$

in which $b^{(i)}(\mathbf{x}, t)$, $\mathbf{x}: \{x, y\} \in \Omega$, $t > 0$ $(i = 1, 2, 3)$ represent time-dependent fictitious sources, unknown in the first instance. Note that time is a parameter, i.e., the equations are quasi-static.

In the following we present the AEM for the case where the boundary displacements are specified. The procedure yields the following system of $3 \times M$ nonlinear semi-discretized equations of motion for the fictitious loads $\mathbf{b}^{(i)}$:

$$\mathbf{M}^{(1)}\ddot{\mathbf{b}}^{(1)} + \mathbf{F}^{(1)}\left(\mathbf{b}^{(1)}, \ddot{\mathbf{b}}^{(1)}, \mathbf{b}^{(2)}, \ddot{\mathbf{b}}^{(2)}, \mathbf{b}^{(3)}, \ddot{\mathbf{b}}^{(3)}\right) = -\mathbf{f} \tag{5.37a}$$

$$\mathbf{M}^{(2)}\ddot{\mathbf{b}}^{(2)} + \mathbf{A}_{11}\mathbf{b}^{(2)} + \mathbf{A}_{12}\mathbf{b}^{(3)} + \mathbf{F}^{(2)}\left(\mathbf{b}^{(1)}\right) = 0 \tag{5.37b}$$

$$\mathbf{M}^{(3)}\ddot{\mathbf{b}}^{(3)} + \mathbf{A}_{21}\mathbf{b}^{(2)} + \mathbf{A}_{22}\mathbf{b}^{(3)} + \mathbf{F}^{(3)}\left(\mathbf{b}^{(1)}\right) = 0 \tag{5.37c}$$

where $\mathbf{M}^{(k)}$ $(k = 1, 2, 3)$ are $M \times M$ known mass matrices, $\mathbf{A}_{ij}$ $(i, j = 1, 2)$ are $M \times M$ known matrices, and $\mathbf{F}^{(k)}$ $(k = 1, 2, 3)$ are vectors nonlinearly depending on the components of $\mathbf{b}^{(i)}$. The initial conditions for $\mathbf{b}^{(i)}$ result from Eqs. (5.33) using Eqs. (5.26) with $p, q = 0$

$$\mathbf{b}^{(i)}(0) = \left[\mathbf{U}^{(i)}\right]^{-1}(\mathbf{g}_i - \mathbf{e}_i), \quad \dot{\mathbf{b}}^{(i)}(0) = \left[\mathbf{U}^{(i)}\right]^{-1}\mathbf{h}_i, \quad (i = 1, 2, 3) \tag{5.38}$$

Equations (5.37) are solved using the numerical solution presented in Appendix C. However, the use of all the degrees of freedom, i.e., $3M$, may be computationally costly and in some cases inefficient due to the large number of time-dependent variables $b_k^{(i)}(t)$, $(k = 1, 2, \ldots, M)$. To overcome this difficulty the number of degrees of freedom is reduced using the Ritz method. Good numerical results are obtained if the mode shapes of the linear membrane problem are employed as Ritz vectors.

5.3.3 Examples

EXAMPLE 5.3

The dynamic response of the membrane of Fig. 5.11 is studied in this example. The boundary of the domain is defined by the curve $r=(ab)^{1/2}/\left[(\cos\theta/a)^2+(\sin\theta/b)^2\right]^{(1/4)}/\left[(\cos\theta/b)^2+(\sin\theta/a)^2\right]^{(1/4)}$, $0\leq\theta\leq 2\pi$. The membrane is prestressed by the imposed displacement $u_n=0.2\text{m}$ normal to the boundary, while the tangential displacement is zero, $u_t=0$. The employed data are: $a=3\text{m}$, $b=1.3\text{m}$, $h=0.002\text{m}$, $\rho=5000\text{kg/m}^3$, $\nu=0.3$, $E=1.1\times10^5\,\text{kN/m}^2$. The results have been obtained with $N=210$ boundary elements and $M=106$ domain nodal points resulting from 164 linear triangular elements. The membrane is subjected to the transverse load $p(t)=2H(t)\,\text{kN/m}^2$. Fig. 5.12 presents the time history of the transverse displacement $w(0{,}0;\,t)$, the membrane displacement $u(-1.741,0;t)$, and the membrane force $N_x(-1.741,0;t)$, when the membrane inertia forces are taken into account and when they are neglected. Evidently, the influence of the membrane inertia forces is very small and hence they can be neglected in the dynamic analysis of the membranes.

The solution of the nonlinear system of the semi-discretized equation of motion was obtained using the method presented in Appendix C after reducing the number of the degrees of freedom via the Ritz method. The mode shapes of the linear vibrations of the membrane were used as Ritz vectors. Fig. 5.13 shows the time history of the transverse displacement $w(0{,}0;\,t)$, and the membrane displacement $u(-1.741,0;t)$ for different number of Ritz vectors as compared with that obtained with no reduction, namely, with 106 degrees of freedom. It is noteworthy that the use of more than 20 modes changes the results negligibly. The Ritz reduction is very important in nonlinear structural dynamics, because it allows obtaining accurate results using a small number of degrees of freedom.

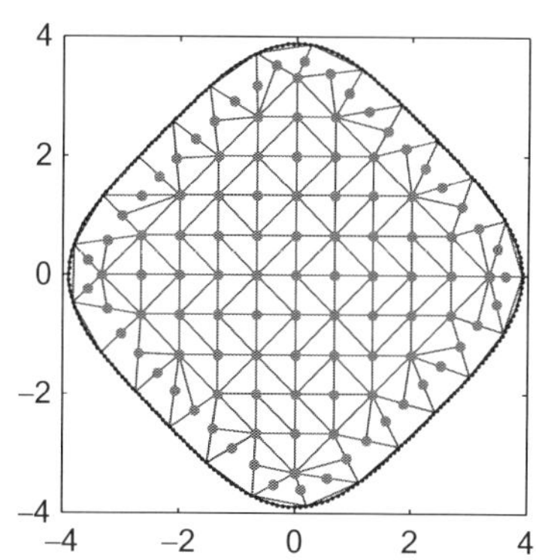

FIGURE 5.11 Boundary and domain nodal points of the membrane in Example 5.3.

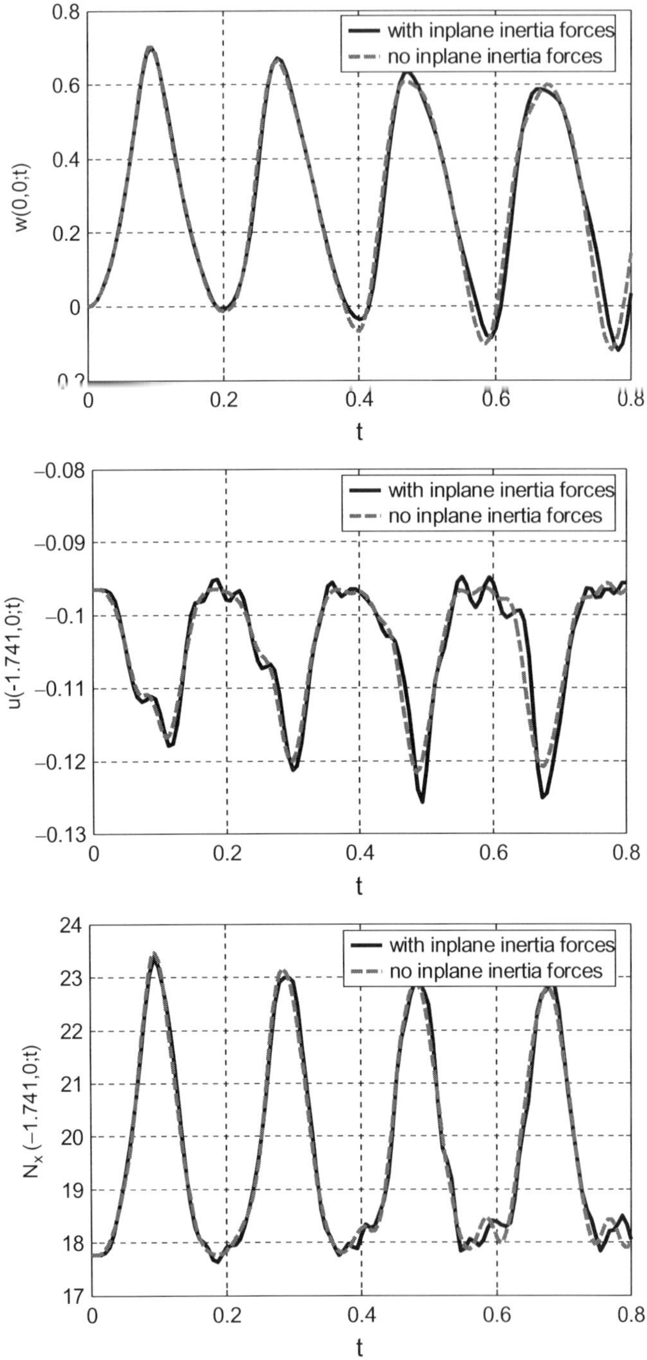

FIGURE 5.12 Time history of the displacements $w(0{,}0;t)$, $u(-1.741,0;t)$ and membrane force $N_x(-1.741,0;t)$ in Example 5.3.

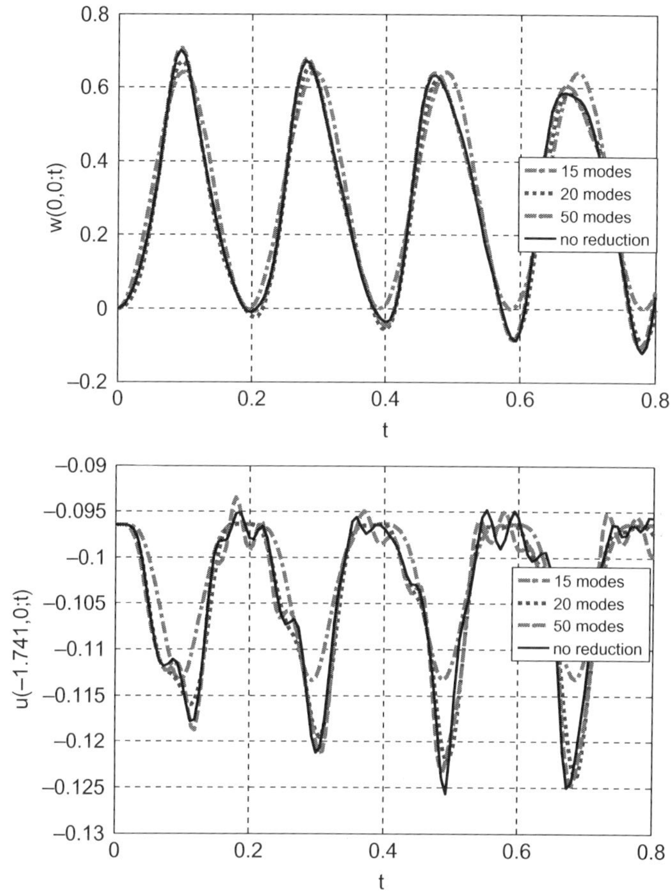

FIGURE 5.13 Time history of the displacements $w(0,0;t)$, $u(-1.741, 0; t)$ for different number of Ritz vectors in Example 5.3.

5.4 VISCOELASTIC MEMBRANES

5.4.1 Differential viscoelastic models of fractional order

Membranes made of linear viscoelastic materials are extensively used in modern engineering applications. Viscoelastic models of differential or integral form have been proposed to describe the mechanical behavior of such materials. It has been shown recently that viscoelastic models of differential form with fractional derivatives are in better agreement with the experimental results than the integer derivative models [24–26]. The dynamic response of pure elastic and viscoelastic membranes, using differential constitutive equations of integer-order derivative or hereditary integral type models, have been examined by several investigators [27–31]. The use of fractional differential models is very limited and is restricted to viscoelastic beams [32], three-dimensional viscoelastic bodies [33], viscoelastic plates (see Chapter 4), and viscoelastic membranes [34].

The response of such membranes is described by a system of nonlinear partial fractional differential equations, for which an analytical solution is out of the question. Therefore, the recourse to numerical solutions is inevitable. The BEM in conjunction with the AEM provides an efficient computational tool for solving linear and nonlinear partial fractional differential equations [35]. The equations of integer-order result as special cases. This method is employed in the following for the static and dynamic analysis of viscoelastic membranes.

Without excluding other fractional differential models [36], the employed herein viscoelastic material is described by the Kelvin-Voigt type model with fractional-order derivative

$$\begin{Bmatrix} \sigma_x \\ \sigma_y \\ \tau_{xy} \end{Bmatrix} = \frac{E}{1-\nu^2} \begin{bmatrix} 1 & \nu & 0 \\ \nu & 1 & 0 \\ 0 & 0 & \dfrac{1-\nu}{2} \end{bmatrix} \begin{Bmatrix} \varepsilon_x + \eta D_c^a \varepsilon_x \\ \varepsilon_y + \eta D_c^a \varepsilon_y \\ \gamma_{xy} + \eta D_c^a \gamma_{xy} \end{Bmatrix} \tag{5.39}$$

where E, ν are the elastic material constants, η the viscoelastic parameter, and D_c^a the Caputo fractional derivative of order α defined as

$$D_c^\alpha u(t) = \begin{cases} \dfrac{1}{\Gamma(m-\alpha)} \displaystyle\int_0^t \dfrac{u^{(m)}(\tau)}{(t-\tau)^{\alpha+1-m}} d\tau, & m-1<\alpha<m \\ \dfrac{d^m}{dt^m} u(t), & m=\alpha \end{cases} \tag{5.40}$$

where m is a positive integer. The advantage of this definition of the fractional derivative is that it permits the assignment of initial conditions that have direct physical significance [37]. Apparently, the classical derivatives result from Eq. (5.40) for integer values of α. Thus, Eq. (5.40) for $\alpha=1$ gives the constitutive equation for the conventional Kelvin-Voigt model, while for $\alpha=0$ the pure elastic model with elastic modulus $E^* = E(1+\eta)$. The use of fractional differential models, besides their simplicity to formulate the equations of structural viscoelastic systems, exhibit the significant advantage of describing the viscoelastic response using much fewer parameters than the integer-order differential models. Nevertheless, due to lack of efficient numerical methods to solve the FDEs, the use of such models, despite their advantages, is very limited to date. The efficient numerical methods developed recently bring the fractional derivative models in the center of the arena.

5.4.2 Differential equations for the viscoelastic membrane

We consider a thin flexible initially flat membrane of thickness h and mass density ρ consisting of homogeneous linearly viscoelastic material occupying the two-dimensional, in general multiply connected, domain Ω in x, y plane. The membrane is prestressed either by imposed displacement u_n^*, v_t^* or by external forces N_n^*, N_t^* acting along the boundary Γ. Moderately large deflections are examined with nonlinear kinematic relations given by Eqs. (5.5). The membrane is subjected to the transverse load $f = f(x, y, t)$ and the inplane loads $p_x = p_x(x, y, t)$ and $p_y = p_y(x, y, t)$.

On the basis of Eqs. (5.39) and (5.40) the stress resultants are written as

$$\overline{N}_x = N_x + \eta D_c^\alpha N_x \tag{5.41a}$$

$$\overline{N}_y = N_y + \eta D_c^\alpha N_y \tag{5.41b}$$

$$\overline{N}_{xy} = N_{xy} + \eta D_c^\alpha N_{xy} \tag{5.41c}$$

where N_x, N_y, N_{xy} are given by Eqs. (5.12).

The governing equations result by taking the equilibrium of the membrane element in the slightly deformed configuration. This yields the following equations in Ω:

$$\begin{aligned}\overline{N}_x w_{,xx} + 2\overline{N}_{xy} w_{,xy} + \overline{N}_y w_{,yy} - p_x w_{,x} - p_y w_{,y} \\ -\rho h\ddot{w} + \rho h\ddot{u} w_{,x} + \rho h\ddot{v} w_{,y} = -f\end{aligned} \tag{5.42a}$$

$$\overline{N}_{x,x} + \overline{N}_{xy,y} - \rho h\ddot{u} = -p_x \tag{5.42b}$$

$$\overline{N}_{y,y} + \overline{N}_{xy,x} - \rho h\ddot{v} = -p_y \tag{5.42c}$$

Without restricting the generality, we consider displacement boundary conditions on Γ:

$$u_n = u_n^*, \qquad u_t = v_t^*, \qquad w = w^* \tag{5.43a,b,c}$$

Inserting Eqs. (5.41) in Eqs. (5.42) we obtain the equations governing the response of the viscoelastic membrane in terms of the displacements in Ω:

$$\begin{aligned}C\left\{\left[u_{,x} + \nu v_{,y} + \frac{1}{2}\left(w_{,x}^2 + \nu w_{,y}^2\right) + \eta D_c^\alpha\left[u_{,x} + \nu v_{,y} + \frac{1}{2}\left(w_{,x}^2 + \nu w_{,y}^2\right)\right\}w_{,xx}\right.\right. \\ + C(1-\nu)\left\{\left(u_{,y} + v_{,x} + w_{,x}w_{,y}\right) + \eta D_c^\alpha\left(u_{,y} + v_{,x} + w_{,x}w_{,y}\right)\right\}w_{,xy} \\ + C\left\{\left[\nu u_{,x} + v_{,y} + \frac{1}{2}\left(\nu w_{,x}^2 + w_{,y}^2\right)\right] + \eta D_c^\alpha\left[\nu u_{,x} + v_{,y} + \frac{1}{2}\left(\nu w_{,x}^2 + w_{,y}^2\right)\right]\right\}w_{,yy} \\ -p_x w_{,x} - p_y w_{,y} - \rho h\ddot{w} + \rho h\ddot{u} w_{,x} + \rho h\ddot{v} w_{,y} = -f\end{aligned} \tag{5.44a}$$

$$\begin{aligned}C\left\{\left[u_{,x} + \nu v_{,y} + \frac{1}{2}\left(w_{,x}^2 + \nu w_{,y}^2\right) + \eta D_c^\alpha\left[u_{,x} + \nu v_{,y} + \frac{1}{2}\left(w_{,x}^2 + \nu w_{,y}^2\right)\right\}_{,x}\right.\right. \\ + C\frac{1-\nu}{2}\left\{\left(u_{,y} + v_{,x} + w_{,x}w_{,y}\right) + \eta D_c^\alpha\left(u_{,y} + v_{,x} + w_{,x}w_{,y}\right)\right\}_{,y} - \rho h\ddot{u} = -p_x\end{aligned} \tag{5.44b}$$

$$C\frac{1-\nu}{2}\left\{\left(u_{,y}+v_{,x}+w_{,x}w_{,y}\right)+\eta D_c^{\alpha}\left(u_{,y}+v_{,x}+w_{,x}w_{,y}\right)\right\}_{,x}$$
$$+C\left\{\left[\nu u_{,x}+v_{,y}+\frac{1}{2}\left(\nu w_{,x}^2+w_{,y}^2\right)\right]+\eta D_c^{\alpha}\left[\nu u_{,x}+v_{,y}+\frac{1}{2}\left(\nu w_{,x}^2+w_{,y}^2\right)\right]\right\}_{,y}-\rho h\ddot{v}=-p_y \tag{5.44c}$$

Note that indicated differentiations in the above equations have not been performed for the conciseness of the expressions.

Equations (5.44) are subjected to the boundary conditions (5.43a,b,c) and to the initial conditions (5.33).

Equations (5.44) constitute a system of three coupled nonlinear partial fractional differential equations of evolution type that are solved using the BEM presented in [35], which for the problem at hand is adjusted as follows.

5.4.3 AEM solution

Equations (5.44) are of the second-order with regard to the spatial derivatives. Thus, according to the principle of the analog equation they can be replaced by Eqs. (5.36). Working as in Section 5.2, we obtain the following semi-discretized equations of motion:

$$\mathbf{M}^{(i)}\ddot{\mathbf{b}}^{(i)}+\mathbf{F}^{(i)}\left(\mathbf{b}^{(1)},\ddot{\mathbf{b}}^{(1)},\mathbf{b}^{(2)},\ddot{\mathbf{b}}^{(2)},\mathbf{b}^{(3)},\ddot{\mathbf{b}}^{(3)},D_c^{\alpha}\mathbf{b}^{(1)},D_c^{\alpha}\mathbf{b}^{(2)},D_c^{\alpha}\mathbf{b}^{(3)}\right)$$
$$=\mathbf{f}^{(i)},\quad (i=1,2,3) \tag{5.45}$$

with initial conditions given by Eqs. (5.38). $\mathbf{M}^{(i)}$ are $M\times M$ consistent mass matrices, $\mathbf{F}^{(i)}$ are stiffness vectors whose elements depend nonlinearly on the components of $\mathbf{b}^{(i)}$, $D_c^{\alpha}\mathbf{b}^{(i)}$, and are $\mathbf{f}^{(i)}$ known load vectors.

Equations (5.45) constitute a system of $3M$ nonlinear fractional ordinary differential equations, which are solved using the time-step numerical method presented in [38].

5.4.4 Examples

EXAMPLE 5.4 FREE AND FORCED VIBRATIONS

The dynamic response of the membrane shown in Fig. 5.11 is studied when the material is viscoelastic, described by the fractional Kelvin-Voigt type model. The employed data and discretization are the same as in Example 5.3. First, the free vibrations of the elastic ($\eta=0$) and the viscoelastic membrane ($\eta=0.2$) are examined. The initial conditions are $w(x,y;0)=$ the deflection of the membrane due to a uniform static load $f=1.0\text{kN/m}^2$ and $\dot{w}(x,y;0)=0$. The results were obtained using 20 linear modes for the Ritz reduction. Fig. 5.14 shows the time history of the deflection $w(0,0;t)$, the membrane displacement $u(-1.741,0;t)$, and the membrane force $N_x(-1.741,0;t)$ for elastic ($\eta=0$) and viscoelastic ($\alpha=1$, $\eta=0.2$) material.

Subsequently, the forced vibrations of the viscoelastic membrane are studied when the membrane is subjected to the suddenly applied transverse load

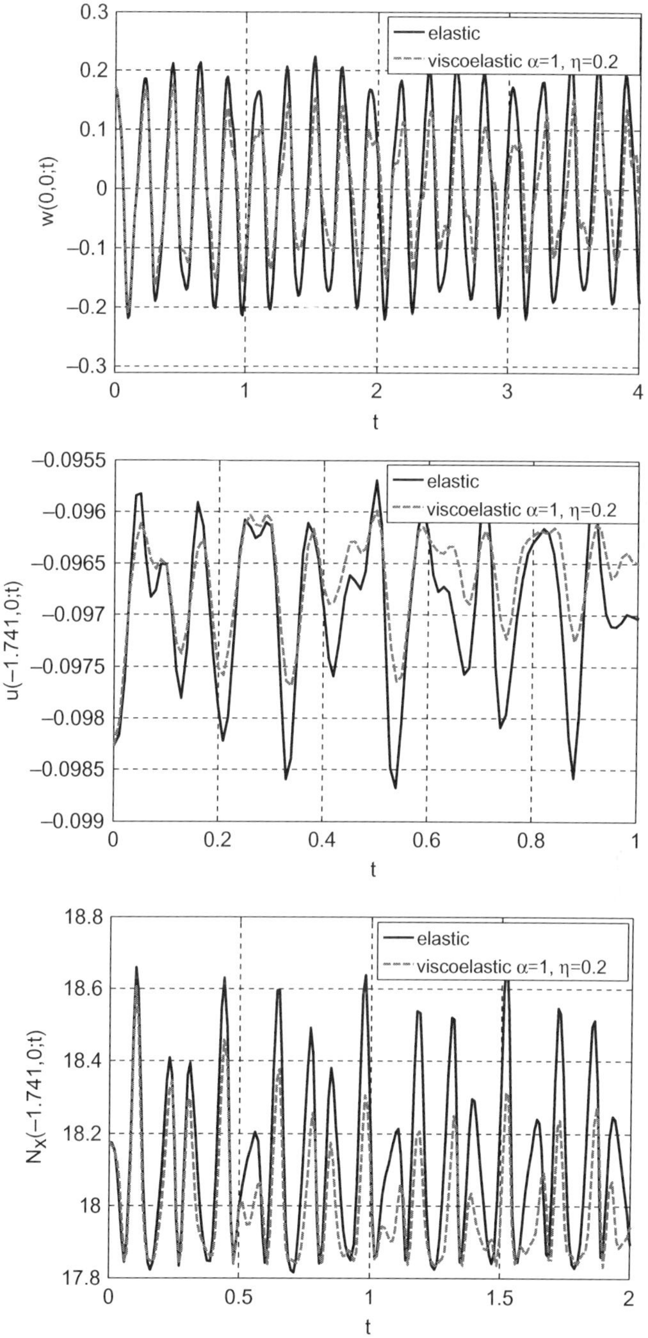

FIGURE 5.14 Time history of $w(0,0;t)$, $u(-1.741,0;t)$, $N_x(-1.741,0;t)$ for elastic ($\eta=0$) and viscoelastic ($\alpha=1$, $\eta=0.2$) material in Example 5.4.

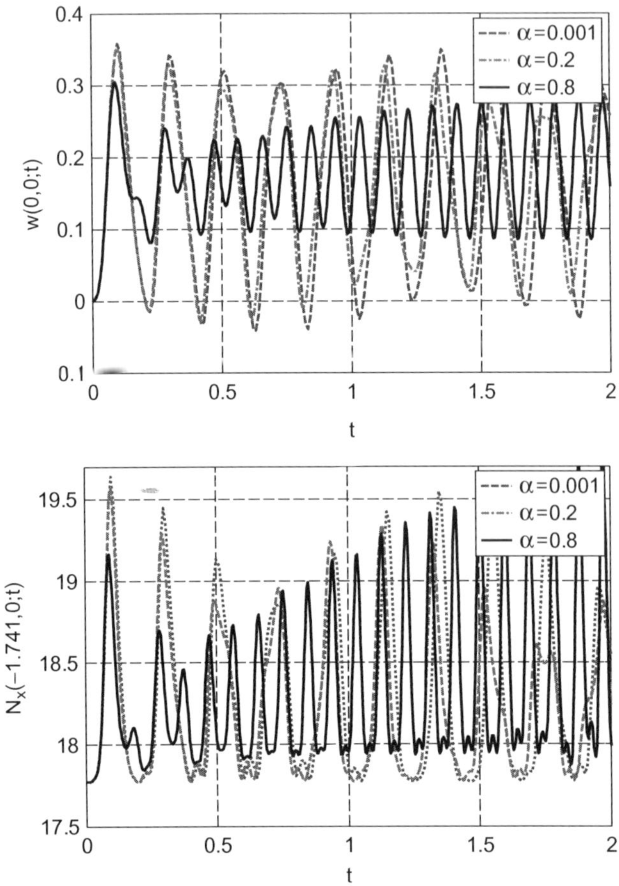

FIGURE 5.15 Time history of $w(0, 0; t)$ and $N_x(-1.741, 0; t)$ for various values of the order α ($\eta = 0.5$) in Example 5.4.

$f = 1.0H(t)\,\mathrm{kN/m^2}$. The results were obtained using 20 linear modes for the Ritz reduction. Fig. 5.15 presents the time history of the deflection $w(0, 0; t)$ and the membrane force $N_x(-1.741, 0; t)$ for various values of the order of the fractional derivative. This figure shows how the fractional order influences the viscoelastic response. Fig. 5.16 shows the deflection $w(0, 0; t)$ for elastic and viscoelastic material as compared with the static deflection. Finally, Fig. 5.17 presents the phase plane of the deflection $w(0, 0; t)$ for $\alpha = 0.5$, $\eta = 0.5$.

EXAMPLE 5.5 FORCED VIBRATIONS UNDER HARMONIC LOAD; RESONANCE

The dynamic response of the viscoelastic membrane of Example 5.4 under the transverse harmonic excitation $f = f_0 \sin \Omega t$ is studied. The results were obtained using $N = 210$ boundary elements and $M = 122$ internal nodal points resulting from 193 linear triangular elements and 10 linear mode shapes for the Ritz reduction of the degrees of freedom. First, the response of the

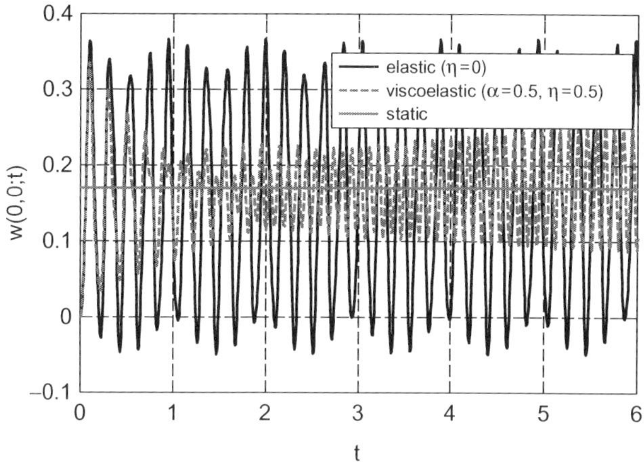

FIGURE 5.16 Time history of deflection $w(0, 0, t)$ of a viscoelastic membrane ($\alpha = 0.5$, $\eta = 0.5$) in Example 5.4.

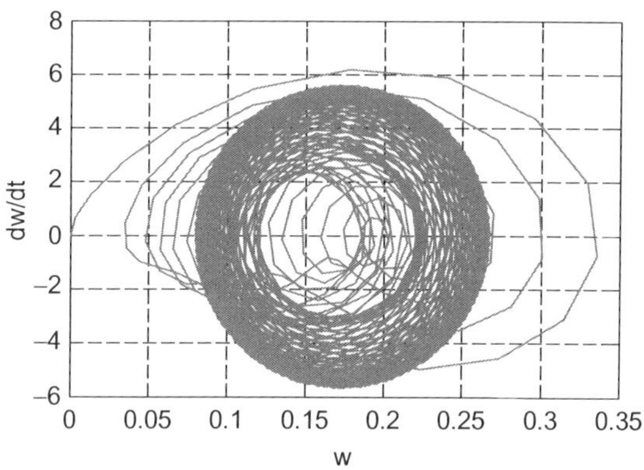

FIGURE 5.17 Phase plane of deflection $w(0, 0, t)$ of the viscoelastic membrane ($\alpha = 0.5$, $\eta = 0.5$) in Example 5.4.

membrane for the conventional Kelvin-Voigt viscoelastic model ($\alpha = 1$) is studied when the external frequency Ω is close to the lower eigenfrequency, which was found to be $\omega = 0.916$. Fig. 5.18 shows the amplitude-frequency curve of the vibration (steady state response) at the center of the membrane for the nonlinear problem together with that of the linear undamped membrane, i.e., when the nonlinear terms are neglected.

It is observed that the amplitude of the vibrations increases as the external frequency Ω approaches the natural frequency. However, the amplitude curves of the nonlinear problem move to the right, indicating a hardening nonlinear behavior, while two stable solutions are observed for some values of the excitation frequency. Moreover, due to the viscoelastic material, the vibrations are bounded contrary to the linear membrane problem. Fig. 5.19 presents the

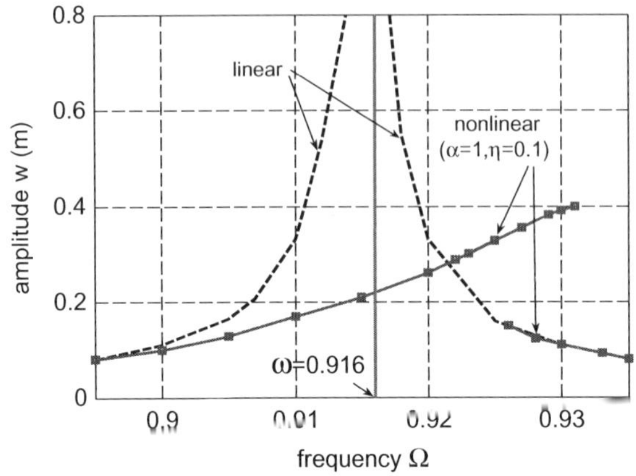

FIGURE 5.18 Amplitude-frequency curves at the center of the membrane for the linear and nonlinear problems in Example 5.5.

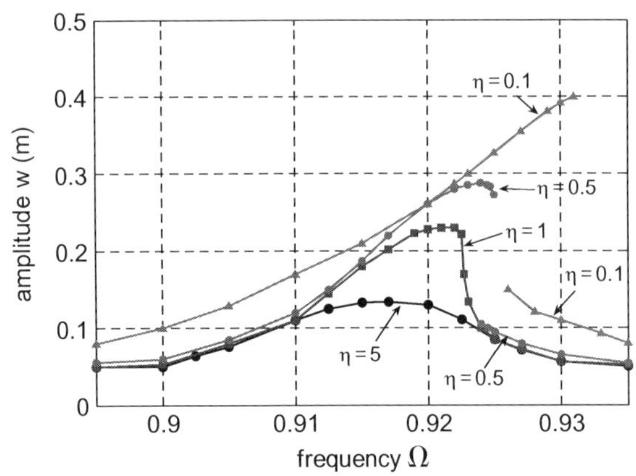

FIGURE 5.19 Amplitude-frequency curves at the center of the membrane for various values of the viscous parameter for the classical Voigt model ($\alpha = 1$) in Example 5.5.

amplitude-frequency curves for various values of the viscous parameter η. The amplitude of the vibrations increases as the viscous parameter decreases. For $\eta < 1$ a jump phenomenon appears and two stable solutions are observed. Fig. 5.20 shows the time history of the deflection at the center of the membrane for two values of the external frequency, just before and after the jump. The frequency changes from $\Omega = 0.925$ to $\Omega = 0.922$ at instant $t = 2400\,\text{sec}$. Next, we study the response of the membrane for the fractional Kelvin-Voigt model. Fig 5.21 presents the amplitude of the steady state response at the center of the membrane for $\eta = 5$ and various values of the order of the fractional derivative α.

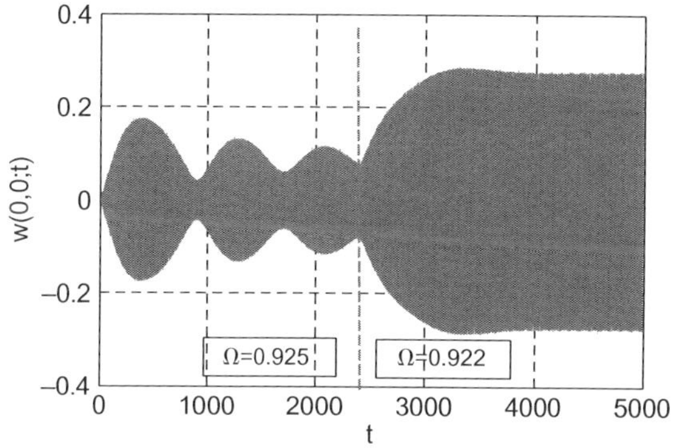

FIGURE 5.20 Time history of the central deflection for two values of the external frequency ($\alpha = 1$, $\eta = 0.5$, $f_0 = 0.01$) in Example 5.5.

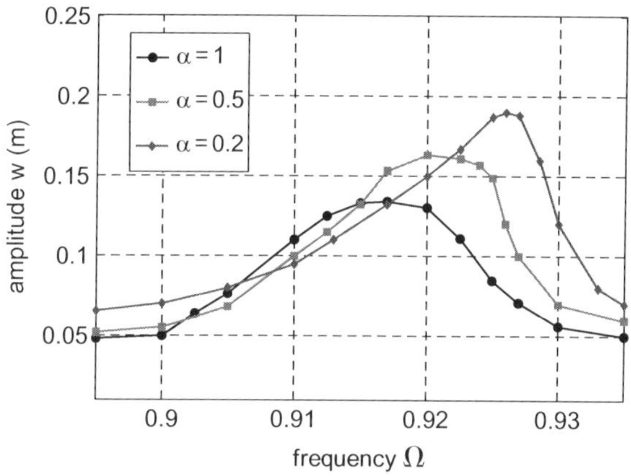

FIGURE 5.21 Amplitude-frequency curves for various values of the fractional derivative α ($\eta = 5$) in Example 5.5.

5.5 REFERENCES

[1] J.T. Katsikadelis, Boundary Elements. Theory and Applications, Elsevier, Oxford, UK, 2002.

[2] H. Hencky, Über den Spannungszustand in kreisrunden Platten mit verschwindender Biegefestigkeit, Zeitschrift für Mathematik und Physik 63 (1915) 311–317.

[3] R. Kao, N. Perrone, Large defections of axisymmetric circular membranes, Int. J. Solid. Struct. 7 (1971) 1601–1612.

[4] B. Storåkers, Small defections of linear elastic circular membranes under lateral pressure, ASME, J. Appl. Mech. 50 (1983) 735–739.

[5] A. Föppl, L. Föppl, Drang und Zwang, 1, R. Oldenburg, Munich, 1920.

[6] M. Neubert, A. Sommer, Rectangular Shell Plating under Uniformly Distributed Hydrostatic Pressure, N.A.C.A, 1940 Technical Memorandum, N. 965.

[7] B. Storåkers, Variation principles and bounds for the approximate analysis of plane membranes under lateral pressure, ASME, J. Appl. Mech. 50 (1983) 743–749.

[8] H. Hencky, Die Berechnung dunner rechteckiger Platten mit verschwindender Biegungsteifigkeit, Zeitschrift fur Angewandte Mathematik und Mechanik Vol. 1, (1921) 81–89.

[9] R. Kao, N. Perrone, Large defections of flat arbitrary membranes, Comput. Struct. 2 (1972) 535–546.
[10] J.W. Leonard, Tension Structures, McGraw-Hill, New York, 1988.
[11] J.T. Katsikadelis, M.S. Nerantzaki, Non linear structural problems by the analog equation method, in: Proceedings, 2nd National Congress of Computational Mechanics, 1996, , pp. 841–850.
[12] J.T. Katsikadelis, M.S. Nerantzaki, G.C. Tsiatas, The analog equation method for large deflection analysis of membranes. A boundary-only solution, Comput. Mech. 27 (2001) 513–523.
[13] J.T. Katsikadelis, M.S. Nerantzaki, The boundary element method for nonlinear problems, Eng. Anal. Bound. Elem. 23 (5) (1999) 365–373.
[14] J.T. Katsikadelis, C.G. Tsiatas, The analog equation method for large deflection analysis of heterogeneous orthotropic membrane: a boundary-only solution, Eng. Anal. Bound. Elem. 25 (8) (2001) 655–667.
[15] J.T. Katsikadelis, Dynamic analysis of nonlinear membranes by the analog equation method. A boundary-only solution, Comput. Mech. 29 (2) (2002) 170–177.
[16] J.T. Katsikadelis, C.G. Tsiatas, Nonlinear dynamic analysis of heterogeneous orthotropic membranes, Eng. Anal. Bound. Elem. 27 (2003) 115–124.
[17] J.T. Katsikadelis, M. Nerantzaki, The ponding problem on membranes. An analog equation solution, Comput. Mech. 28 (2) (2002) 122–128.
[18] M.S. Nerantzaki, J.T. Katsikadelis, Ponding on floating membranes, Eng. Anal. Bound. Elem. 27 (6) (2003) 589–596.
[19] J.T. Katsikadelis, M.S. Nerantzaki, A boundary element solution to the soap bubble, problem, Comput. Mech. 27 (2) (2001) 154–159.
[20] C.G. Tsiatas, J.T. Katsikadelis, Large deflection analysis of elastic space membranes, Int. J. Numer. Meth. Eng. 65 (2) (2006) 264–294.
[21] C.G. Tsiatas, J.T. Katsikadelis, A BEM based domain decomposition method for nonlinear analysis of elastic membranes, Comput. Mech. 38 (2) (2006) 119–131.
[22] C.G. Tsiatas, J.T. Katsikadelis, Nonlinear analysis of elastic cable-supported membranes, Eng. Anal. Bound. Elem. 35 (2011) 1149–1158.
[23] S.G. Mikhlin (Ed.), Linear Equations of Mathematical Physics, Holt. Rinehart und Winston, New York, 1967.
[24] M. Stiassnie, On the application of fractional calculus for the formulation of viscoelastic models, Appl. Math. Model. 3 (1979) 300–302.
[25] K. Adolfsson, M. Enelund, P. Olsson, On the fractional order model of viscoelasticity, Mech. Time-Depend. Mat. 9 (2005) 15–34.
[26] F.C. Meral, T.J. Royston, R. Magin, Fractional calculus in viscoelasticity: an experimental study, Commun. Nonlinear Sci. Numer. Simulat. 15 (2010) 939–945.
[27] R. Plaut, Parametric excitation of an inextensible, air-inflated cylindrical membrane, Int. J. Nonlin. Mech. 25 (1990) 253–262.
[28] H. Koivurova, A. Pramila, Nonlinear vibration of axially moving membrane by finite element method, Comput. Mech. 20 (1997) 573–581.
[29] A.J. Levine, F.C. Mackintosh, Dynamics of viscoelastic membranes, Phys. Rev. E 66 (2002) 061606.
[30] A. Wineman, Nonlinear viscoelastic membranes, Comput. Math. Appl. 53 (2007) 168–181.
[31] P.B. Goncales, R.M. Soares, D. Pamplona, Nonlinear vibrations of a radially stretched circular hyperelastic membrane, J. Sound Vib. 327 (2009) 231–248.
[32] A.C. Galucio, J.F. Deu, R. Ohayon, Finite element formulation of viscoelastic sandwich beams using fractional derivative operators, Comput. Mech. 33 (2004) 282–291.
[33] A. Schmidt, L. Gaul, Finite element formulation of viscoelastic constitutive equations using fractional time derivatives, Nonlinear Dynam. 29 (1–4) (2002) 37–55.

[34] J.T. Katsikadelis, Nonlinear dynamic analysis of viscoelastic plates described with fractional differential models, J. Theor. Appl. Mech. 50 (3) (2012) 743–753, 50th Anniversary Issue.
[35] J.T. Katsikadelis, The BEM for numerical solution of partial fractional differential equations, Comput. Math. Appl. 62 (2011) 891–901.
[36] M.S. Nerantzaki, Babouskos, Analysis of inhomogeneous anisotropic viscoelastic bodies described by multi-parameter fractional differential constitutive models, Comput. Math. Appl. 62 (2011) 891–901.
[37] I. Podlubny, Fractional Differential Equations, Academic Press, New York, 1999.
[38] J.T. Katsikadelis, Numerical solution of multi-term fractional differential equations, ZAMM Zeitschrift für Angewandte Mathematik und Mechanik 89 (7) (2009) 593–608.

PROBLEMS

5.1 Using the procedure described in Section 5.2.2 write a computer code and analyze the static response of the membrane of Fig. P5.1. The boundary is described by the curve $r = a\left(|\sin\theta|^3 + |\cos\theta|^3\right)$, $0 \leq \theta \leq 2\pi$. The membrane is prestressed by imposed displacements $u_n^* = 0.10\text{m}$ in the direction normal to the boundary, while $u_t^* = 0$ in the tangential direction. The other data are $a = 5.0\text{m}$, $h = 0.002\text{m}$, $E = 1.1 \times 10^5\text{kN/m}^2$, and $\nu = 0.3$, $f = 10\text{kN/m}^2$.

5.2 Using the procedure described in Section 5.3.2 write a computer code and analyze the dynamic response of the square membrane of Fig. P5.2. The membrane is prestressed by imposed displacements $u_n^* = 0.10\text{m}$ in the direction normal to the boundary, while $u_t^* = 0$ in the tangential direction. Data: $a = 10.0\text{m}$, $h = 0.002\text{m}$, $E = 1.1 \times 10^5\text{kN/m}^2$, and $\nu = 0.3$, $f = 10\text{kN/m}^2$, $R = a/5$, $\rho = 5000\text{kg/m}^3$.

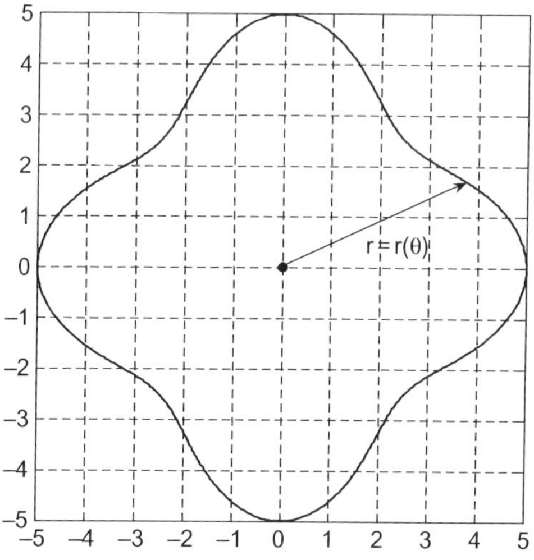

FIGURE P5.1 Cross-shaped membrane.

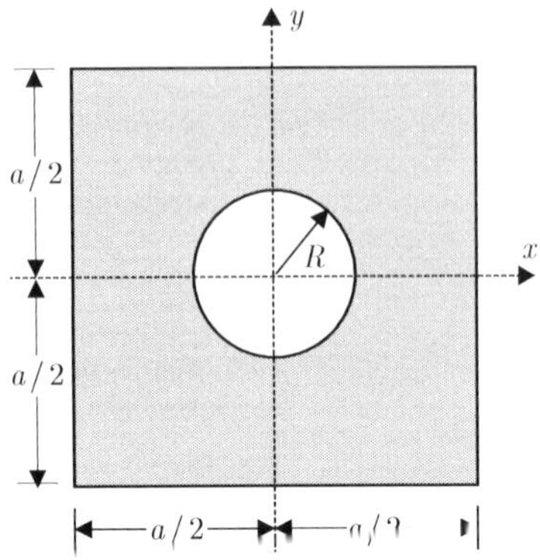

FIGURE P5.2 Square membrane with a hole.

5.3 Derive the differential equations governing the dynamic response of an orthotropic membrane.

5.4 Derive the differential equations governing the dynamic response of an isotropic viscoelastic membrane described with the three-parameter (solid) fractional differential model.

$$\begin{Bmatrix} \sigma_x \\ \sigma_y \\ \tau_{xy} \end{Bmatrix} + \frac{\eta b}{1+b} \begin{Bmatrix} D_c^a \sigma_x \\ D_c^a \sigma_y \\ D_c^a \tau_{xy} \end{Bmatrix} = \frac{b}{1+b} \frac{E}{1-\nu^2} \begin{bmatrix} 1 & \nu & 0 \\ \nu & 1 & 0 \\ 0 & 0 & \dfrac{1-\nu}{2} \end{bmatrix} \begin{Bmatrix} \varepsilon_x + \eta D_c^a \varepsilon_x \\ \varepsilon_y + \eta D_c^a \varepsilon_y \\ \gamma_{xy} + \eta D_c^a \gamma_{xy} \end{Bmatrix}$$

Appendix | A

Derivatives of r and Kernels, Particular Solutions and Tangential Derivatives

A.1 DERIVATIVES OF r

We present certain relations that facilitate the differentiation of the kernels of the integral equations. Points inside the domain Ω are denoted by uppercase letters, e.g., $P(x, y)$, and points on the boundary by lowercase letters, e.g., $q(\xi, \eta)$. The angle between the x axis and the vector r is denoted by α and the angle between the x axis and the unit vector $\mathbf{n}$ normal to the boundary at point q by β (see Fig. A.1). Using these two angles, we also define the angle ϕ as

$$\phi = \text{angle}(\mathbf{r}, \mathbf{n}) = \beta - \alpha \tag{A.1}$$

Referring to Fig. A.1, it is

$$\cos\alpha = \frac{\xi - x}{r} \tag{A.2}$$

$$\sin\alpha = \frac{\eta - y}{r} \tag{A.3}$$

where

$$r = \sqrt{(\xi - x)^2 + (\eta - y)^2} \tag{A.4}$$

Differentiation of Eq. (A.4) yields

The Boundary Element Method for Plate Analysis

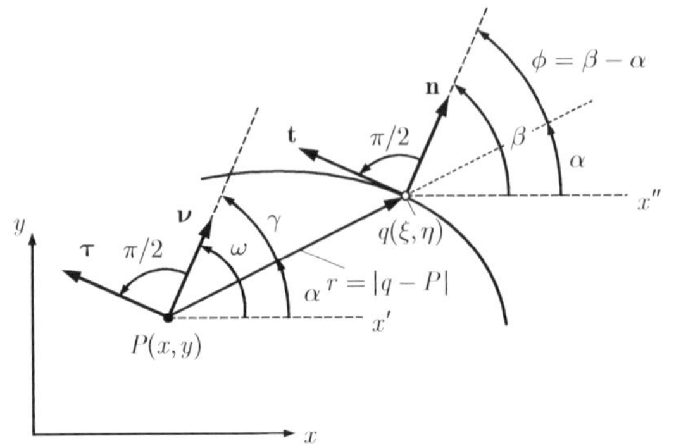

FIGURE A.1 Geometric definitions related to the relative position of a field point P and a boundary point q.

$$r_{,x} = -r_{,\xi} = -\frac{\xi - x}{r} = -\cos\alpha \tag{A.5}$$

$$r_{,y} = -r_{,\eta} = -\frac{\eta - y}{r} = -\sin\alpha \tag{A.6}$$

where the subscripts x, y, ξ, and η preceded by a comma denote differentiation with respect to the corresponding independent variable.

Noting that

$$\cos\beta = n_x$$
$$\sin\beta = n_y$$

we can derive the following expressions for the derivatives of r:

$$\begin{aligned} r_{,n} &= r_{,\xi} n_x + r_{,\eta} n_y \\ &= r_{,\xi} \cos\beta + r_{,\eta} \sin\beta \\ &= \cos\alpha\cos\beta + \sin\alpha\sin\beta \\ &= \cos(\beta - \alpha) \\ &= \cos\phi \end{aligned} \tag{A.7}$$

$$\begin{aligned} r_{,t} &= -r_{,\xi} n_y + r_{,\eta} n_x \\ &= -r_{,\xi} \sin\beta + r_{,\eta} \cos\beta \\ &= -\cos\alpha\sin\beta + \sin\alpha\cos\beta \\ &= -\sin(\beta - \alpha) \\ &= -\sin\phi \end{aligned} \tag{A.8}$$

$$
\begin{aligned}
r_{,xx} &= -\left(\frac{\xi - x}{r}\right)_{,x} \\
&= -\frac{(\xi - x)_{,x} r - (\xi - x) r_{,x}}{r^2} \\
&= \frac{(\eta - y)^2}{r^3} \\
&= \frac{r_{,y}^2}{r}
\end{aligned}
\tag{A.9}
$$

Similarly, we obtain

$$r_{,yy} = \frac{r_{,x}^2}{r} \tag{A.10}$$

$$r_{,xy} = -\frac{r_{,x} r_{,y}}{r} \tag{A.11}$$

$$r_{,\xi\xi} = \frac{r_{,\eta}^2}{r} \tag{A.12}$$

$$r_{,\eta\eta} = \frac{r_{,\xi}^2}{r} \tag{A.13}$$

$$r_{,\xi\eta} = -\frac{r_{,\xi} r_{,\eta}}{r} \tag{A.14}$$

$$r_{,\xi x} = -r_{,xx} = -\frac{r_{,y}^2}{r} \tag{A.15}$$

$$r_{,\xi y} = -r_{,xy} = \frac{r_{,x} r_{,y}}{r} \tag{A.16}$$

$$r_{,\eta x} = -r_{,yx} = \frac{r_{,x} r_{,y}}{r} \tag{A.17}$$

$$r_{,\eta y} = -r_{,yy} = -\frac{r_{,x}^2}{r} \tag{A.18}$$

$$
\begin{aligned}
r_{,nx} &= \left(r_{,\xi} \cos\beta + r_{,\eta} \sin\beta\right)_{,x} \\
&= r_{,\xi x} \cos\beta + r_{,\eta x} \sin\beta \\
&= \frac{r_{,y}}{r}\left(r_{,\eta} \cos\beta - r_{,\xi} \sin\beta\right) \\
&= \frac{r_{,t} r_{,y}}{r}
\end{aligned}
\tag{A.19}
$$

$$
\begin{aligned}
r_{,ny} &= \left(r_{,\xi} \cos\beta + r_{,\eta} \sin\beta\right)_{,y} \\
&= r_{,\xi y} \cos\beta + r_{,\eta y} \sin\beta \\
&= \frac{r_{,x}}{r}\left(-r_{,\eta} \cos\beta + r_{,\xi} \sin\beta\right) \\
&= -\frac{r_{,t} r_{,x}}{r}
\end{aligned}
\tag{A.20}
$$

$$\begin{aligned} r_{,tx} &= \left(-r_{,\xi}\sin\beta + r_{,\eta}\cos\beta\right)_{,x} \\ &= -r_{,\xi x}\sin\beta + r_{,\eta x}\cos\beta \\ &= -\frac{r_{,y}}{r}\left(r_{,\xi}\cos\beta + r_{,\eta}\sin\beta\right) \\ &= -\frac{r_{,y}r_{,n}}{r} \end{aligned} \tag{A.21}$$

$$\begin{aligned} r_{,ty} &= \left(-r_{,\xi}\sin\beta + r_{,\eta}\cos\beta\right)_{,y} \\ &= -r_{,\xi\eta}\sin\beta + r_{,\eta y}\cos\beta \\ &= \frac{r_{,x}}{r}\left(r_{,\xi}\cos\beta + r_{,\eta}\sin\beta\right) \\ &= \frac{r_{,x}r_{,n}}{r} \end{aligned} \tag{A.22}$$

$$\begin{aligned} r_{,nt} &= \left(r_{,t\xi}\cos\beta + r_{,t\eta}\sin\beta\right) \\ &= \frac{r_{,n}}{r}\left(r_{,\eta}\cos\beta - r_{,\xi}\sin\beta\right) \\ &= \frac{r_{,n}r_{,t}}{r} \end{aligned} \tag{A.23}$$

$$\begin{aligned} r_{,\nu} &= r_{,x}\cos\omega + r_{,y}\sin\omega \\ &= -(\cos\alpha\cos\omega + \sin\alpha\sin\omega) \\ &= \cos(\omega - \alpha) \end{aligned} \tag{A.24}$$

$$\begin{aligned} r_{,\tau} &= \left(-r_{,x}\sin\omega + r_{,y}\cos\omega\right) \\ &= -(-\cos\alpha\sin\omega + \sin\alpha\cos\omega) \\ &= \sin(\omega - \alpha) \end{aligned} \tag{A.25}$$

$$\begin{aligned} r_{,\nu x} &= \left(r_{,x}\cos\omega + r_{,y}\sin\omega\right)_{,x} \\ &= r_{,xx}\cos\omega + r_{,yx}\sin\omega \\ &= \frac{r_{,y}}{r}\left(r_{,y}\cos\omega - r_{,x}\sin\omega\right) \\ &= \frac{r_{,y}r_{,\tau}}{r} \end{aligned} \tag{A.26}$$

$$\begin{aligned} r_{,\nu y} &= \left(r_{,x}\cos\omega + r_{,y}\sin\omega\right)_{,y} \\ &= r_{,xy}\cos\omega + r_{,yy}\sin\omega \\ &= \frac{r_{,x}}{r}\left(-r_{,y}\cos\omega + r_{,x}\sin\omega\right) \\ &= -\frac{r_{,x}r_{,\tau}}{r} \end{aligned} \tag{A.27}$$

$$\begin{aligned} r_{,\tau x} &= \left(-r_{,x}\sin\omega + r_{,y}\cos\omega\right)_{,x} \\ &= -r_{,xx}\sin\omega + r_{,yx}\cos\omega \\ &= -\frac{r_{,y}}{r}\left(r_{,y}\sin\omega + r_{,x}\cos\omega\right) \\ &= \frac{r_{,y}r_{,\nu}}{r} \end{aligned} \tag{A.28}$$

$$\begin{aligned} r_{,\tau y} &= \left(-r_{,x}\sin\omega + r_{,y}\cos\omega\right)_{,y} \\ &= -r_{,xy}\sin\omega + r_{,yy}\cos\omega \\ &= \frac{r_{,x}}{r}\left(r_{,y}\sin\omega + r_{,x}\cos\omega\right) \\ &= \frac{r_{,x}r_{,\nu}}{r} \end{aligned} \tag{A.29}$$

A.2 DERIVATIVES OF KERNELS

In this section, the derivatives of the kernels of the integral representation (2.84) are given, when the stress resultants are obtained by direct differentiation (see Section 2.3.5).

$$v_{,xx} = \frac{1}{8\pi D}\left(2\ \ln r + 2r_{,x}^2 + 1\right) \tag{A.30}$$

$$v_{,yy} = \frac{1}{8\pi D}\left(2\ \ln r + 2r_{,y}^2 + 1\right) \tag{A.31}$$

$$v_{,xy} = \frac{1}{4\pi D} r_{,x} r_{,y} \tag{A.32}$$

$$\left(\frac{\partial v}{\partial n}\right)_{,xx} = \frac{1}{4\pi D}\frac{1}{r}\left(r_{,n} + 2r_{,x}r_{,y}r_{,t}\right) \tag{A.33}$$

$$\left(\frac{\partial v}{\partial n}\right)_{,yy} = \frac{1}{4\pi D}\frac{1}{r}\left(r_{,n} - 2r_{,x}r_{,y}r_{,t}\right) \tag{A.34}$$

$$\left(\frac{\partial v}{\partial n}\right)_{,xy} = \frac{1}{4\pi D}\frac{1}{r}\left(r_{,y}^2 - r_{,x}^2\right) r_{,t} \tag{A.35}$$

$$(Mv)_{,xx} = \frac{1}{4\pi D r^2}\left[-(1+\nu)r_{,y}^2 + 2(1-\nu)r_{,y}^2\left(r_{,n}^2 - r_{,t}^2\right) + 2(1-\nu)r_{,x}r_{,y}r_{,n}r_{,t}\right] \tag{A.36}$$

$$(Mv)_{,yy} = \frac{1}{4\pi D r^2}\left[-(1+\nu)r_{,x}^2 + 2(1-\nu)r_{,x}^2\left(r_{,n}^2 - r_{,t}^2\right) - 2(1-\nu)r_{,x}r_{,y}r_{,n}r_{,t}\right] \tag{A.37}$$

$$(Mv)_{,xy} = \frac{1}{4\pi D r^2}\left[(1+\nu)r_{,x}r_{,y} - 2(1-\nu)r_{,x}r_{,y}\left(r_{,n}^2 - r_{,t}^2\right) - 2(1+\nu)r_{,n}r_{,t}r_{,x}^2\right] \tag{A.38}$$

$$\begin{aligned} (Vv)_{,xx} = \frac{1}{2\pi D r^3}\Big\{&(1-\nu)\Big(-7r_{,t}^2 r_{,y}^2 r_{,n} + 3r_{,y}^2 r_{,n}^3 + 10r_{,x}r_{,t}r_{,y}r_{,n}^2 \\ &-2r_{,x}r_{,t}^3 r_{,y} + r_{,n}r_{,x}^2 r_{,t}^2 - r_{,n}^3 r_{,x}^2\Big) + 2r_{,n}\left(r_{,y}^2 - r_{,x}^2\right) \\ &+ 4r_{,x}r_{,t}r_{,y} - \kappa(1-\nu)r\left[2\left(r_{,n}^2 - r_{,t}^2\right)r_{,y}^2 + 4r_{,x}r_{,y}r_{,t}r_{,n}\right]\Big\} \end{aligned} \tag{A.39}$$

$$(Vv)_{,yy} = \frac{1}{2\pi D r^3}\{(1-\nu)\left(-7r_{,t}^2 r_{,x}^2 r_{,n} + 3r_{,x}^2 r_{,n}^3 - 10r_{,x}r_{,t}r_{,y}r_{,n}^2 + 2r_{,x}r_{,t}^3 r_{,y} + r_{,n}r_{,y}^2 r_{,t}^2 - r_{,n}^3 r_{,y}^2\right) + 2r_{,n}\left(r_{,x}^2 - r_{,y}^2\right) - 4r_{,x}r_{,t}r_{,y} - \kappa(1-\nu)r\left[2\left(r_{,n}^2 - r_{,t}^2\right)r_{,x}^2 - 4r_{,x}r_{,y}r_{,t}r_{,n}\right]\} \tag{A.40}$$

$$(Vv)_{,xy} = -\frac{1}{2\pi D r^3}\left\{(1-\nu)\left[-8r_{,x}r_{,y}r_{,n}r_{,t}^2 + 4r_{,x}r_{,y}r_{,n}^3 + 5r_{,t}\left(r_{,x}^2 - r_{,y}^2\right)r_{,n}^2\right] + \left[2r_{,t} - r_{,t}^3(1-\nu)\right]\left(r_{,x}^2 - r_{,y}^2\right) + 4r_{,x}r_{,y}r_{,n} + 2\kappa r(1-\nu)\left[r_{,x}r_{,y}\left(r_{,t}^2 - r_{,n}^2\right) + r_{,t}r_{,n}\left(r_{,y}^2 - r_{,x}^2\right)\right]\right\} \tag{A.41}$$

$$Tv_{,xx} = -\frac{1-\nu}{2\pi D r^2}r_{,y}\left[r_{,x}\left(r_{,t}^2 - r_{,n}^2\right) + 2r_{,y}r_{,n}r_{,t}\right] \tag{A.42}$$

$$Tv_{,yy} = \frac{1-\nu}{2\pi D r^2}r_{,x}\left[r_{,y}\left(r_{,t}^2 - r_{,n}^2\right) - 2r_{,x}r_{,n}r_{,t}\right] \tag{A.43}$$

$$Tv_{,xy} = -\frac{1-\nu}{4\pi D r^2}\left[\left(r_{,y}^2 - r_{,x}^2\right)\left(r_{,t}^2 - r_{,n}^2\right) - 4r_{,x}r_{,y}r_{,n}r_{,t}\right] \tag{A.44}$$

$$\nabla^2 v_{,x} = \frac{1}{2\pi D}\frac{r_{,x}}{r} \tag{A.45}$$

$$\nabla^2 v_{,y} = \frac{1}{2\pi D}\frac{r_{,y}}{r} \tag{A.46}$$

$$\nabla^2\left(\frac{\partial v}{\partial n}\right)_{,x} = \frac{r_{,t}r_{,y} - r_{,n}r_{,x}}{2\pi D r^2} \tag{A.47}$$

$$\nabla^2\left(\frac{\partial v}{\partial n}\right)_{,y} = -\frac{r_{,t}r_{,x} + r_{,n}r_{,y}}{2\pi D r^2} \tag{A.48}$$

$$\nabla^2(Mv)_{,x} = \frac{1}{2\pi D}\frac{1}{r^3}\left[4r_{,y}r_{,t}r_{,n}(1-\nu) - 2r_{,x}(1-\nu)\left(r_{,n}^2 - r_{,t}^2\right) + r_{,x}(1+\nu)\right] \tag{A.49}$$

$$\nabla^2(Mv)_{,y} = \frac{1}{2\pi D}\frac{1}{r^3}\left[-4r_{,x}r_{,t}r_{,n}(1-\nu) + 2r_{,y}(1-\nu)\left(r_{,n}^2 - r_{,t}^2\right) + r_{,y}(1+\nu)\right] \tag{A.50}$$

$$\nabla^2(Vv)_{,x} = \frac{\nu-1}{\pi r^4}\left[-9r_{,t}r_{,n}\left(r_{,x}r_{,t} + r_{,y}r_{,n}\right) + 3\left(r_{,x}r_{n}^3 + r_{,y}r_{,t}^3\right) + 2\kappa r r_{,x}\left(r_t^2 - r_{,n}^2\right) + 4\kappa r r_{,y}r_{,n}r_{,t}\right] \tag{A.51}$$

$$\nabla^2(Vv)_{,y} = \frac{\nu - 1}{\pi r^4}\left[9r_{,t}r_{,n}\left(r_{,x}r_{,n} - r_{,y}r_{,t}\right) + 3\left(r_{,y}r_{n}^{3} - r_{,x}r_{,t}^{3}\right) + 2\kappa r r_{,y}\left(r_{t}^{2} - r_{,n}^{2}\right) - 4\kappa r r_{,x}r_{,n}r_{,t}\right] \tag{A.52}$$

$$\nabla^2(Tv)_{,x} = -\frac{1-\nu}{2\pi D r^3}\left[r_{,y}\left(r_{,t}^{2} - r_{,n}^{2}\right) - 2r_{,x}r_{,n}r_{,t}\right] \tag{A.53}$$

$$\nabla^2(Tv)_{,y} = \frac{1-\nu}{2\pi D r^3}\left[r_{,x}\left(r_{,t}^{2} - r_{,n}^{2}\right) + 2r_{,y}r_{,n}r_{,t}\right] \tag{A.54}$$

A.3 PARTICULAR SOLUTIONS OF THE POISSON EQUATION (3.57)

Table A.1 gives the particular solutions $\mathcal{K}_n(r)$ $(n = 1, 2, \ldots, 10)$ of Eq. (3.57), which are employed in Eq. (3.58) to convert the domain integrals into boundary line integrals.

TABLE A.1 Particular solutions of the equation $\nabla^2\mathcal{K}_n(r) = K_n(r)$

Kernel function $K_n(r)$	Particular solution $\mathcal{K}_n(r)$
$K_1(r) = v = \frac{1}{8\pi}r^2 \ln r$	$\mathcal{K}_1(r) = \frac{r^4}{256\pi D}(2 \ln r - 1)$
$K_2(r) = v_1 = \frac{1}{8\pi}rr_{,\nu}(2 \ln r + 1)$	$\mathcal{K}_2(r) = \frac{r^3 r_{,n}}{128\pi D}(4 \ln r - 1)$
$K_3(r) = v_{,x} = \frac{1}{8\pi}r(2 \ln r + 1)r_{,x}$	$\mathcal{K}_3(r) = \frac{r^3 r_{,x}}{128\pi D}(4 \ln r - 1)$
$K_4(r) = v_{,y} = \frac{1}{8\pi}r(2 \ln r + 1)r_{,y}$	$\mathcal{K}_4(r) = \frac{r^3 r_{,y}}{128\pi D}(4 \ln r - 1)$
$K_5(r) = v_{,xx} = \frac{1}{8\pi}(2 \ln r + 2r_{,x}^2 + 1)$	$\mathcal{K}_5(r) = \frac{r^2}{128\pi D}\left[r_{,x}^2 - r_{,y}^2 + \left(12r_{,x}^2 + 4r_{,y}^2\right) \ln r\right]$
$K_6(r) = v_{,yy} = \frac{1}{8\pi}\left(2 \ln r + 2r_{,y}^2 + 1\right)$	$\mathcal{K}_6(r) = \frac{r^2}{128\pi D}\left[r_{,y}^2 - r_{,x}^2 + \left(12r_{,y}^2 + 4r_{,x}^2\right) \ln r\right]$
$K_7(r) = v_{,xy} = \frac{1}{4\pi}r_{,x}r_{,y}$	$\mathcal{K}_7(r) = \frac{r^2 r_{,x}r_{,y}}{64\pi D}(1 + 4 \ln r)$
$K_8(r) = \nabla^2 v = \frac{1}{2\pi}(\ln r + 1)$	$\mathcal{K}_8(r) = \frac{r^2}{8\pi D} \ln r$
$K_9(r) = \nabla^2 v_{,x} = \frac{1}{2\pi}\frac{r_{,x}}{r}$	$\mathcal{K}_9(r) = \frac{rr_{,x}}{8\pi D}(2 \ln r + 1)$
$K_{10}(r) = \nabla^2 v_{,y} = \frac{1}{2\pi}\frac{r_{,y}}{r}$	$\mathcal{K}_{10}(r) = \frac{rr_{,y}}{8\pi D}(2 \ln r + 1)$

A.4 TANGENTIAL DERIVATIVES $w_{,ntt}$, $w_{,ttt}$, $w_{,tts}$ AND THEIR DIFFERENT APPROXIMATIONS

In this section the tangential derivatives $w_{,ntt}$, $w_{,ttt}$, and $w_{,tts}$ on the boundary, which are involved in Eqs. (3.195), are expressed in terms of the boundary quantities w and $w_{,n}$, which explicitly appear in the boundary integral equations.

If the positive direction of the arc length s coincides with that of the tangential vector (Fig. A.2), we have for the first derivative

$$w_{,t} = w_{,s} \tag{A.55}$$

Therefore, this derivative can be directly approximated with finite difference schemes using the arc length s as the independent variable, regardless of whether the boundary is rectilinear or curvilinear. However, for the second and higher derivatives special care should be given to express them with intrinsic coordinates, namely, the arc length s and the normal derivative n.

Apparently, if the boundary is rectilinear the intrinsic coordinates are identical with the local Cartesian coordinates. Thus, we have

$$w_{,tt} = w_{,ss} \tag{A.56a}$$

$$w_{,nt} = w_{,ns} \tag{A.56b}$$

Therefore, $w_{,tt}$ and $w_{,nt}$ can be expressed in terms of the boundary values of w and $w_{,n}$, respectively, using finite difference schemes. However, for a curvilinear boundary the second derivatives are expressed by Eqs. (2.49b,c), namely,

$$w_{,tt} = w_{,ss} + \kappa w_{,n} \tag{A.57a}$$

$$w_{,nt} = w_{,ns} - \kappa w_{,s} \tag{A.57b}$$

where $\kappa = \kappa(s)$ is the curvature of the boundary at point s, which is evaluated from the relations (2.51a,b). Obviously, for a rectilinear boundary it is $\kappa = 0$ and Eqs. (A.57) become identical to Eqs. (A.56).

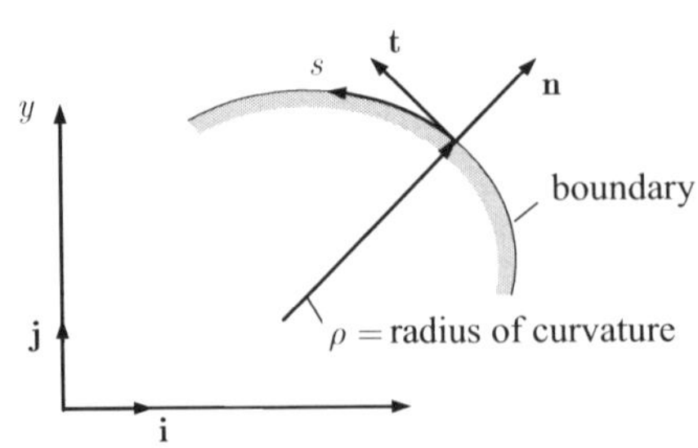

FIGURE A.2 Intrinsic coordinates on the boundary.

The third-order derivatives can be obtained using the procedure developed in [1] or [2].

(a) Expression for $w_{,ttt}$

$$w_{,ttt} = w_{,sss} + \kappa_{,s}\, w_{,n} + 3\kappa w_{,ns} - \kappa^2 w_{,s} \tag{A.58}$$

(b) Expression for $w_{,tts}$

$$w_{,tts} = w_{,sss} + \kappa_{,s}\, w_{,n} + \kappa w_{,ns} + \kappa^2 w_{,s} \tag{A.59}$$

(c) Expression for $w_{,ntt}$

$$w_{,ntt} = w_{,nss} - \kappa_{,s}\, w_{,s} - (2+\nu)\kappa w_{,ss} - \kappa \overline{M}_n - (1+\nu)\kappa^2 w_{,n} \tag{A.60}$$

(d) Expression for $w_{,nnt}$

$$w_{,nnt} = w_{,nns} - 2\kappa w_{,ns} + 2\kappa^2 w_{,s} \tag{A.61}$$

(e) Expression for $\overline{M}_{n,t}$

$$\begin{aligned} \overline{M}_{n,t} &= -(w_{,nnt} + \nu w_{,ttt}) \\ &= \overline{M}_{n,s} + 2\kappa(1-\nu) w_{,ns} + 2\kappa^2(\nu - 1) w_{,s} \end{aligned} \tag{A.62}$$

Note that $\overline{M}_n = -(w_{,nn} + \nu w_{,tt})$, namely it designates the normal bending moment on the boundary of the thin isotropic plate with $D = 1$.

The derivatives with respect to the arc length s are approximated using finite differences. For equidistant nodal points, $\Delta s = h$, the following schemes can be employed. For nodal points adjacent to a corner backward and forward differences are employed depending on whether the nodal point lies before or after the corner.

First derivative

$$\text{Central difference } (f_{,s})_i = \frac{f_{i+1} - f_{i-1}}{2h} \tag{A.63a}$$

$$\text{Backward difference } (f_{,s})_i = \frac{3f_i - 4f_{i-1} + f_{i-2}}{2h} \tag{A.63b}$$

$$\text{Forward difference } (f_{,s})_i = \frac{-3f_i + 4f_{i+1} - f_{i+2}}{2h} \tag{A.63c}$$

Second derivative

$$\text{Central difference } (f_{,ss})_i = \frac{f_{i+1} - 2f_i + f_{i-1}}{h^2} \tag{A.64a}$$

$$\text{Backward difference } (f_{,ss})_i = \frac{-f_{i-3} + 4f_{i-2} - 5f_{i-1} + 2f_i}{h^2} \tag{A.64b}$$

$$\text{Forward difference } (f_{,ss})_i = \frac{2f_i - 5f_{i+1} + 4f_{i+2} - f_{i+3}}{h^2} \tag{A.64c}$$

Third derivative

$$\text{Central difference } (f_{,sss})_i = \frac{f_{i+2} - 2f_{i+1} + 2f_{i-1} - f_{i-2}}{2h^3} \tag{A.65a}$$

$$\text{Backward difference } (f_{,sss})_i = \frac{3f_{i-4} - 14f_{i-3} + 24f_{i-2} - 18f_{i-1} + 5f_i}{2h^3} \tag{A.65b}$$

$$\text{Forward difference } (f_{,sss})_i = \frac{-5f_i + 18f_{i+1} - 24f_{i+2} + 14f_{i+3} - 3f_{i+4}}{2h^3} \tag{A.65c}$$

$$\text{Mixed backward } f_{,sss} = \frac{-f_{i-2} + 3f_{i-1} - 3f_i + f_{i+1}}{h^3} \tag{A.65d}$$

$$\text{Mixed forward } f_{,sss} = \frac{-f_{i-1} + 3f_i - 3f_{i+1} + f_{i+2}}{h^3} \tag{A.65e}$$

A.5 REFERENCES

[1] J.T. Katsikadelis, The Analysis of Plates on Elastic Foundation by the Boundary Integral Equation Method, 1982. Dissertation in partial fulfillment for the Degree of Doctor of Philosophy (PhD) in Applied Mechanics at the Polytechnic University of New York, New York.

[2] V. Slivker, Mechanics of Structural Elements, Theory and Applications, Springer-Verlag Berlin Heidelberg, 2007.

Appendix | B

Gauss Integration

B.1 GAUSS INTEGRATION OF A REGULAR FUNCTION

The success of the BEM as a computational method is contingent upon evaluating accurately line and domain integrals. Any numerical integration method can be utilized, e.g., trapezoidal rule, Simpson rule, Newton-Cotes rule, etc. The numerical methods approximate an integral with a sum of products of the values of the integrand at specific points of the integration interval, usually equidistant, multiplied by known coefficients, the *weight factors* (or *weights*), resulting from the employed integration rule. Hence, an essential criterion for choosing the integration rule is the achievement of the desired degree of accuracy in the approximation of the integral by utilizing the lowest possible number of integration points. The Gauss integration method satisfies this criterion. In this method, the points are not equidistant but they arrange themselves in the optimal pattern. The weight factors are also adjusted in this optimization process. The concept behind the Gauss integration is quite simple, and will be explained in what follows.

Let us consider the integral

$$I = \int_{-1}^{1} f(\xi)\, d\xi \tag{B.1}$$

It is approximated by the sum

$$I \approx I_n = \sum_{k=1}^{n} f(\xi_k)\, w_k \tag{B.2}$$

where $f(\xi_k)$ $(k = 1, 2, \ldots, n)$ are the values of the function $f(\xi)$ at the n points ξ_k $(-1 < \xi_k < +1)$ and w_k are the associated *weights*. The points ξ_k, which are called *Gauss integration points* or simply *Gauss points*, are not equidistant within the integration interval, but, for a given n, their location and the associated weights are chosen so that the error $E_n = I - I_n$ becomes minimum.

We first study the trivial case where the function $f(\xi)$ is a linear polynomial

$$f(\xi) = a_0 + a_1 \xi \tag{B.3}$$

The Boundary Element Method for Plate Analysis

The exact value of the integral is

$$\begin{aligned} I &= \int_{-1}^{+1} (a_0 + a_1\xi)\, d\xi \\ &= \left[a_0\xi + a_1\frac{\xi^2}{2} \right]_{-1}^{+1} \\ &= 2a_0 \end{aligned}$$

while the approximate value obtained from Eq. (B.2) for $n=1$ is

$$I_1 = (a_0 + a_1\xi_1)\, w_1$$

Therefore, the error is

$$E_1(a_0, a_1) = I - I_1 = 2a_0 - (a_0 + a_1\xi_1)\, w_1$$

and is minimized for

$$\frac{\partial E_1}{\partial a_0} = 2 - w_1 = 0 \tag{B.4a}$$

$$\frac{\partial E_1}{\partial a_1} = -w_1\xi_1 = 0 \tag{B.4b}$$

which yield

$$\xi_1 = 0, \quad w_1 = 2$$

and $E_1(a_0, a_1) = 0$.

We notice that Eq. (B.2) gives

$$\begin{aligned} \int_{-1}^{1} f(\xi)\, d\xi &= 2f_0 \\ &= 2a_0 \end{aligned} \tag{B.5}$$

where $f_0 = f(0)$.

Namely, the value of the integral is independent of the slope a_1 of the straight line representing the function. Hence, the definite integrals on the interval $[-1, +1]$ of all the linear functions passing through the point $(0, f_0)$ have the same value (see Fig. B.1).

Consider now the case of a function $f(\xi)$ that is a quadratic polynomial. The exact value of the integral is

$$\begin{aligned} I &= \int_{-1}^{1} \left(a_0 + a_1\xi + a_2\xi^2\right) d\xi \\ &= \left[a_0\xi + a_1\frac{\xi^2}{2} + a_2\frac{\xi^3}{3} \right]_{-1}^{+1} \\ &= 2a_0 + \frac{2}{3}a_2 \end{aligned}$$

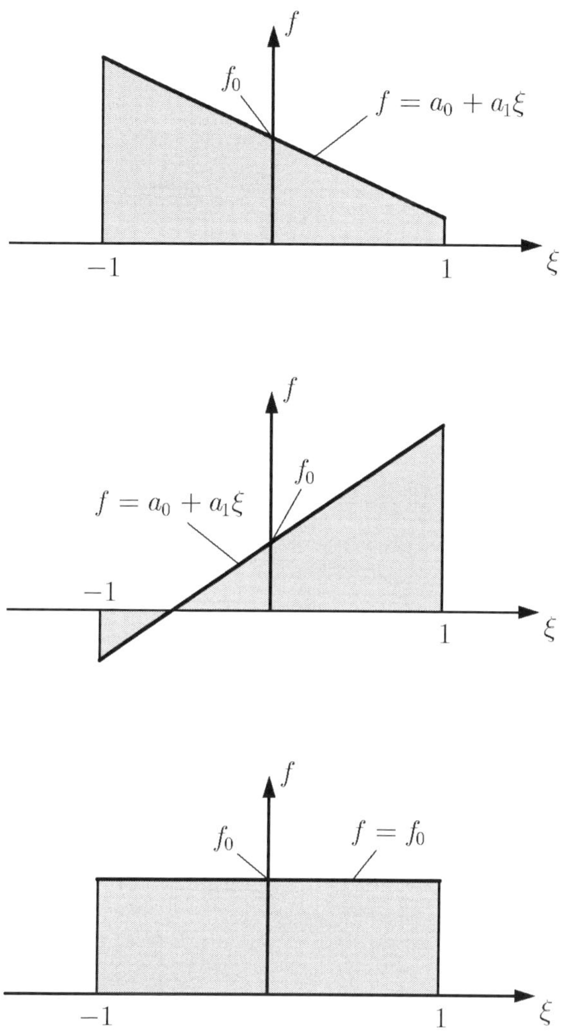

FIGURE B.1 Typical linear polynomials.

while the approximate one for two Gauss points ($n = 2$) is

$$
\begin{aligned}
I_2 &= f(\xi_1)\, w_1 + f(\xi_2)\, w_2 \\
&= \left(a_0 + a_1\,\xi_1 + a_2\,\xi_1^2\right) w_1 + \left(a_0 + a_1\,\xi_2 + a_2\,\xi_2^2\right) w_2
\end{aligned}
$$

The resulting error is

$$
\begin{aligned}
E_2(a_0, a_1, a_2) &= I - I_2 \\
&= 2a_0 + \frac{2}{3}a_2 - \left(a_0 + a_1\,\xi_1 + a_2\,\xi_1^2\right) w_1 \\
&\quad - \left(a_0 + a_1\,\xi_2 + a_2\,\xi_2^2\right) w_2
\end{aligned}
\tag{B.6}
$$

which is minimized under the following conditions:

$$\frac{\partial E_2}{\partial a_0} = 2 - w_1 - w_2 = 0 \tag{B.7}$$

$$\frac{\partial E_2}{\partial a_1} = -\xi_1 w_1 - \xi_2 w_2 = 0 \tag{B.8}$$

$$\frac{\partial E_2}{\partial a_2} = \frac{2}{3} - \xi_1^2 w_1 - \xi_2^2 w_2 = 0 \tag{B.9}$$

The above three simultaneous equations include four unknowns, w_1, w_2, ξ_1, and ξ_2. Therefore, one of them can be chosen arbitrarily, and since it is convenient to place the points ξ_1 and ξ_2 symmetrically with respect to the origin, we set

$$\xi_2 = -\xi_1 \tag{B.10}$$

From Eq. (B.8), we obtain $w_1 = w_2$ and from Eq. (B.7) we find

$$w_1 = w_2 = 1 \tag{B.11}$$

Finally, Eq. (B9) gives

$$\xi_1 = \frac{1}{\sqrt{3}} = 0.577350269189626 \tag{B.12}$$

Introducing the above values of ξ_1, ξ_2, w_1, and w_2 into Eq. (B.6), the error becomes $E_2(a_0, a_1, a_2) = 0$. The approximate value of the integral is computed from the sum

$$I_2 = f(\xi_1) w_1 + f(\xi_2) w_2 \tag{B.13}$$

where ξ_1 and ξ_2 are given by Eqs. (B.10) and (B.12) and w_1, w_2 by Eq. (B.11).

We consider the case where the function $f(\xi)$ is a cubic polynomial and we apply the same procedure. The exact value is

$$\begin{aligned} I_3 &= \int_{-1}^{1} \left(a_0 + a_1 \xi + a_2 \xi^2 + a_3 \xi^3\right) d\xi \\ &= 2a_0 + \frac{2}{3} a_2 \end{aligned}$$

Using again two Gauss points ($n = 2$), the expression for the error becomes

$$\begin{aligned} E_3(a_0, a_1, a_2, a_3) = \left(2a_0 + \frac{2}{3} a_2\right) &- \left(a_0 + a_1 \xi_1 + a_2 \xi_1^2 + a_3 \xi_1^3\right) w_1 \\ &- \left(a_0 + a_1 \xi_2 + a_2 \xi_2^2 + a_3 \xi_2^3\right) w_2 \end{aligned}$$

which is minimized if the following conditions are satisfied:

$$\frac{\partial E_3}{\partial a_0} = 2 - w_1 - w_2 = 0 \tag{B.14a}$$

$$\frac{\partial E_3}{\partial a_1} = -\xi_1 w_1 - \xi_2 w_2 = 0 \tag{B.14b}$$

$$\frac{\partial E_3}{\partial a_2} = \frac{2}{3} - \xi_1^2 w_1 - \xi_2^2 w_2 = 0 \tag{B.14c}$$

$$\frac{\partial E_3}{\partial a_3} = -\xi_1^3 w_1 - \xi_2^3 w_2 = 0 \tag{B.14d}$$

The solution $w_1 = w_2 = 1$ and $\xi_1 = -\xi_2 = 1/\sqrt{3}$, obtained when $f(\xi)$ is a second-order polynomial, is also a solution of the simultaneous equations (B.14) and yields $E_3(a_0, a_1, a_2, a_3) = 0$. Hence, Eq. (B.2) renders exact values for the integrals of the second- and third-order polynomials using the same Gauss integration rule, i.e., the same integration points and weights.

This procedure can also be applied for higher-order polynomials. In general, it can be proved that the formula of Eq. (B.2) with n terms gives the exact value for the integral of a polynomial $f(\xi)$ of order less than or equal to $2n - 1$. However, this procedure can hardly be employed to determine the coordinates of the Gauss points and the weight factors, because it becomes complicated and laborious for increasing degree of the polynomial. The Gauss points can be obtained by a simpler method, which uses Legendre's orthogonal polynomials [1,2] to represent the function $f(\xi)$ and that is why it is known as *Gauss-Legendre integration.*

Legendre's polynomials are defined as

$$P_n(\xi) = \frac{1}{2^n n!} \frac{d^n}{d\xi^n} (\xi^2 - 1)^n \tag{B.15}$$

For example, the first-, second- and third-order polynomials are

$$\text{for } n = 1: \quad P_1(\xi) = \xi \tag{B.16a}$$

$$\text{for } n = 2: \quad P_2(\xi) = \frac{1}{2}(3\xi^2 - 1) \tag{B.16b}$$

$$\text{for } n = 3: \quad P_3(\xi) = \frac{1}{2}(5\xi^3 - 3\xi) \tag{B.16c}$$

It is proven that the coordinates ξ_k of the integration points are the zeros of these polynomials. The weight factors w_k are computed by the expression

$$w_k = \frac{2(1 - \xi_k^2)}{n^2 [P_{n-1}(\xi_k)]^2} \tag{B.17}$$

Table B.1 provides the coordinates of the integration points and the weights for various values of n.

If $f(\xi)$ is not a polynomial, the integral can be evaluated approximately as

$$\int_{-1}^{1} f(\xi)\, d\xi \approx \sum_{k=1}^{n} f(\xi_k)\, w_k \tag{B.18}$$

where the function is actually approximated by a polynomial of degree $2n-1$. The error associated with the Gauss-Legendre integration is given by Lanczos in the following form [1]:

$$E_n = I - I_n = \frac{1}{2n+1}\left[f(1) + f(-1) - I_n - \sum_{k=1}^{n} w_k \xi_k f'(\xi_k)\right] \tag{B.19}$$

in which n is the number of integration points. This estimation is very good for smooth functions $f(\xi)$.

Usually, the interval over which a function is integrated differs from the interval $[-1,\ +1]$. Suppose that we have to evaluate the integral

$$I = \int_{a}^{b} f(x)\, dx \tag{B.20}$$

Using the transformation

$$x = \frac{b-a}{2}\xi + \frac{b+a}{2} \tag{B.21}$$

the interval of Eq. (B.20) is mapped onto the interval $[-1,\ +1]$. Thus,

$$I = \int_{a}^{b} f(x)\, dx = \frac{b-a}{2}\int_{-1}^{+1} f(\xi)\, d\xi \tag{B.22}$$

Example B.1

Determine the value of the integral

$$I = \int_{1}^{3} \frac{\sin^2 x}{x}\, dx$$

The transformation $x = \xi + 2$ yields the expression of the above integral in the interval $[-1,\ +1]$, which is of the following form:

$$I = \int_{-1}^{+1} \frac{\sin^2(\xi + 2)}{\xi + 2}\, d\xi$$

TABLE B.1 Abscissas and weights for the Gauss-Legendre integration

$$I=\int_{-1}^{1} f(\xi)\,d\xi \approx \sum_{k=1}^{n} f(\xi_k)\,w_k$$

n	$\pm\xi_k$	w_k
$n=2$	0.57735 02691 89626	1.00000 00000 00000
$n=3$	0.00000 00000 00000	0.88888 88888 88889
	0.77459 66692 41483	0.55555 55555 55556
$n=4$	0.33998 10435 84856	0.65214 51548 62546
	0.86113 63115 94053	0.34785 48451 37454
$n=5$	0.00000 00000 00000	0.56888 88888 88889
	0.53846 93101 05683	0.47862 86704 99366
	0.90617 98459 38664	0.23692 68850 56189
$n=6$	0.23861 91860 83197	0.46791 39345 72691
	0.66120 93864 66265	0.36076 15730 48139
	0.93246 95142 03152	0.17132 44923 79170
$n=7$	0.00000 00000 00000	0.41795 91836 73469
	0.40584 51513 77397	0.38183 00505 05119
	0.74153 11855 99394	0.27970 53914 89277
	0.94910 79123 42759	0.12948 49661 68870
$n=8$	0.18343 46424 95650	0.36268 37833 78362
	0.52553 24099 16329	0.31370 66458 77887
	0.79666 64774 13627	0.22238 10344 53374
	0.96028 98564 97536	0.10122 85362 90376
$n=9$	0.00000 00000 00000	0.33023 93550 01260
	0.32425 34234 03809	0.31234 70770 40003
	0.61337 14327 00590	0.26061 06964 02935
	0.83603 11073 26636	0.18064 81606 94857
	0.96816 02395 07626	0.08127 43883 61574
$n=10$	0.14887 43389 81631	0.29552 42247 14753
	0.43339 53941 29247	0.26926 67193 09996
	0.67940 95682 99024	0.21908 63625 15982
	0.86506 33666 88985	0.14945 13491 50581
	0.97390 65285 17172	0.06667 13443 08688
$n=12$	0.12523 34085 11469	0.24914 70458 13403
	0.36783 14989 98180	0.23349 25365 38355
	0.58731 79542 86617	0.20316 74267 23066
	0.76990 26741 94305	0.16007 83285 43346
	0.90411 72563 70475	0.10693 93259 95318
	0.98156 06342 46719	0.04717 53363 86512
$n=16$	0.09501 25098 37637	0.18945 06104 55068
	0.28160 35507 79258	0.18260 34150 44923
	0.45801 67776 57227	0.16915 65193 95002
	0.61787 62444 02643	0.14959 59888 16576
	0.75540 44083 55003	0.12462 89712 55533
	0.86563 12023 87831	0.09515 85116 82492
	0.94457 50230 73232	0.06225 35239 38647
	0.98940 09349 91649	0.02715 24594 11754

TABLE B.2 Computed values of the integral $I = \int_1^3 \frac{\sin^2 x}{x}\, dx$ for various numbers of Gauss points, n

n	ξ_k	w_k	$\int_1^3 \frac{\sin^2 x}{x}\, dx$
1	$\xi_1 = 0$	$w_1 = 2.0000000000$	0.82682181
2	$\xi_1 = -0.5773502692$ $\xi_2 = -\xi_1$	$w_1 = 1.0000000000$ $w_2 = w_1$	0.79856002
3	$\xi_1 = -0.7745966692$ $\xi_2 = 0$ $\xi_3 = -\xi_1$	$w_1 = 0.5555555556$ $w_2 = 0.8888888889$ $w_3 = w_1$	0.79465269
4	$\xi_1 = -0.8611363116$ $\xi_2 = -0.3399810436$ $\xi_3 = -\xi_2$ $\xi_4 = -\xi_1$	$w_1 = 0.3478548451$ $w_2 = 0.6521451549$ $w_3 = w_2$ $w_4 = w_1$	0.79482835
5	$\xi_1 = -0.9061798459$ $\xi_2 = -0.5384693101$ $\xi_3 = 0$ $\xi_4 = -\xi_2$ $\xi_5 = -\xi_1$	$w_1 = 0.2369268851$ $w_2 = 0.4786286705$ $w_3 = 0.5688888889$ $w_4 = w_2$ $w_5 = w_1$	0.79482516
Exact value			0.79482518

The integral is computed using Gaussian integration. Results for various values of n are presented in Table B.2. We notice that a five-point Gaussian quadrature is sufficient to approximate the value of the integral to seven significant figures.

B.2 INTEGRALS WITH A LOGARITHMIC SINGULARITY

The Gaussian integration gives good results when a function $f(\xi)$ varies smoothly within the integration interval. If the integrand is singular the Gauss-Legendre integration cannot be employed and for this purpose special integration rules have been developed that depend on the type of singularity. Integrals whose integrand includes a logarithmic singularity are approximated by the formula [3]

$$\int_0^1 f(\xi) \ln \xi \, d\xi \approx -\sum_{k=1}^{n} f(\xi_k)\, w_k \tag{B.23}$$

The integration points ξ_k and the weights w_k are given in Table B.3. Singular integrals in BEM are discussed in [9].

TABLE B.3 Abscissas and weights for the Gaussian quadrature of a function with logarithmic singularity

$$\int_0^1 f(\xi)\ln\xi\, d\xi \approx -\sum_{k=1}^{n} f(\xi_k)\, w_k$$

n	ξ_k	w_k
$n=2$	0.11200 88061 66976	0.71853 93190 30384
	0.60227 69081 18738	0.28146 06809 69615
$n=3$	0.06389 07930 873254	0.51340 45522 32363
	0.36899 70637 15618	0.39198 00412 01487
	0.76688 03039 38941	0.09461 54065 661491
$n=4$	0.04144 84801 993832	0.38346 40681 45135
	0.24547 49143 20602	0.38687 53177 74762
	0.55616 54535 60278	0.19043 51269 50142
	0.84898 23945 32985	0.03922 54871 299598
$n=5$	0.02913 44721 519720	0.29789 34717 82894
	0.17397 72133 20897	0.34977 62265 13224
	0.41170 25202 84902	0.23448 82900 44052
	0.67731 41745 82820	0.09893 04595 166331
	0.89477 13610 31008	0.01891 15521 431957
$n=6$	0.02163 40058 441169	0.23876 36625 78547
	0.12958 33911 54950	0.30828 65732 73946
	0.31402 04499 14765	0.24531 74265 63210
	0.53865 72173 51802	0.14200 87565 66476
	0.75691 53373 77402	0.05545 46223 248862
	0.92266 88513 72120	0.01016 89586 929322
$n=7$	0.01671 93554 082581	0.19616 93894 25248
	0.10018 56779 15675	0.27030 26442 47272
	0.24629 42462 07930	0.23968 18730 07690
	0.43346 34932 57033	0.16577 57748 10432
	0.63235 09880 47766	0.08894 32271 376579
	0.81111 86267 40105	0.03319 43043 565710
	0.94084 81667 43347	0.05932 78701 512592
$n=8$	0.01332 02441 608924	0.16441 66047 28002
	0.07975 04290 138949	0.23752 56100 23306
	0.19787 10293 26188	0.22684 19844 31919
	0.35415 39943 51909	0.17575 40790 06070
	0.52945 85752 34917	0.11292 40302 46759
	0.70181 45299 39099	0.05787 22107 177820
	0.84937 93204 41106	0.02097 90737 421329
	0.95332 64500 56359	0.03686 40710 402761

B.3 DOUBLE INTEGRALS OF A REGULAR FUNCTION

The Gaussian integration can also be employed to evaluate double integrals. In this case the integration rule depends on the geometry of the domain over which the integration is performed. Various Gauss integration rules have been developed for domains of specific geometry, such as rectangles, triangles, circles as well as domains that can be transformed into one of those geometries. Efforts have been made to develop Gauss integration rules for domains of arbitrary geometry, e.g., the method of finite sectors [4]. In what follows, we will limit our discussion to Gaussian quadratures for rectangular and triangular domains, which are the common geometries for two-dimensional discretizations.

B.3.1 Gauss integration for rectangular domains

For domains resulting as Cartesian products of lower dimensions, e.g., squares, cubes, cylinders, etc., integration rules can always be formulated by *multiplying* rules of lower dimensions [2,5]. For example, if

$$\int_{-1}^{+1} f(\xi)\, d\xi \approx \sum_{k=1}^{n} f(\xi_k)\, w_k \tag{B.24}$$

is a one-dimensional rule, then

$$\int_{-1}^{+1}\int_{-1}^{+1} f(\xi,\eta)\, d\xi\, d\eta \approx \sum_{j=1}^{n}\sum_{i=1}^{n} f(\xi_j,\eta_i)\, w_i w_j \tag{B.25}$$

is the corresponding two-dimensional rule. These rules, however, are not necessarily on the side of economy, with regard to the number of integration points. For square domains Ω: $|\mathrm{x}| < h$, $|\mathrm{y}| < h$, the following simple rule can be employed [2]:

$$\frac{1}{4h^2}\iint_{\Omega} f(x,y)\, dx\, dy \approx \sum_{k=1}^{n} f(x_k, y_k)\, w_k \tag{B.26}$$

The coordinates of the Gauss points x_k and y_k and the associated weight factors are given in Table B.4. The order of the error is $E = O(h^4)$ for the first two cases and $E = O(h^6)$ for the third one.

B.3.2 Gauss integration for triangular domains

The integration over a triangular domain is simplified by employing a natural coordinate system, which here is referred to as a *triangular coordinate system* (see Fig. B.2). The sides a_1, a_2, a_3 are designated by the number of the opposite vertices. The triangular coordinates $\xi_i (i = 1,2,3)$ for an interior point P are defined as the ratios of the areas A_i to the total area A of the triangle:

$$\xi_1 = \frac{A_1}{A},\quad \xi_2 = \frac{A_2}{A},\quad \xi_3 = \frac{A_3}{A} \tag{B.27}$$

TABLE B.4 Coordinates and weights for Gauss integration on square domains of side $2h$

Number and position of the Gauss points	x_k	y_k	w_k
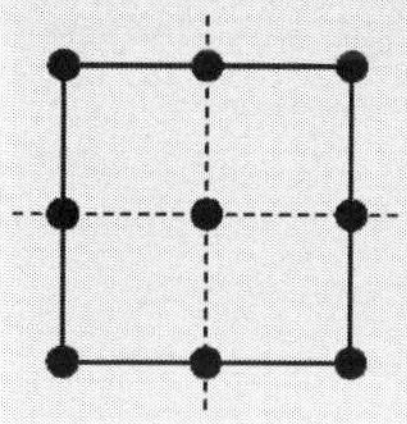 $n=9$	0	0	$^4/_9$
	$\pm h$	$\pm h$	$^1/_{36}$
	$\pm h$	0	$^1/_9$
	0	$\pm h$	$^1/_9$
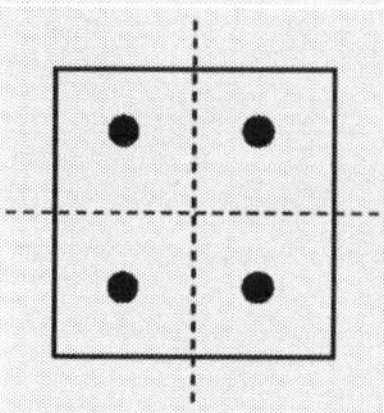 $n=4$	$\pm h\frac{\sqrt{3}}{3}$	$\pm h\frac{\sqrt{3}}{3}$	$^1/_4$
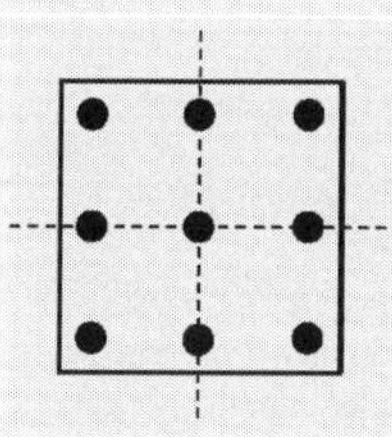 $n=9$	0	0	$^{16}/_{81}$
	$\pm h\sqrt{\frac{3}{5}}$	$\pm h\sqrt{\frac{3}{5}}$	$^{25}/_{324}$
	0	$\pm h\sqrt{\frac{3}{5}}$	$^{10}/_{81}$
	$\pm h\sqrt{\frac{3}{5}}$	0	$^{10}/_{81}$

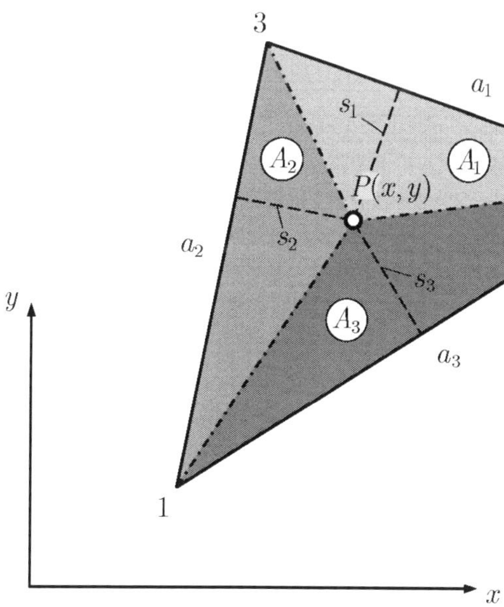

FIGURE B.2 Triangular coordinate system.

Since the sum of the three areas is

$$A_1 + A_2 + A_3 = A$$

it is evident that

$$\xi_1 + \xi_2 + \xi_3 = 1 \tag{B.28}$$

The Cartesian coordinates x and y of point P are related to the triangular coordinates as

$$x = \xi_1 x_1 + \xi_2 x_2 + \xi_3 x_3 \tag{B.29a}$$

$$y = \xi_1 y_1 + \xi_2 y_2 + \xi_3 y_3 \tag{B.29b}$$

where x_i, y_i $(i = 1, 2, 3)$ are the coordinates of the triangle's vertices. Equations (B.29) can be checked at a few specific points. For example, at the centroid of the triangle it is

$$A_1 = A_2 = A_3 = \frac{A}{3},$$

$$x = \frac{x_1 + x_2 + x_3}{3} \quad \text{and} \quad y = \frac{y_1 + y_2 + y_3}{3}$$

while at point 2 (see Fig. B.2): $A_2 = A$, $A_1 = A_3 = 0$, $x = x_2$, and $y = y_2$.

Equations (B.28) and (B.29) yield the relation between triangular and orthogonal coordinates, which can be written in matrix form as

$$\begin{Bmatrix} 1 \\ x \\ y \end{Bmatrix} = \begin{bmatrix} 1 & 1 & 1 \\ x_1 & x_2 & x_3 \\ y_1 & y_2 & y_3 \end{bmatrix} \begin{Bmatrix} \xi_1 \\ \xi_2 \\ \xi_3 \end{Bmatrix} \tag{B.30}$$

The integration over the triangular domain can be performed by considering the infinitesimal parallelogram shown in Fig. B.3. Denoting by s_1, s_2, and s_3 the distances of point P from the corresponding opposite sides and by h_1, h_2, and h_3 the respective heights of the triangle, we can write

$$A = \frac{a_i h_i}{2} \quad \text{and} \quad A_i = \frac{a_i s_i}{2} (i = 1,2,3)$$

which yield according to Eq. (B.27)

$$\frac{s_i}{h_i} = \frac{A_i}{A} = \xi_i$$

and hence

$$s_i = h_i \xi_i$$

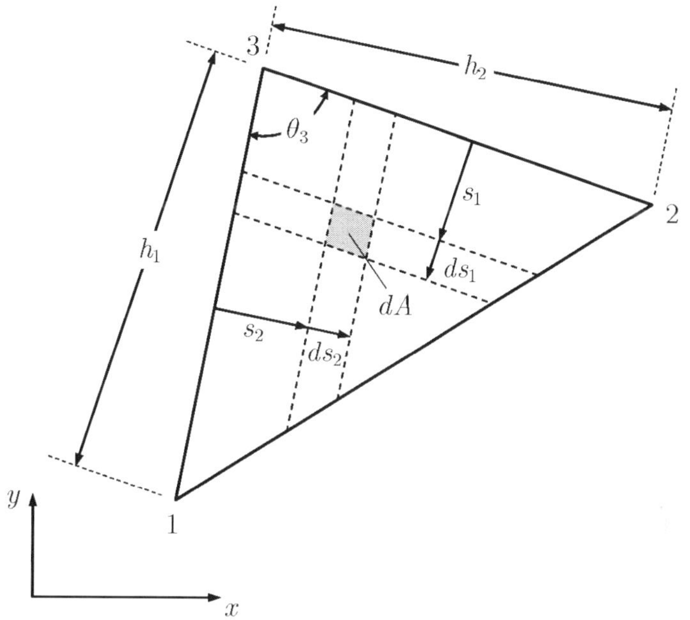

FIGURE B.3 Integration over a triangular domain.

So, the surface element dA can be expressed as

$$\begin{aligned} dA &= \frac{ds_1\, ds_2}{\sin\theta_3} \\ &= \frac{(h_1\, d\xi_1)(h_2\, d\xi_2)}{\sin\theta_3} \\ &= 2A\, d\xi_1\, d\xi_2 \end{aligned} \tag{B.31}$$

Consequently, a domain integral may be written as

$$\begin{aligned} \int_A f(x,y)\, dA &= 2A \int_0^1 \int_0^{1-\xi_1} f(\xi_1, \xi_2, \xi_3)\, d\xi_2\, d\xi_1 \\ &= 2A \int_0^1 \int_0^{1-\xi_1} f[\xi_1, \xi_2, (1-\xi_1-\xi_2)]\, d\xi_2\, d\xi_1 \end{aligned} \tag{B.32}$$

If $f(\xi_1, \xi_2, \xi_3)$ is a polynomial term of the form $\xi_1^a\, \xi_2^b\, \xi_3^c$, where a, b, and c are non-negative integers, then we have [6]

$$\int_A \xi_1^a \xi_2^b \xi_3^c\, dA = \frac{a!\, b!\, c!}{(a+b+c+2)!}\, 2A \tag{B.33}$$

A comprehensive introduction to triangular coordinates can be found in the book by Gallagher [7].

Gauss integration rules over triangles have been derived by means of triangular coordinates and have the form

$$
\begin{aligned}
I &= \int_0^1 \left[\int_0^{1-\xi_1} f(\xi_1, \xi_2, \xi_3)\, d\xi_2 \right] d\xi_1 \\
&\approx \sum_{k=1}^{n} f\left(\xi_1^k, \xi_2^k, \xi_3^k\right) w_k
\end{aligned}
\tag{B.34}
$$

The coordinates ξ_1^k, ξ_2^k, ξ_3^k and the weight factors w_k for integration rules, which are accurate for polynomials of order 1 to 5, are given in Table B.5. These values have been derived by Hammer *et al.* [8].

TABLE B.5 Coordinates and weights for Gauss integration over triangular domains

Number and position of the Gauss points	k	ξ_1^k	ξ_2^k	ξ_3^k	w_k/A
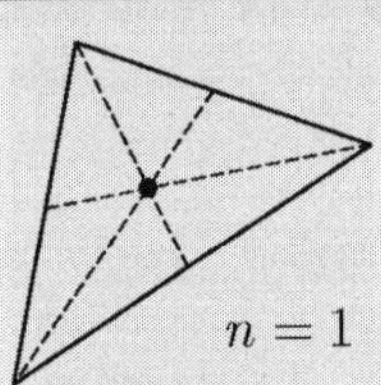 $n = 1$ Degree of accuracy: 1	1	$1/3$	$1/3$	$1/3$	1
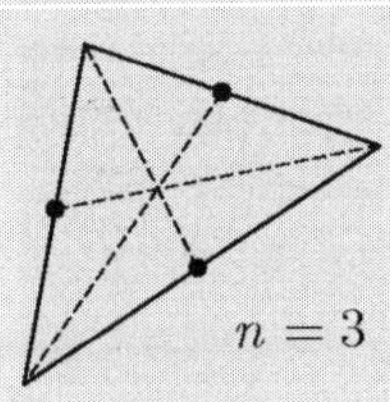 $n = 3$ Degree of accuracy: 2	1	$1/2$	$1/2$	0	$1/3$
	2	0	$1/2$	$1/2$	$1/3$
	3	$1/2$	0	$1/2$	$1/3$
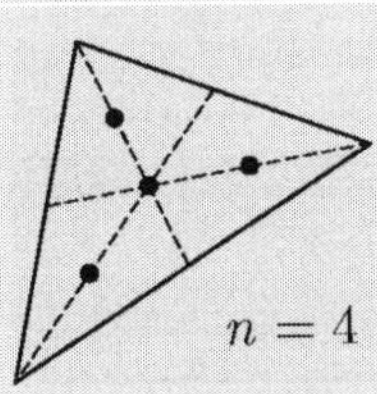 $n = 4$ Degree of accuracy: 3	1	$1/3$	$1/3$	$1/3$	$-27/48$
	2	$3/5$	$1/5$	$1/5$	$25/48$
	3	$1/5$	$3/5$	$1/5$	$25/48$
	4	$1/5$	$1/5$	$3/5$	$25/48$
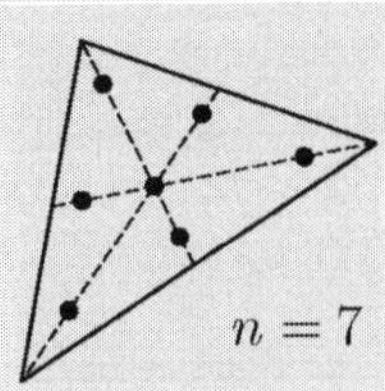 $n = 7$ Degree of accuracy: 5	1	0.33333333	0.33333333	0.33333333	0.22500000
	2	0.79742699	0.10128651	0.10128651	0.12593918
	3	0.10128651	0.79742699	0.10128651	0.12593918
	4	0.10128651	0.10128651	0.79742699	0.12593918
	5	0.05971587	0.47014206	0.47014206	0.13239415
	6	0.47014206	0.05971587	0.47014206	0.13239415
	7	0.47014206	0.47014206	0.05971587	0.13239415

Example B.2

The integrals below are computed using a four-point Gaussian integration over the triangle shown in Fig. B.4.

Equations (B.29) applied to the triangle of Fig. B.4 become

$$x = 2\xi_1 + 3\xi_2 + \xi_3$$

$$y = \xi_1 + 2\xi_2 + 3\xi_3$$

and formula (B.34) yields

$$\int_A (1)\, dA = \sum_{k=1}^{4} (1)_k w_k$$

$$= (1)\left(-\frac{27A}{48}\right) + (1)\left(\frac{25A}{48}\right) + (1)\left(\frac{25A}{48}\right) + (1)\left(\frac{25A}{48}\right) = A$$

$$\int_A x\, dA = \sum_{k=1}^{4} \left(2\xi_1^k + 3\xi_2^k + \xi_3^k\right) w_k$$

$$= \left(2\times\frac{1}{3} + 3\times\frac{1}{3} + \frac{1}{3}\right)\left(-\frac{27}{48}A\right) + \left(2\times\frac{3}{5} + 3\times\frac{1}{5} + \frac{1}{5}\right)\left(\frac{25}{48}A\right)$$

$$+ \left(2\times\frac{1}{5} + 3\times\frac{3}{5} + \frac{1}{5}\right)\left(\frac{25}{48}A\right) + \left(2\times\frac{1}{5} + 3\times\frac{1}{5} + \frac{3}{5}\right)\left(\frac{25}{48}A\right)$$

$$= 2A$$

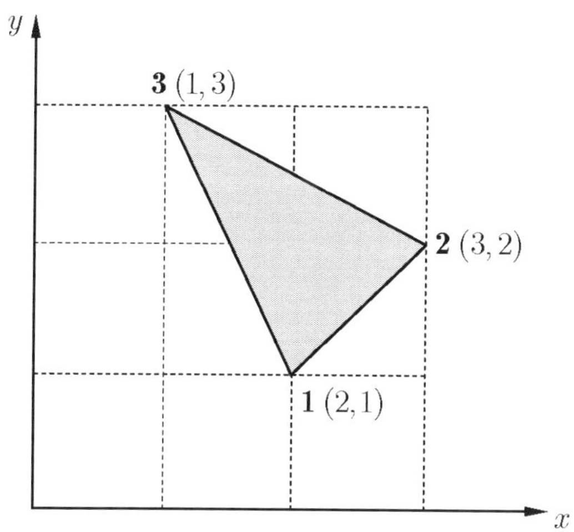

FIGURE B.4 Triangular domain of integration.

$$
\begin{aligned}
\int_A y^2\,dA &= \sum_{k=1}^{4}\left(\xi_1^k+2\xi_2^k+3\xi_3^k\right)^2 w_k \\
&= \left(\frac{1}{3}+2\times\frac{1}{3}+3\times\frac{1}{3}\right)^2\left(-\frac{27}{48}A\right)+\left(\frac{3}{5}+2\times\frac{1}{5}+3\times\frac{1}{5}\right)^2\left(\frac{25}{48}A\right) \\
&+\left(\frac{1}{5}+2\times\frac{3}{5}+3\times\frac{1}{5}\right)^2\left(\frac{25}{48}A\right)+\left(\frac{1}{5}+2\times\frac{1}{5}+3\times\frac{3}{5}\right)^2\left(\frac{25}{48}A\right) \\
&= \frac{25}{6}A
\end{aligned}
$$

The computed values are exact because the four-point integration rule is exact for polynomials up to the third degree.

B.5 REFERENCES

[1] F. Scheid, Numerical Analysis, in: Schaum's Outline SeriesMcGraw-Hill Book Company, New York, 1968.

[2] M. Abramowitz, I. Stegun (Eds.), Handbook of Mathematical Functions, 10th ed., Dover Publications, New York, 1972.

[3] A.H. Stroud, D. Secrest, Gaussian Quadrature Formulas, Prentice Hall, Inc., Englewood Cliffs, New Jersey, 1966

[4] J.T. Katsikadelis, A boundary element solution to the vibration problem of plates, J. Sound Vib. 141 (1990) 313–322.

[5] P. Davis, P. Rabinowitz, Methods of Numerical Integration, Academic Press, New York, 1975.

[6] M.A. Eisenberg, L.E. Malvern, On finite element integration in natural coordinates, Int. J. Numer. Meth. Eng. 7 (1973) 574–575.

[7] R.H. Gallagher, Finite Element Analysis Fundamentals, Prentice Hall, Englewood Cliffs, New Jersey, 1975.

[8] P.C. Hammer, O.J. Marlowe, A.H. Stroud, Numerical integration over simplexes and cones, Math. Tables Other Aids Comput. X (1956) 130–137.

[9] V. Sladek, J. Sladek (Eds.), Singular Integrals in Boundary Element Methods, Computational Mechanics Publications, Southampton, 1998.

Appendix | C

Numerical Integration of the Equations of Motion

C.1 INTRODUCTION

In this Appendix a direct time integration method for the numerical solution of the equations of motion in structural dynamics is presented. The method has been used to obtain the numerical results in the examples of Chapters 4 and 5. It applies to systems of linear and nonlinear equations of motion. It is simple to implement. It is self-starting, second-order accurate, unconditionally stable, and it does not exhibit amplitude decay or period elongation. It performs well when large deformations and longtime durations are considered in the transient response analysis of structures. It can be used as a practical method for the integration of the equations of motion in cases where widely used time integration procedures, e.g., Newmark's, do not conserve energy and momentum and become unstable. Another important advantage of this method is that it can be adjusted to solve evolution equations of higher order than two, such as those arising from the static or dynamic response of viscoelastic structures described with multi-term differential viscoelastic models (see Sections 4.4 and 5.4). The method is described in detail in [1]. Here, the basic ingredients of the method are presented.

C.2 LINEAR SYSTEMS

C.2.1 One-degree-of-freedom system

The procedure is illustrated with the linear one-degree-of-freedom system. In this case, the initial value problem is stated as

$$m\ddot{u} + c\dot{u} + ku = p(t) \tag{C.1}$$

$$u(0) = u_0,\ \dot{u}(0) = \dot{u}_0 \tag{C.2a,b}$$

where m, c and k are the mass, damping and stiffness of the system, respectively.

The Boundary Element Method for Plate Analysis

The problem is solved using the principle of the analog equation. Hence, if $u = u(t)$ is the sought solution and the operator d^2/dt^2 is applied to it we have

$$\ddot{u} = q(t) \tag{C.3}$$

where $q(t)$ is a fictitious source, unknown in the first instance. Eq. (C.3) is the analog equation of (C.1). It indicates that the solution of Eq. (C.1) can be obtained by solving Eq. (C.3) with the initial conditions (C.2). This is achieved as follows.

Taking the Laplace transform of Eq. (C.3) we obtain

$$U(s) = \frac{1}{s}u(0) + \frac{1}{s^2}\dot{u}(0) + \frac{1}{s^2}Q(s) \tag{C.4}$$

where $U(s)$, $Q(s)$ are the Laplace transforms of $u(t)$, $q(t)$. The inverse Laplace transform of the above expression yields the solution of Eq. (C.3) in integral from as

$$u(t) = u(0) + \dot{u}(0)\, t + \int_0^t q(\tau)(t-\tau)\, d\tau \tag{C.5}$$

Thus, the initial value problem is transformed into the equivalent Volterra integral equation for $q(t)$.

Eq. (C.5) is solved numerically within a time interval $[0, T]$. The interval $[0, T]$ is divided into N equal intervals $\Delta t = h$, $h = T/N$ (Fig. C.1), in which $q(t)$ is assumed to vary according to a certain law, e.g. constant, linear, etc. In this analysis $q(t)$ is assumed to be constant and equal to the mean value in the interval h, i.e.,

$$q_r^m = \frac{q_{r-1} + q_r}{2} \tag{C.6}$$

Hence, Eq. (C.5) at instant $t = nh$ can be written as

$$\begin{aligned} u_n = {} & u_0 + nh\dot{u}_0 \\ & + \left[q_1^m \int_0^h (nh-\tau)\, d\tau + q_2^m \int_h^{2h} (nh-\tau)\, d\tau + \cdots + q_n^m \int_{(n-1)h}^{nh} (nh-\tau)\, d\tau \right] \end{aligned} \tag{C.7}$$

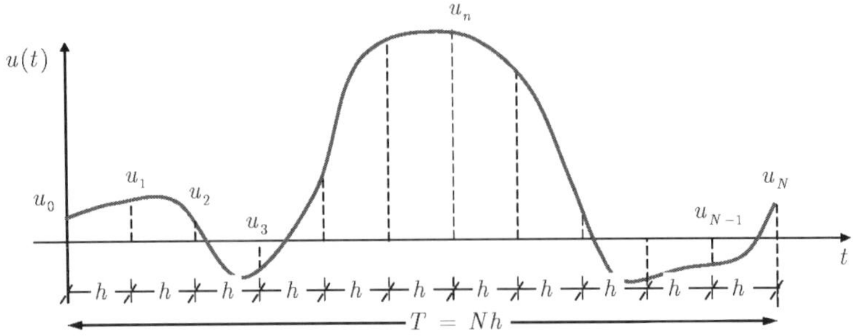

FIGURE C.1 Discretization of the interval $[0, T]$ into N equal intervals $h = T/N$.

which after evaluation of the integrals yields

$$\begin{aligned} u_n &= u_0 + nh\dot{u}_0 + c_1\sum_{r=1}^{n-1}[2(n-r)+1]q_r^m + c_1 q_n^m \\ &= u_{n-1} + h\dot{u}_0 + 2c_1\sum_{r=1}^{n-1} q_r^m + c_1 q_n^m \end{aligned} \tag{C.8}$$

where

$$c_1 = \frac{h^2}{2} \tag{C.9}$$

The velocity is obtained by direct differentiation of Eq. (C.5) making use of the Leibniz rule. Thus, we have

$$\dot{u}(t) = \dot{u}(0) + \int_0^t q(\tau)\, d\tau \tag{C.10}$$

Using the same discretization for the interval $[0,\ T]$ to approximate the integral in Eq. (C.10), we have

$$\begin{aligned} \dot{u}_n &= \dot{u}_0 + c_2\sum_{r=1}^{n-1} q_r^m + c_2 q_n^m \\ &= \dot{u}_{n-1} + c_2 q_n^m \end{aligned} \tag{C.11}$$

where

$$c_2 = h \tag{C.12}$$

Solving Eq. (C.11) for $\sum_{r=1}^{n-1} q_r^m$ and substituting in Eq. (C.8) gives

$$u_n = u_{n-1} + h\dot{u}_n - c_1 q_n^m \tag{C.13}$$

Taking into account Eq. (C.6), Eqs. (C.13) and (C.11) can be further written as

$$\frac{c_1}{2} q_n - h\dot{u}_n + u_n = -\frac{c_1}{2} q_{n-1} + u_{n-1} \tag{C.14}$$

$$-\frac{c_2}{2} q_n + \dot{u}_n = \dot{u}_{n-1} + \frac{c_2}{2} q_{n-1} \tag{C.15}$$

Moreover, Eq. (C.1) at time $t = nh$ is written as

$$mq_n + c\dot{u}_n + ku_n = p_n \tag{C.16}$$

Equations (C.14), (C.15), and (C.16) can be combined as

$$\begin{bmatrix} m & c & k \\ \frac{1}{2}c_1 & -h & 1 \\ -\frac{1}{2}c_2 & 1 & 0 \end{bmatrix} \begin{Bmatrix} q_n \\ \dot{u}_n \\ u_n \end{Bmatrix} = \begin{bmatrix} 0 & 0 & 0 \\ -\frac{1}{2}c_1 & 0 & 1 \\ \frac{1}{2}c_2 & 1 & 0 \end{bmatrix} \begin{Bmatrix} q_{n-1} \\ \dot{u}_{n-1} \\ u_{n-1} \end{Bmatrix} + \begin{Bmatrix} 1 \\ 0 \\ 0 \end{Bmatrix} p_n \tag{C.17}$$

The coefficient matrix in Eq. (C.17) is not singular for sufficient small h and hence the system can be solved successively for $n = 1, 2, \ldots$ to yield the solution u_n and the derivatives $\dot{u}_n$, $\ddot{u}_n = q_n$ at instant $t = nh \leq T$. For $n = 1$, the value q_0 appears in the right-hand side of Eq. (C.17). This quantity can be readily obtained from Eq. (C.1) for $t = 0$. This yields

$$q_0 = (p_0 - c\dot{u}_0 - ku_0)/m \tag{C.18}$$

Equation (C.17) is written in matrix form

$$\mathbf{U}_n = \mathbf{A}\mathbf{U}_{n-1} + \mathbf{b}\overline{p}_n, \quad n = 1, 2, \ldots N \tag{C.19}$$

in which

$$\mathbf{A} = \begin{bmatrix} 1 & 2\xi\omega & \omega^2 \\ \frac{1}{2}c_1 & -h & 1 \\ -\frac{1}{2}c_2 & 1 & 0 \end{bmatrix}^{-1} \begin{bmatrix} 0 & 0 & 0 \\ -\frac{1}{2}c_1 & 0 & 1 \\ \frac{1}{2}c_2 & 1 & 0 \end{bmatrix} \tag{C.20a}$$

$$\mathbf{U}_n = \begin{Bmatrix} q_n \\ \dot{u}_n \\ u_n \end{Bmatrix} \tag{C.20b}$$

$$\mathbf{b} = \begin{bmatrix} 1 & 2\xi\omega & \omega^2 \\ \frac{1}{2}c_1 & -h & 1 \\ -\frac{1}{2}c_2 & 1 & 0 \end{bmatrix}^{-1} \begin{Bmatrix} 1 \\ 0 \\ 0 \end{Bmatrix}, \quad \overline{p}_n = \frac{p_n}{m} \tag{C.20c}$$

$\omega = \sqrt{k/m}$ is the eigenfrequency and $\xi = c/2m\omega$ the damping ratio. The recurrence formula (C.19) can be employed to construct the solution algorithm. However, the solution procedure can be further simplified. Thus, applying Eq. (C.19) for $n = 1, 2, \ldots$ we have

$$\begin{aligned} \mathbf{U}_1 &= \mathbf{A}\mathbf{U}_o + \mathbf{b}\overline{p}_1 \\ \mathbf{U}_2 &= \mathbf{A}\mathbf{U}_1 + \mathbf{b}\overline{p}_2 \\ &= \mathbf{A}(\mathbf{A}\mathbf{U}_o + \mathbf{b}\overline{p}_1) + \mathbf{b}\overline{p}_2 \\ &= \mathbf{A}^2\mathbf{U}_o + \mathbf{A}\mathbf{b}\overline{p}_1 + \mathbf{b}\overline{p}_2 \\ \ldots &= \ldots\ldots\ldots\ldots\ldots\ldots \\ \mathbf{U}_n &= \mathbf{A}^n\mathbf{U}_o + \left(\mathbf{A}^{n-1}\overline{p}_1 + \mathbf{A}^{n-2}\overline{p}_2 + \ldots \mathbf{A}^0\overline{p}_n\right)\mathbf{b} \end{aligned} \tag{C.21}$$

Apparently, the last of Eqs. (C.21) gives the solution vector $\mathbf{U}_n$ at instant $t_n = nh$ using only the known vector $\mathbf{U}_0$ at $t=0$. The matrix $\mathbf{A}$ and the vector $\mathbf{b}$ are computed only once. It can be shown that for the spectral radius it is $\rho(\mathbf{A}) \leq 1$, which implies that the method is unconditionally stable. Moreover, it is proved that the method is second-order accurate and exhibits no amplitude decay and period elongation. For detailed proofs see in [1].

C.2.2 Multi-degree-of-freedom system

The initial value problem for the multiple-degree-of-freedom system is stated as

$$\mathbf{M}\ddot{\mathbf{u}} + \mathbf{C}\dot{\mathbf{u}} + \mathbf{K}\mathbf{u} = \mathbf{p}(t) \tag{C.22}$$

$$\mathbf{u}(0) = \mathbf{u}_0, \quad \dot{\mathbf{u}}(0) = \dot{\mathbf{u}}_0 \tag{C.23}$$

where $\mathbf{M}$, $\mathbf{C}$, and $\mathbf{K}$ are the mass, damping, and stiffness matrix of the structure, respectively.

The solution procedure described in Section C.2.1 may also be applied to this case provided that the coefficients m, c, k and the quantities $u_0, \dot{u}_0, u_n, \dot{u}_n, q_n, p_n$ are replaced with the coefficients matrices $\mathbf{M}$, $\mathbf{C}$, $\mathbf{K}$ and the vectors $\mathbf{u}_0, \dot{\mathbf{u}}_0, \mathbf{u}_n, \dot{\mathbf{u}}_n, \mathbf{q}_n, \mathbf{p}_n$, respectively, and the scalar operations with matrix operations. Thus, Eqs. (C.17) and (C.18) read

$$\begin{bmatrix} \mathbf{M} & \mathbf{C} & \mathbf{K} \\ \frac{c_1}{2}\mathbf{I} & -h\mathbf{I} & \mathbf{I} \\ -\frac{c_2}{2}\mathbf{I} & \mathbf{I} & \mathbf{0} \end{bmatrix} \begin{Bmatrix} \mathbf{q}_n \\ \dot{\mathbf{u}}_n \\ \mathbf{u}_n \end{Bmatrix} = \begin{bmatrix} \mathbf{0} & \mathbf{0} & \mathbf{0} \\ -\frac{c_1}{2}\mathbf{I} & \mathbf{0} & \mathbf{I} \\ \frac{c_2}{2}\mathbf{I} & \mathbf{I} & \mathbf{0} \end{bmatrix} \begin{Bmatrix} \mathbf{q}_{n-1} \\ \dot{\mathbf{u}}_{n-1} \\ \mathbf{u}_{n-1} \end{Bmatrix} + \begin{Bmatrix} \mathbf{I} \\ \mathbf{0} \\ \mathbf{0} \end{Bmatrix} \mathbf{p}_n \tag{C.24}$$

$$\mathbf{q}_0 = \mathbf{M}^{-1}(\mathbf{p}_0 - \mathbf{C}\dot{\mathbf{u}}_0 - \mathbf{K}\mathbf{u}_0), \quad \det(\mathbf{M}) \neq 0 \tag{C.25}$$

Eq. (C.24) is solved successively for $n = 1, 2, \ldots$.

The solution algorithm is stated in a pseudo-code type notation in Table C.1. The reader interested in implementing the method can easily do so in the computer language of his preference.

Example C.1. Forced damped vibrations

Equation (C.1) has been solved with the data $m=1$, $\xi=0.07$, $\omega=2$, $u_0=1$, $\dot{u}_0=20$, $T=25$, $h=1/500$, $p(t)=p_0 H(t)$, $p_0=10$. Fig. C.2 shows the obtained solution together with the error.

Example C.2.

Equation (C.22) has been solved for various values of the number of the degrees of freedom N and the total time of the motion t_{tot}. The employed matrices $\mathbf{M}$, $\mathbf{C}$, $\mathbf{K}$ are symmetric and fully populated and they are randomly generated for each set of values (N, t_{tot}). The computation time for each case is shown

TABLE C.1 Algorithm 1: Linear Equations of Motion

A. Data

Read: $\mathbf{M}, \mathbf{C}, \mathbf{K}, \mathbf{u}_0, \dot{\mathbf{u}}_0, \mathbf{p}(t), t_{tot}$

B. Initial computations

Choose: $h := \Delta t$ and compute n_{tot}

Compute: $c_1 := h^2/2 \quad c_2 := h \quad \mathbf{q}_0 := \mathbf{M}^{-1}(\mathbf{p}_0 - \mathbf{C}\dot{\mathbf{u}}_0 - \mathbf{K}\mathbf{u}_0)$

Formulate: $\mathbf{U}_0 := \{\mathbf{q}_0 \quad \dot{\mathbf{u}}_0 \quad \mathbf{u}_0\}^T$

Compute:

$$\mathbf{A} := \begin{bmatrix} \mathbf{M} & \mathbf{C} & \mathbf{K} \\ \frac{c_1}{2}\mathbf{I} & -h\mathbf{I} & \mathbf{I} \\ -\frac{c_2}{2}\mathbf{I} & \mathbf{I} & \mathbf{0} \end{bmatrix}^{-1} \begin{bmatrix} \mathbf{0} & \mathbf{0} & \mathbf{0} \\ -\frac{c_1}{2}\mathbf{I} & \mathbf{0} & \mathbf{I} \\ \frac{c_2}{2}\mathbf{I} & \mathbf{I} & \mathbf{0} \end{bmatrix}, \quad \mathbf{b} := \begin{bmatrix} \mathbf{M} & \mathbf{C} & \mathbf{K} \\ \frac{c_1}{2}\mathbf{I} & -h\mathbf{I} & \mathbf{I} \\ -\frac{c_2}{2}\mathbf{I} & \mathbf{I} & \mathbf{0} \end{bmatrix}^{-1} \begin{Bmatrix} \mathbf{I} \\ \mathbf{0} \\ \mathbf{0} \end{Bmatrix}$$

C. Compute solution

for $n := 1$ to n_{tot}

$\mathbf{U}_n = \mathbf{A}\mathbf{U}_{n-1} + \mathbf{b}\mathbf{p}_n$

end

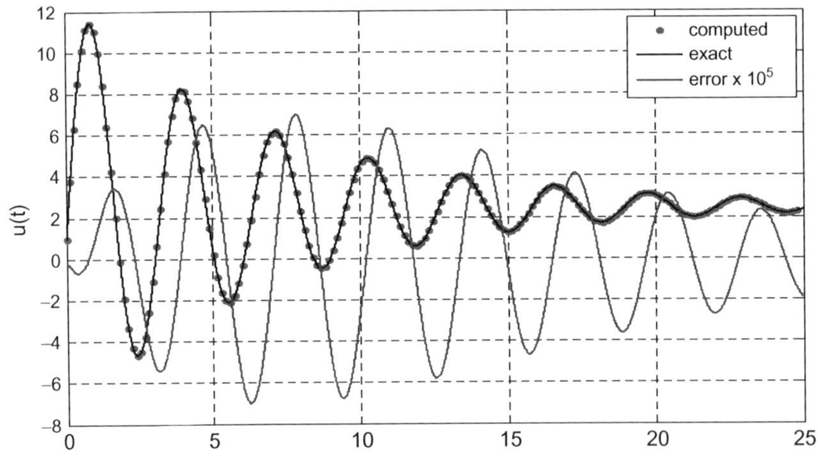

FIGURE C.2 Displacement and error in Example C.1.

in Table C.2 as compared with Newmark's trapezoidal rule. The scheme was programmed in MATLAB (R2009a) and the solution with $\Delta t = 0.01$ was obtained on a Fujitsu Celsius Series 700 notebook. These results show efficiency of the scheme with regard to the time cost.

C.3 NONLINEAR EQUATIONS OF MOTION

The nonlinear initial value problem for multi-degree-of-freedom systems is described as

TABLE C.2 Computation time in Example C.2. Upper row: Newmark's trapezoidal rule; Lower row: present scheme

t_{tot} sec	N 20	40	80	100	120	160	180	200
25	0.439	1.091	2.937	3.667	4.532	5.958	6.752	7.510
	0.020	0.031	0.146	0.219	0.249	0.402	0.527	0.652
50	2.138	5.471	11.901	14.845	17.904	25.507	27.583	30.667
	0.039	0.108	0.260	0.366	0.434	0.781	0.971	1.148
100	10.786	23.365	48.101	60.808	72.76	98.61	116.044	121.544
	0.091	0.142	0.435	0.623	0.817	1.477	1.801	2.11

$$\mathbf{M}\ddot{\mathbf{u}} + \mathbf{F}(\dot{\mathbf{u}}, \mathbf{u}) = \mathbf{p}(t) \tag{C.26}$$

$$\mathbf{u}(0) = \mathbf{u}_0,\ \dot{\mathbf{u}}(0) = \dot{\mathbf{u}}_0 \tag{C.27}$$

where $\mathbf{M}$ is the $N \times N$ known coefficient matrix with $\det(\mathbf{M}) \neq 0$, $\mathbf{F}(\dot{\mathbf{u}}, \mathbf{u})$ is an $N \times 1$ vector, whose elements are nonlinear functions of the components of $\mathbf{u}$, $\dot{\mathbf{u}}$, $\mathbf{p}(t)$ is the vector of the N given load functions, and $\mathbf{u}_0$, $\dot{\mathbf{u}}_0$ the given constant vectors.

The solution procedure is similar to that for the linear systems. Thus, Eq. (C.26) for $t = 0$ gives the initial acceleration vector

$$\mathbf{q}_0 = \mathbf{M}^{-1}[\mathbf{p}_0 - \mathbf{F}(\dot{\mathbf{u}}_0, \mathbf{u}_0)], \quad \mathbf{q}_0 = \ddot{\mathbf{u}}_0 \tag{C.28}$$

Subsequently, we apply Eq. (C.26) for $t = t_n$

$$\mathbf{M}\mathbf{q}_n + \mathbf{F}(\dot{\mathbf{u}}_n, \mathbf{u}_n) = \mathbf{p}_n \tag{C.29}$$

Apparently, Eqs. (C.14) and (C.15) are valid for the multi-degree-of-freedom system (see Eq. (C.24)), too, and can be written as

$$\begin{bmatrix} -h\mathbf{I} & \mathbf{I} \\ \mathbf{I} & \mathbf{0} \end{bmatrix} \begin{Bmatrix} \dot{\mathbf{u}}_n \\ \mathbf{u}_n \end{Bmatrix} = \begin{bmatrix} \mathbf{0} & \mathbf{I} \\ \mathbf{I} & \mathbf{0} \end{bmatrix} \begin{Bmatrix} \dot{\mathbf{u}}_{n-1} \\ \mathbf{u}_{n-1} \end{Bmatrix} + \begin{bmatrix} -\dfrac{c_1}{2}\mathbf{I} \\ \dfrac{c_2}{2}\mathbf{I} \end{bmatrix} \mathbf{q}_n + \begin{bmatrix} -\dfrac{c_1}{2}\mathbf{I} \\ \dfrac{c_2}{2}\mathbf{I} \end{bmatrix} \mathbf{q}_{n-1} \tag{C.30}$$

Eqs. (C.29) and (C.30) are combined and solved for $\mathbf{q}_n$, $\dot{\mathbf{u}}_n$, $\mathbf{u}_n$ with $n = 1, 2, \ldots$. Note that Eq. (C.30) is linear and can be solved for $\dot{\mathbf{u}}_n$, $\mathbf{u}_n$. Then substitution into Eq. (C.29) results in a nonlinear equation, which can be solved to yield $\mathbf{q}_n$. In our examples the function *fsolve* in MATLAB has been employed to obtain the numerical results. The solution algorithm is stated in a pseudo-code type notation in Table C.3.

TABLE C.3 Algorithm 2: Nonlinear Equations of Motion

A. Data
Read: $\mathbf{M}$, $\mathbf{F}(\dot{\mathbf{u}}, \mathbf{u})$, $\mathbf{u}_0$, $\dot{\mathbf{u}}_0$, $\mathbf{p}(t)$, t_{tot}

B. Initial computations
Choose: $h := \Delta t$ and compute n_{tot}
Compute: $c_1 := h^2/2 \; c_2 := h \; \mathbf{q}_0 := \mathbf{M}^{-1}[\mathbf{p}_0 - \mathbf{F}(\dot{\mathbf{u}}_0, \mathbf{u}_0)]$

C. Compute solution
for $n := 1$ to n_{tol} solve for $\{\mathbf{q}_n \; \dot{\mathbf{u}}_n \; \mathbf{u}_n\}^T$ the system of the nonlinear algebraic equations:

$$\mathbf{M}\mathbf{q}_n + \mathbf{F}(\dot{\mathbf{u}}_n, \mathbf{u}_n) = \mathbf{p}_n$$

$$\begin{bmatrix} -h\mathbf{I} & \mathbf{I} \\ \mathbf{I} & \mathbf{0} \end{bmatrix} \begin{Bmatrix} \dot{\mathbf{u}}_n \\ \mathbf{u}_n \end{Bmatrix} = \begin{bmatrix} \mathbf{0} & \mathbf{I} \\ \mathbf{I} & \mathbf{0} \end{bmatrix} \begin{Bmatrix} \dot{\mathbf{u}}_{n-1} \\ \mathbf{u}_{n-1} \end{Bmatrix} + \begin{bmatrix} \frac{c_1}{2}\mathbf{I} \\ \frac{c_2}{2}\mathbf{I} \end{bmatrix} \mathbf{q}_n + \begin{bmatrix} -\frac{c_1}{2}\mathbf{I} \\ \frac{c_2}{2}\mathbf{I} \end{bmatrix} \mathbf{q}_{n-1}$$

end

Example C.3. The elastic pendulum

The elastic pendulum, also called the "soft pendulum" (Fig. C.3) is chosen to illustrate the efficiency of the scheme in solving nonlinear equations of motion. In this pendulum the rod is assumed elastically extensible with a stiffness $k = EA/l$; A is the area of the cross-section and E the modulus of elasticity. The system has two degrees of freedom and its motion is described by the equations [2]:

$$m\ddot{x} + \frac{EA}{L}\left(1 - \frac{L}{\sqrt{x^2 + y^2}}\right)x = 0 \tag{C.31a}$$

$$m\ddot{y} + \frac{EA}{L}\left(1 - \frac{L}{\sqrt{x^2 + y^2}}\right)y = -mg \tag{C.31b}$$

with the initial conditions

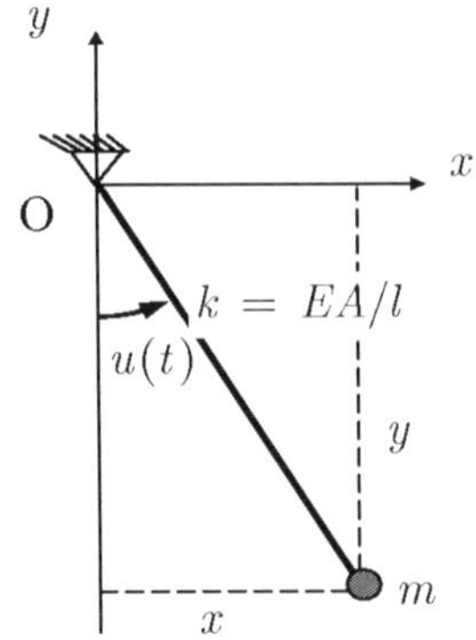

FIGURE C.3 Elastic pendulum

$$x(0) = x_0,\ \dot{x}(0) = \dot{x}_0,\ y(0) = y_0,\ \dot{y}(0) = \dot{y}_0 \tag{C.32}$$

This problem, in absence of gravity ($g = 0$), has been used as a benchmark problem by earlier investigators [3,4] to check the performance of their method in an effort to overcome the instability of Newmark's method arising when long duration motions are considered in nonlinear structural dynamics. The pendulum is studied using the numerical scheme with $l = 3.0443\text{m}$, $EA = 10^4\text{N}$, $x_0 = 0$, $\dot{x}_0 = 7.72\,\text{ms}^{-1}$, $y_0 = \dot{y}_0 = 0$, and $m = 6.667\,\text{kg}$. The rod has been treated as a truss element with consistent mass having $\rho A = 6.57\text{kg/m}$. The response of the system obtained with $\Delta t = 0.01$ is presented in Fig. C.4, which is identical

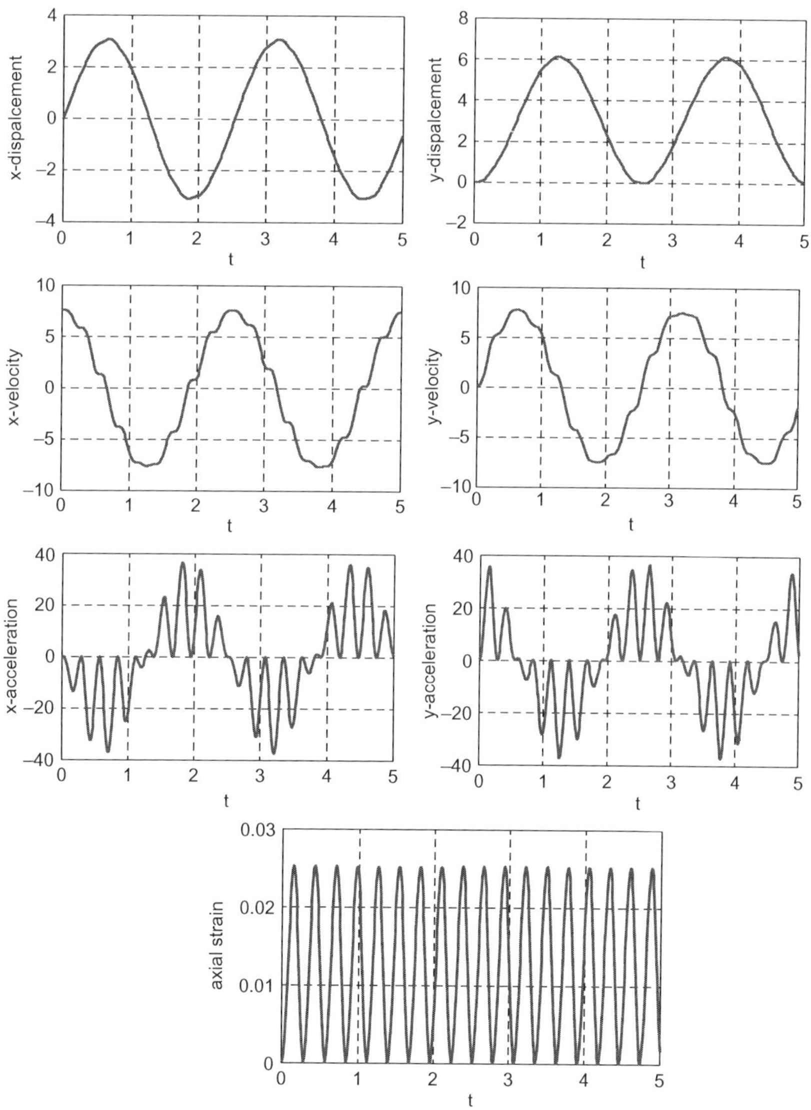

FIGURE C.4 Response of the elastic pendulum using the scheme with $\Delta t = 0.01$ in Example C.3.

with that in [4]. In Fig. C.5 the x-displacement in the intervals $0 \leq t \leq 5$ and $990.71 \leq t \leq 995.71$ has been plotted. This demonstrates that the response remains unchanged after a long duration of motion. Fig. C.6 shows that the total energy of the system is conserved. Apparently, the proposed scheme exhibits no period elongation or amplitude decay in analyzing nonlinear dynamic systems. Finally, Fig. C.7 shows the response of the elastic pendulum obtained using the

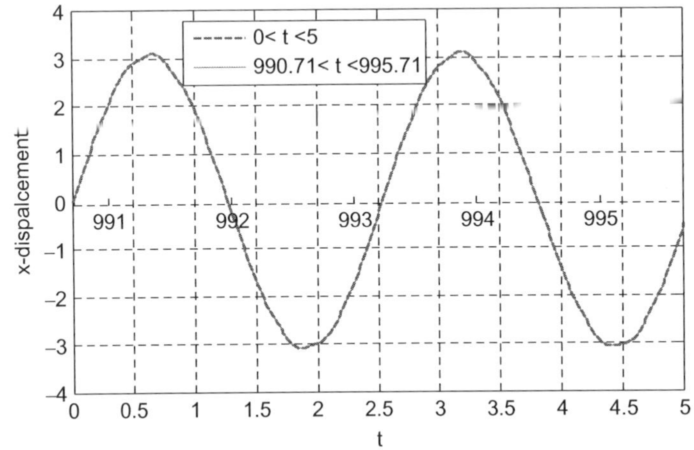

FIGURE C.5 Elastic pendulum. x-displacement in the time intervals $0 < t < 5$ and $990.71 < t < 995.71$ with $\Delta t = 0.01$ in Example C.3.

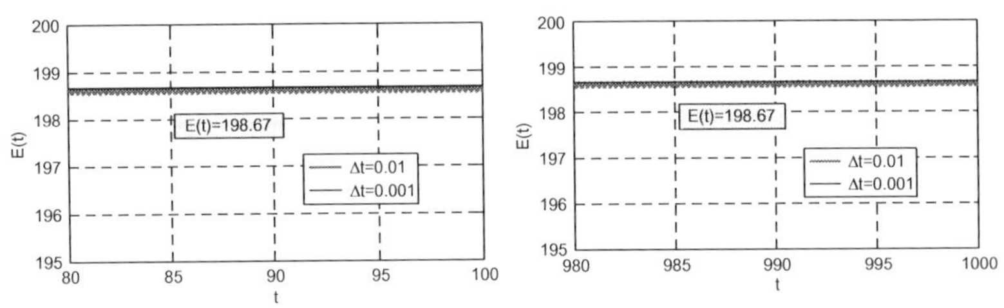

FIGURE C.6 Elastic pendulum. Energy variation

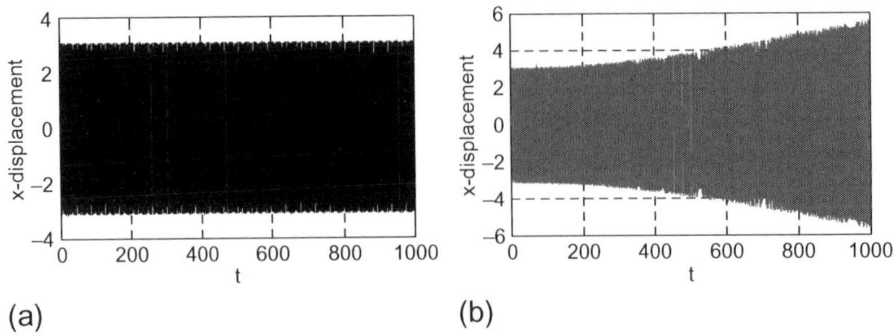

FIGURE C.7 Elastic pendulum. x-displacement (a) current scheme with $\Delta t = 0.01$. (b) Newmark's trapezoidal rule with $\Delta t = 0.00001$.

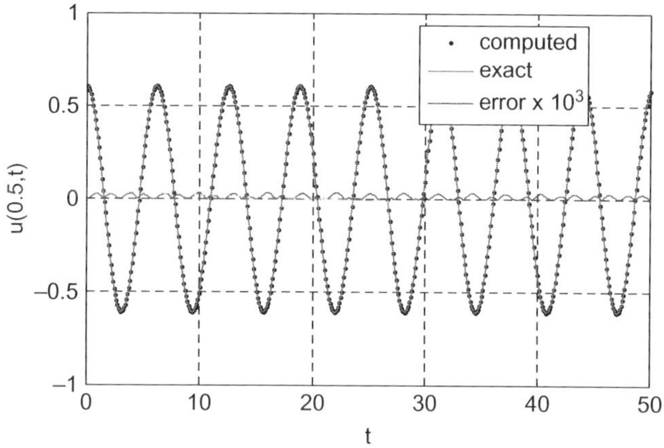

FIGURE C.8 Solution at $x=0.5$, $\Delta t=0.1$ in Example C.4.

current scheme with $\Delta t=0.01$ and Newmark's trapezoidal scheme with $\Delta t=0.00001$. Obviously, the present scheme performs well for a relatively large time step, while Newmark's scheme becomes unstable even for a very small time step.

Example C.4.

In this example the numerical scheme is employed to solve a large system of nonlinear equations of motion, which represents the semi-discretized equations of motion resulting from the analysis of the nonlinear wave equation

$$u_{,xx}-uu_{,x}-\ddot{u}=e^{-2x}\cos^2 t+2e^{-x}\cos t,\quad 0\le x\le 1,\quad 0<t \tag{C.33}$$

$$u(x,0)=e^{-x},\quad \dot{u}(x,0)=0 \tag{C.34a,b}$$

$$u(0,0)=1,\quad u(1,0)=e^{-1} \tag{C.35a,b}$$

The AEM is employed with 31 constant elements [5]. This procedure yields a system of 31 nonlinear equations of motion of the form (C.26) in terms of the nodal values, which are solved using the present numerical scheme. The problem admits an exact solution $u_{ex}=e^{-x}\cos t$. Fig. C.8 shows the solution at $x=0.5$ as compared with the exact one together with the error $=|u(0.5,t)-u_{ex}(0.5,t)|$.

C.4 VARIABLE COEFFICIENTS

The method applies also when the coefficient matrices in Eq. (C.22) are variable, i.e., they are functions of the independent variable t. Obviously, the previously described solution procedure remains the same except that the elements $\mathbf{M}$, $\mathbf{C}$, $\mathbf{K}$ in the first row of the coefficient matrix in the left-hand side of

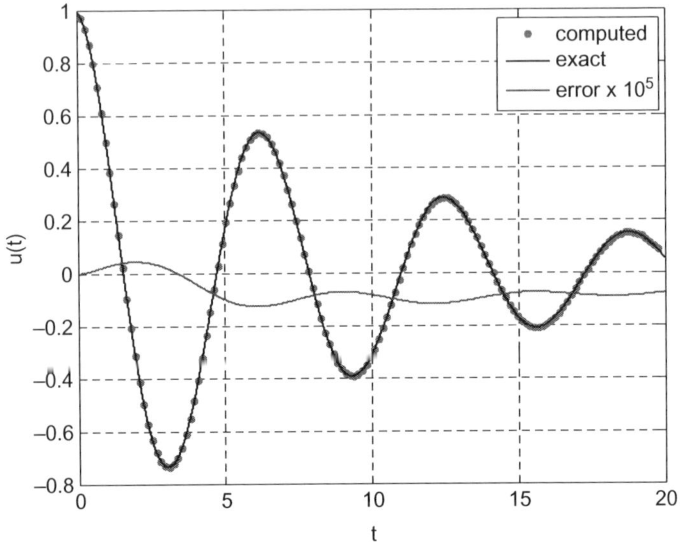

FIGURE C.9 Solution in Example C.4.

Eq. (C.24) must be reevaluated in each time step. The efficiency of the method is demonstrated by solving the following initial value problem:

$$(1+t^2)\ddot{u}+t\dot{u}+e^{1/(1+t)}u=-0.01e^{-0.1t}[(99+10t+99t^2)\cos t \\ +\left(-20+100t-20t^2\right)\sin t-100e^{1/(1+t)}\cos t)] \quad (36a)$$

$$u_0=1,\ \dot{u}_0=-0.1 \quad (36b)$$

The equation admits an exact solution $u_{ex}(t)=e^{-0.1t}\text{cost}$. The computed solution for $T=20$ is shown in Fig. C.9 as compared with the exact one.

C.5 REFERENCES

[1] J.T. Katsikadelis, A New direct time integration method for the equations of motion in structural dynamics, ZAMM, Zeitschrft fur Angewandte Mathematik und Mechanik (2013) 1–18.

[2] J.T. Katsikadelis, Dynamic Analysis of Structures. Theory and Applications, second ed., Symmetria Publications, Athens, Greece, 2012 (in Greek).

[3] D. Kuhl, M.A. Crisfield, Energy-conserving and decaying algorithms, Int. J. Numer. Meth. Eng. 45 (1999) 569–599.

[4] K.J. Bathe, Conserving energy and momentum in nonlinear dynamics: a simple implicit time integration scheme, Comput. Struct. 85 (2007) 437–445.

[5] J.T. Katsikadelis, C.G. Tsiatas, Nonlinear dynamic analysis of beams with variable stiffness, J. Sound Vib. 270 (4–5) (2004) 847–863.

Index

Note: Page numbers followed by *f* indicate figures, *t* indicate tables and *np* indicate footnote.

F

G

H

I

K

L

V

W

CPI Antony Rowe
Eastbourne, UK
July 23, 2014

CPI Antony Rowe
Eastbourne, UK
July 23, 2014